A Moral Climate

A Moral Climate

The Ethics of Global Warming

MICHAEL S. NORTHCOTT

Maryknoll, New York 10545

Founded in 1970, Orbis Books endeavors to publish works that enlighten the mind, nourish the spirit, and challenge the conscience. The publishing arm of the Maryknoll Fathers and Brothers, Orbis seeks to explore the global dimensions of the Christian faith and mission, to invite dialogue with diverse cultures and religious traditions, and to serve the cause of reconciliation and peace. The books published reflect the views of their authors and do not represent the official position of the Maryknoll Society. To learn more about Maryknoll and Orbis Books, please visit our website at www.maryknoll.org.

First published in Great Britain in 2007 by
Darton, Longman and Todd Ltd
1 Spencer Court
140-142 Wandsworth High Street
London SW18 4JJ
Great Britain

First published in the USA in 2007 by
Orbis Books
P.O. Box 308
Maryknoll, New York 10545-0308
U.S.A.

FSC
Mixed Sources
Product group from well-managed forests and other controlled sources
Cert no. SGS-COC-2953
www.fsc.org
© 1996 Forest Stewardship Council

Orbis ISBN-13: 978-1-57075-711-2

Printed and bound in Great Britain.

Library of Congress Cataloging-in-Publication Data

Northcott, Michael S.
A moral climate : the ethics of global warming / Michael Northcott.
p. cm.
ISBN-13: 978-1-57075-711-2
1. Human ecology--Religious aspects--Christianity. 2. Christian ethics. 3. Global warming--Moral and ethical aspects. 4. Climatic changes--Moral and ethical aspects. 5. Environmental ethics. 6. Environmental responsibility. I. Title.

BT695.5.N69 2007
241'.691--dc22 2006032382

Contents

Foreword

Over the last decade, global warming has become arguably the most talked about issue of our time. At the G8 meeting in June 2007 in Germany it took centre stage. I am writing this Foreword just as the G8 has closed with its compromise statement on climate change that has at least shown agreement about an international framework for action.

The media has been full of debate about the science, technology and economics of the issue. My main involvement has been with the science, especially from 1988 to 2002 when I was privileged to chair or co-chair the scientific assessment for the Intergovernmental Panel on Climate Change (IPCC) and work with many hundreds of scientists from around the world from different cultures and backgrounds. What was remarkable was that under the honest and balanced discipline of science together with the sense of shared responsibility for conveying accurate information, it was possible to come to consensus agreement. Not of course an agreement about everything, but identifying what we knew with reasonable certainty as distinct from those areas where a great deal of uncertainty remained. Arguments about the basic science are largely over. Now some of the big questions are about the technology (what is the best action to take?), and the economics (can we afford to act or can we afford not to act?).

The biggest challenge the world faces, however, is how allowable carbon emissions are to be shared between industrialised nations who have already benefited enormously from cheap fossil fuel energy and developing nations who are determined to grow their economies to a more acceptable level. That challenge raises deeper considerations that are the subject of this book – considerations of morality, equity (both international and intergenerational), justice, attitudes and motivation – qualities that make up the *moral climate* that need to be put alongside the physics, chemistry, biology and dynamics that govern the equations describing the *physical climate*.

Faced with the complication of this challenge, two extreme attitudes are commonly held. The first is *Denial* – doomsters are hyping it up and things are just not that serious. The second is *Despair* – the doomsters are right and we are already beyond the point of no return. Neither demands action. With the first, action is not necessary, with the second it is too late. Inaction always seems a comfortable position!

Response to the challenge requires very large change – locally and globally, individually and corporately, not superficially but fundamentally. Northcott addresses all aspects of that change. The world needs to set its priorities very clearly. The writer C. S. Lewis, in an essay 'First and Second Things'[1] written in 1942, pointed out that by concentrating on *second* things, *first* things are likely to be lost. In prioritising as first the trappings of civilisation – higher living standards, for instance - there was great danger of losing civilisation itself. Jesus put over the same principle when he said 'Seek first the kingdom of God and his righteousness and all these things (food, drink, clothing) will be added on to you'.[2] In facing climate change, putting selfish interests ahead of the common good will likely lead to the loss of both.

The realisation of change requires both motivation and application. Motivation for action and change generally comes from what we believe, bringing spiritual and religious considerations to the fore – another of Northcott's strong themes. Given motivation, application has to be effective. Earlier this year I visited New Orleans and saw some of the awful devastation caused by Hurricane Katrina in 2005. Resources had been allocated towards its remediation but hardly spent. Application, amazingly, seemed almost wholly absent. The most cheering part of the visit was to see parties of young groups from churches, sports clubs and the like from all over the US who had come in to try to do their bit, inadequate though it was bound to be.

Northcott engages his readers by skilfully weaving together the *moral climate* and the *physical climate* illustrated by real life stories from the Arctic and the Tropics. He also challenges established attitudes and demonstrates that the problems of climate change can only be solved by bringing together all our intellectual, moral and practical capabilities in an integrated way (lots of joined up thinking required!). Finally he leaves us with hope (not despair) and points out that the benefits that ensue from concerted action are not just those of relieving the worst damage from climate change, essential though that is. But along with that action comes the opportunity to face and solve other world problems – world poverty, the rich/poor divide, over use of resources and the appreciation and preservation of the non-human creation. Further, as nations and communities learn to share and work together much more closely, the goal comes in sight of achieving much more understanding, friendship and cooperation between all the world's nations and peoples.

Northcott's book is very readable and informative but also challenging. Allow it not only to inform your mind but to touch your conscience. You will then be able to chart a path for the future that quietly, radically and effectively will not only save us from the worst ravages of anthropogenic climate change but also bring about change towards a more sustainable, fairer, safer and happier world.

JOHN HOUGHTON

Preface

This book has been something of an odyssey. It began as a sabbatical research project in the Lilly library on Duke University's belle époque East Campus in North Carolina in January 2005. I am very grateful to Dean Gregory Jones of the Duke Divinity School and Dean William Schlesinger of the Nicholas School of the Environment and Earth Sciences for inviting me to take up a visiting professorship there. I also thank Ellen Davis, Stanley Hauerwas, Amy Laura Hall, John Utz, Stuart Pimm, Eric and Laura Pritchard and Kyle and Kelly Van Houtan for their hospitality and friendship during my sabbatical there, and for their responses to the first glimmerings of my findings in this project. Since then I have presented ideas from this book at the Hong Kong Baptist University, Oxford Brookes University, the University of Nottingham, the Centre International Reformé at John Knox House, Geneva, Trondheim University, Liverpool Cathedral, Christ Church, Oxford, and the Greenbelt Festival. I am grateful to my hosts at these occasions for their hospitality, and to the audiences for helping me think through the issues in this book with greater clarity.

When I write down the journeys undertaken while writing this book it becomes evident that the book already has a carbon footprint even before being published. I can only say in mitigation that within the United Kingdom I travelled by train, and in the course of writing I scrapped my car and did not replace it. Offsetting the carbon used in flying is no substitute for not flying and yet it is hard to see how the required international exchange of academic life can be sustained without flying. But in the past scholars exchanged letters in which they debated their ideas and challenged one another across continents and over many years, often without meeting face-to-face. And such conversations, because they were recorded in the form of letters, are a valuable scholarly resource many centuries later. Unless we can find a low-carbon form of flying – space frames attached to helium balloons offer the only form of air travel that is as low-carbon as rail or coach travel, but no airlines are proposing to build them – scholars will need to return to writing letters and exchanging papers at a distance, or else adopt new versions of international exchange on the internet instead of attending inter-continental meetings.

Well, enough of confessions. I also want to thank those friends and colleagues who have read through the whole of this work before publication, including Bryan Brock, Tim Foulger, Elizabeth Vander Meer, Joslyn Ogden, William Schlesinger, Wilf Wilde and the anonymous reader engaged by Darton, Longman and Todd. I am very grateful for their perceptive criticism and generous advice. I am also grateful to colleagues in the School of Divinity in the University of Edinburgh for granting me the research leave in which this book began, and for bearing with me in its latter stages as I devoted time to bringing it to completion. I thank the librarians at New College library, Duke University Library and the National Library of Scotland, where much of the research was undertaken. I also thank the unsung digital librarians both at Duke and Edinburgh whose genius in linking up the scholarly universe through digital access to journals and other research resources made the construction of such a cross-disciplinary work so much easier. Without their assistance it would not have been possible to write this book in the interstices of the normal demands of academic life. I thank Virginia Hearn, who commissioned this book for Darton, Longman and Todd. And finally I thank my wife Jill and my daughter Rebecca, who have borne with the frequent travels this book has involved. Rebecca is the youngest of our three children and as she now begins to make her way in the world she reminds me that there are new agents and new possibilities and the young can behave differently.

I completed this manuscript in the week in which the world celebrated the 200th anniversary of the abolition of slavery. This is a propitious moment to finish a book on global warming. The industrial revolution began by enslaving human beings to its imperatives, expropriating their common lands and wealth and enslaving their labour power and then even their bodies as tradeable chattels. Industrialism is now enslaving the climate of the planet to the same imperatives of market values and property accumulation. And the international arrangements so far constructed to resolve this crisis are turning the climate, and carbon dioxide, into traded chattels through carbon trades and offsets. What empowered many individuals around the world in the struggle against the evils of slavery and the coercive injustices of early industrialism was a spiritual vision of divine grace. Only such a vision can save the present generation from subjecting the peoples and species of the earth to a future of dangerous climate change. Mary, the Mother of Jesus, had an overwhelming vision of grace at the annunciation of the birth of Christ by the Angel Gabriel. Her response was to submit to its authority – 'let it be to me according to your word'. The earth is an authority more present to us than the angels.

MICHAEL NORTHCOTT
Feast of the Annunciation, 2007

Abbreviations

ANWR: Artic National Wildlife Refuge

CITES: Convention on International Trade in Endangered Species

CO_2: Carbon dioxide

COP: Conference of the Parties

ECEN: European Churches Environmental Network

FAO: Food and Agriculture Organization, United Nations

G8: Group of Eight: international forum for the governments of Canada, France, Germany, Italy, Japan, Russia, the United Kingdom and the United States

GCM: General Circulation Model

GDP: Gross Domestic Product

IEA: International Energy Authority

IMF: International Monetary Fund

IPCC: Intergovernmental Panel on Climate Change

IWC: International Whaling Commission

NGO: Non-governmental organisation

ppm: parts per million

UNEP: United Nations Environment Programme

UNFCCC: United Nations Framework Convention on Climate Change

WCC: World Council of Churches

WHO: World Health Organization

WMO: World Meteorological Organization

WTO: World Trade Organization

For they sow the wind,
and they shall reap the whirlwind.

Hosea 8:7

When it is evening, you say, 'It will be fair weather for the sky is red.' And in the morning, 'It will be stormy today, for the sky is red and threatening.' You know how to interpret the appearance of the sky, but you cannot interpret the signs of the times.

Matthew 16:2–3

Take away justice and what are kingdoms but large-scale criminal enterprises.

Augustine, *De Civitate Dei*

People will not look towards posterity when they fail to reflect on past generations.

Edmund Burke, *Reflections on the Revolution in France*

A change in the weather is sufficient to recreate the world and ourselves.

Marcel Proust, *À La Recherche du Temps Perdu*

Introduction

Take up weeping and wailing for the mountains,
and a lamentation for the pastures of the wilderness,
because they are laid waste so that no one passes through,
and the lowing of cattle is not heard;
both the birds of the air and the animals
have fled and are gone.
I will make Jerusalem a heap of ruins,
a lair of jackals;
and I will make the cities of Judah a desolation
without inhabitant.

Who is the man so wise that he can understand this? To whom has the mouth of the LORD *spoken, that he may declare it? Why is the land ruined and laid waste like a wilderness, so that no one passes through? And the* LORD *says: Because they have forsaken my law which I set before them, and have not obeyed my voice, or walked in accord with it, but have stubbornly followed their own hearts and have gone after the Baals, as their ancestors taught them.*

Jeremiah 9:10–14

The beautiful island of Tasmania is home to the last great eucalyptus rainforest in Australasia. The *Eucalyptus regnans* is the earth's tallest hardwood species and mature trees reach 90m into the air and are up to 20m in girth. The rich, moist forest floor and the tree canopy is home to thousands of species of insects, birds and small mammals and to numerous kinds of mosses, lichen and tree ferns, some of which, like *Diksonia antarctica*, the Tasmanian tree fern, are unique to the island. Gunns, a large timber company, has been clear-cutting this rare forest for many years. In the next decade it plans to work its way through much of the rest of the island, in the teeth of strong local opposition, but with the support of the Australian government.[1] Gunns clears 30 hectares of old growth rainforest a day to feed its woodchip and paper pulp factories, despite extensive local opposition. Pro-logging lobby groups, and Gunns itself, have considerable influence over the political process in Australia and receive far more media coverage than conservation groups.[2] Those who publicly oppose logging do so at considerable risk: the homes and businesses of individuals who have spoken out against the wood chip industry in Tasmania have often been burned to the ground.[3] Despite protests from Australian scientists, citizens and faith groups, the timber of this wondrous island is being turned into wood chip and paper pulp while the remaining forest – once the big trees are harvested – is burned to the ground with napalm and then replanted with a monocrop of eucalyptus trees.[4] Around the tree nurseries the company plants blue carrots treated with a lethal toxin in the ground so that any small mammals that survive the burning of the forest are killed to prevent them eating the young trees.

In 2007 the government of New South Wales turned off the irrigation supply to a large part of the Murray-Darling basin, the principal agricultural region of Australia. The Darling River is dying from over use of its waters for irrigation. And the problem has been exacerbated by a record ten-year drought which has seen large areas of former rich agricultural farmland turn to near desert. Australia is facing an ecological meltdown from climate change which is partly linked to regional land-use changes and partly to anthropogenic global warming, or the warming of the planet consequent upon industrial greenhouse gas emissions. Greenhouse gases naturally trap a proportion of the heat of the sun emitted from the surface of the earth inside the earth system, while letting another proportion out into space. Industrial greenhouse gas emissions are increasing the atmospheric quantity of these gases and so increasing the amount of heat retained by the earth's atmosphere. Australia is among the heaviest emitters per head of greenhouse gases on the planet, after the United States, Canada, Finland and Iceland.[5] Australia is also the most fossil-fuel dependent for its energy production of all developed industrial economies, with renewable sources of energy providing less than 2 per cent of its energy, despite the considerable potential of the interior for solar energy production. And still the Australian government promotes and subsidises the corporate burning and clear felling of the region's remaining ancient forests.[6]

Tropical and subtropical forests in their natural state are rich and wondrous ecosystems which act like natural air conditioning. The microclimate of the forest draws moisture from the ocean and atmosphere and creates updrafts which form rain clouds. As the rain falls it is taken up by tree canopies, fallen leaves, mosses and tree roots, and seeps deep into the soil before reaching streams and rivers. When the trees are gone the rain at first continues to fall, but when it does it falls on bare earth and washes it into streams and rivers.

Over time, with more deforestation, the rains began to dissipate, the clouds no longer form and the land becomes parched. Eventually it is at risk of turning into bare desert, something that is happening all over Australia.

Deforestation is a major contributor to human greenhouse gas emissions. Since colonial times Australian loss of above-ground biomass from deforestation has contributed approximately 757 million tonnes of CO_2 to the industrial legacy of excess greenhouse gas emissions that scientists now believe are changing the atmosphere of the planet.[7] The total of greenhouse gas emissions from deforestation far exceeds this figure since disturbance of the forest floor and peaty soils emits quantities of methane equivalent to the global warming potential of the above-ground carbon, since methane has a greenhouse effect twenty times stronger than CO_2. Consequently, Australian ecosystems are now net emitters of greenhouse gases, and contribute approximately 7 per cent to the global total of greenhouse gases from terrestrial sources.[8] The climate of Australia is changing and its mainly urban residents will experience a worsening water crisis in decades to come. But the government and corporations of Australia continue to view the forest as a commodity to be turned into bankable wealth, despite the effects of deforestation on the climate and hydrology of the region.

Australia is a microcosm for what is happening to planet earth in the present ecological crisis. The last fifty years of industrial development in Australia has seen the growth of ten large air-conditioned cities, connected by highways and airports, the destruction of ancient forests and the draining of underground aquifers to irrigate vineyards, citrus and cereal crops and supply the growing urban demand for fresh water. Industrial Australia is now hitting against the biological limits of this once verdant and lush continent, and the Australian government is fast approaching a biopolitical crisis. Soon its people will no longer be able to feed themselves, and yet the government continues to permit the life-sustaining and climate-regulating forests to be destroyed. The Australian government has for long argued that there is no relationship between climate change and humanly-produced carbon dioxide. It has refused to sign the Kyoto Protocol which limits the greenhouse gas emissions of industrial countries and it rejects such limits as an undue restraint on economic growth. Recently, and in the light of increasing local climate change, there has been some change of heart and the Australian government is proposing to ban the sale of incandescent light bulbs to domestic consumers. But such measures may be too little, too late. The warning signs have for too long been wilfully ignored and the country's ecosystems destroyed by a voracious consumer and corporate economy. With global warming the Great Barrier Reef is now threatened with coral bleaching, raised land temperatures have exacerbated the decade-long drought, the rivers from the red and increasingly desertified heart of Australia are beginning to run dry, and the land lies fissured and exhausted.[9]

The Hebrew prophet Jeremiah reflected on a similar geopolitical crisis in the land of Israel in the sixth century BCE. Jeremiah writes in exile from the ancient city of Babylon. In 587 BCE the Babylonians had invaded Jerusalem, destroyed Solomon's great temple, and exiled the leaders of Jerusalem to Babylon to prevent them organising a revolt against their occupiers. These momentous events created a political and theological conundrum for the people of Israel.[10] They believed themselves to be the chosen people of Yahweh, their society protected by the covenant their ancestors had made with Yahweh before they entered the Promised Land. This covenant set the Israelite state apart from the other states of the region since it provided them with a unique standard of justice and a socio-political vision very different from the imperial nations of Egypt, Assyria and Persia. According to the terms of the covenant, Yahweh would richly bless the people when they fulfilled the commandments of Moses, but he would curse them if they were unfaithful to the commandments. And the Hebrews moreover believed that Yahweh was not just a tribal deity but instead the creator of all the earth, and hence of all the nations. This meant that from an Israelite perspective Yahweh directed and sustained not only Israel, but the whole course of history, and hence even those who had occupied the land of Israel.

When Babylon sacked Jerusalem and the House of David fell, this was not for the Hebrew prophets just a geopolitical event but divinely ordained.[11] If Babylon's star was rising and Israel's fading this could only mean one thing: Yahweh was judging Israel for her unfaithfulness to the Mosaic covenant. The central *reason* for the fall of Israel that the prophets discerned was that in her imperial projects Israel had abandoned the terms of the Mosaic Covenant. Israel had become too powerful, her rulers too successful in their military enterprises, her elites too wealthy, and her smallholders and the soil were being enslaved in the project of a Greater Israel. Walter Brüggemann draws an analogy between this prophetic reading of the fall of the House of David and the rise and fall of nations and imperial powers in modern history.[12] Drawing on Paul Kennedy's analysis of great power collapse, Brüggemann argues that the reason for the fall of the modern great powers – such as the British Empire and the Soviet Union – is that empires reach heights of political power and territorial expansion which take them beyond biopolitical limits and into a condition of imperial 'overstretch'.[13] Just as Jeremiah reads the fall of the House of David, and ecological problems in the land of Israel, as divine judgment on the imperial ambitions of David's successor kings, so the collapse of modern imperial domains and the threatened ecological collapse of industrial civilisation, in Australia and right around the world, can be traced to an imperious refusal of biopolitical limits.

The world is faced with a biopolitical crisis which is more momentous than the geopolitical crisis of ancient Israel, or even the fall of the Soviet Union. At the heart of the present crisis is not a conventional empire but the global market empire fashioned by the United States and Europe in the last fifty years, as governments have deregulated money and trade and freed up economic actors and financial markets to enable maximal wealth accumulation by banks and corporations without regard to political sovereignty or territorial limits.[14] This has involved an expansion in monetary values of unprecedented proportions in the form of bank credits, paper money, stocks and such financial instruments as derivatives, futures, private equity funds, and hedge funds. The consequent expansion in capital investment and industrial development has been achieved at great social and ecological cost. Traditional working-class communities have been torn apart in the North, while jobs exported to the poorly-regulated South have engineered a new Dickensian enslavement of millions to sweat shops and factories which create pollutants threatening the health of factory workers and of whole regions in India, South China, Indonesia, Mexico and elsewhere.

These spatial displacements of the ecological and social costs of a deregulated and growing market empire are now implicated in a potential global collapse in the earth system. The ten-fold increase in greenhouse gas emissions that this economy without limits has sustained in the last fifty years has sent the planet's climate into overdrive. As I will show in the next chapter, scientific data indicate that human greenhouse gas emissions have displaced the sun and the ocean as the primary drivers of the earth's climate. Anthropogenic global warming threatens the collapse of whole ecosystems, from the Poles to the heart of Africa, and the extinction of up to 40 per cent of the earth's species.

The rate of growth in capital accumulation and greenhouse gas emissions are indicative of the extent to which human making has been transformed by industrial capitalism and the fossil-fuelled global supply chains which sustain increasing consumption and wealth accumulation. Through technology and economic and political artifice, and because of growth in the human population, the powers of modern humanity have grown to the point that humans are now the strongest biological force on earth. But these new powers have not been accompanied by a growth in moral responsibility for the condition of the planet, or in relations between distant peoples. On the contrary, as technology has heightened human power over nature, modern humans are increasingly alienated from the earth and their fellow creatures. And as in Australia, so in many other parts of the world politicians and citizen-consumers defer their capacities for moral and political deliberation to the autonomous procedures of the market, and to the automatic machines and monitoring systems which govern so many of the processes of industrial making. People are therefore increasingly poorly equipped – ecologically, morally and politically – to deal with the consequences and dangers of these enlarged powers both for the earth and for human wellbeing.

The refusal to recognise moral or ecological constraints to mercantile and technological power is also a part of the legacy of the European Enlightenment and the modern struggle for emancipation from the authority of the Church, tradition and faith, as well as from nature. But the paradox is that in the course of this emancipation, nature *and* society are increasingly dominated by technological power. Dietrich Bonhoeffer suggested that this is because modern rationalism, science and technology train modern humans not to see the earth as divine creation. Consequently 'the earth is no longer our earth, and then we become strangers on earth', and from strangers we finally become earth's subjects: through the power of technology 'the earth grips man and subdues him'.[15] Global warming represents the greatest instance of this subduing, since if industrial humans do not find ways of reducing their present forcing of the climate future generations will be committed to dangerous and irreversible climate change.

Since I commenced research on this book in 2005 the political tide has in some ways begun to turn on global warming. Political leaders in Europe, and most recently even in Australia and North America, now acknowledge that global warming is happening, and that it is caused by humanly-generated greenhouse gases. Most people no longer treat the science as a hypothesis but as a convincing explanation of changes which are already observable as climatic events move beyond the realm of natural variability. There is growing recognition in Europe that there is a conflict between fossil fuel consumption and the health of the planet, and that future generations, and some present generations in tropical and sub-tropical regions, will experience great hardship if the present generation does not take steps to curtail its addiction to fossil fuel. But few voices in the public debate over global warming are prepared to suggest that the problem runs deeper than that. The argument goes that once the engines of industrial making and the global market empire are decoupled from growth in fossil fuel emissions and hooked up to more efficient technologies and more renewable sources of energy the problem of global warming will have been solved, and the engine can go on working its autonomous magic to deliver a free trade utopia to all of humanity.

In this book I suggest that this is a serious misreading of the nature of the ecological crisis. Global warming is the earth's judgment on the global market empire, and on the heedless consumption it fosters. The neoliberal claim is that the market, combined with technological power, can redeem the peoples of the world from pain and suffering through the autonomous, self-regulating market system. Those who direct the neoliberal project of economic globalisation presume that the welfare of people and planet is advanced when the most powerful economic actors – multinational corporations, bankers, investors, engineers – are freed from regulation and taxation to make and sell more 'consumer objects' and accumulate more wealth. But in reality this collective project of global wealth accumulation disempowers people in communities of place, and so provokes enormous destruction in the welfare of human communities.[16] There are presently around 163 million internally-displaced persons in the world, more than 100 million of them displaced by 'development' projects such as giant dams, forest plantations, agribusiness, roads, factories, mining and quarrying.[17] At the same time it presages the greatest ecological collapse in the history of the human species. Global warming, in other words, is the global market empire hitting its biopolitical limits.

Those nation-states that seek to direct the market empire – and the United States and the United Kingdom in particular – present terrorists and other non-state actors who oppose the globalisation project as the enemies of freedom and democracy. And they suggest that the best protection against these enemies is a war of all against all called the 'war on terror'.[18] Some also read the problem of global warming in this frame and suggest that it is an equivalent security threat to that of terrorism. National or 'homeland' security are the watchwords of this political discourse. But the irony is that the global market empire makes no homeland safe since it requires the constant exchange of great quantities of goods and materials through ports and airports and along highways. These exchanges often involve coercive and immoral contracts, new forms of human slavery, and grave biological and ecological risks. But corporations and consumers know, or choose to know, little about the destructive and immoral conditions in which the goods they commission or purchase are produced. The biggest import–export commodity of the global market empire is not even visible to the naked eye, since it is in the form of carbon dioxide, an invisible gas produced in every breath, and by cars, planes, ships, trucks, factories, furnaces and fossil-fuelled power stations in gargantuan quantities.

Empire and Ecology

The Hebrew Bible is a record of the geopolitical impacts of empire on one minority people, the Hebrews.[19] It is also a record of the infection of empire among the people of God. The author(s) of Jeremiah provide a reading of geopolitical history which is a powerful analogy for the ecological crisis provoked by the present global market empire. There had always been tensions between the covenant vision of a locally-governed society of self-sufficient farmers and the State of Israel sponsored by the House of David. From the days of Solomon Jerusalem was fashioned on a scale to match the grand cities and temples of neighbouring cultures. And so successful did Israel become in its imperial projects that it also took up the expansionist tendencies of empire and ceded lands and natural resources from beyond its early borders.

This imperial drive was as much about the internal as the external political economy of Israel. The covenant laws as developed in the Deuteronomic tradition forbade the Hebrews from exploiting their neighbours, animals or the soil in the way that the Egyptians did. The laws warranted social procedures designed to keep the land fairly distributed and farm sizes small, and so prevent either soil or people from being enslaved to excessive accumulations of material wealth and social power. But powerful elites in Jerusalem began to buy up the countryside, taking the lands of indebted farmers to build themselves larger estates on which they could construct fine palaces and store large surpluses, surpluses which Israel also used to sustain its growing standing army. And by pursuing the imperial dream the merchants and princes of Israel returned many of their own people to the condition of slavery that their ancestors had known in Egypt.

Egypt is on the Western boundary of the region known as the Fertile Crescent, which stretches to Southeast Turkey in the North and the Persian Gulf in the East.[20] The great rivers of the Nile, Jordan, Euphrates and Tigris water the region.[21] The region was blessed with fertile soils which had accumulated beneath the great glaciers that had covered the region and then rapidly melted into the ocean at the end of the last ice age. It was the first region of the world in which substantial numbers of people moved from nomadic hunter-gatherer to agrarian lifestyles, a move which is recorded in the Hebrew Bible in the stories of the Patriarchs, from Abraham through Joseph to Moses and Joshua. Joseph was said to have led the Hebrews from their semi-nomadic existence in the Sinai Peninsula to settle in Egypt, and Moses, and then Joshua, to their own land 'flowing with milk and honey'.

The human move to agriculture made possible the emergence of the world's first imperial civilisations in ancient Mesopotamia. These included not only Egypt but Assyria, Babylon and Persia. And Israel also began to make an imperial mark on the region in the later Hebrew monarchies. The rise of these civilisations depended on a stable climate and it was around 10,000 years ago that the earth entered into the Holocene era, characterised by a relatively warm and stable climate, which Brian Fagan insightfully calls 'the long summer'.[22] A stable climate made it possible for people to acquire the skills in seed cultivation and animal husbandry requisite for settled agriculture. And with settled agriculture it became possible to accumulate surplus food and so release people from food growing for other vocations such as stone masons and lawyers, merchants and musicians, priests and politicians, sword makers and soldiers, travellers and traders. Agricultural surplus and the division of labour enabled the accumulation of material and social power. And hence agriculture, fertile soils and climate stability provided the material and ecological foundations of the empires of the ancient world. These empires spawned human history's first monetary and legal systems and literary and religious traditions. They also sponsored slavery and large-scale warfare as they subjugated large terrains to sustain their monumental projects and ambitions.

The Fertile Crescent was also the location of one of the first great ecological collapses in recorded history. A volcanic eruption caused a general cooling and drying of the region which precipitated the abandonment of many urban settlements and the collapse of the

Akkadian empire on the plains of Syria at around 2200 BCE.[23] People in the region continued to be at risk of agricultural collapses, in part because the soils of the region are fragile, and in part because these first empires put considerable demands on the soil. Jeremiah witnesses to such a collapse in approximately 600 BCE. There is no evidence of a climate event at this time and the likely cause was poor farming practices. As perennial grasses are cleared and the land cropped with cereal plants the soil is at risk of salinisation from mineral salts in the subsoil and bedrock which rise through the soil when the land is ploughed and cropped with non-native plants. Regular ploughing and continuous annual cropping allow the rain to sink deep into the land and the salts to rise up to poison the topsoil. This was likely the pollution of the land of which Jeremiah complained.[24] Once the primeval forests on the high lands were cleared the land also became prone to drought and flash floods, and the precious topsoil which grew the surpluses that sustained empires was gradually washed into the ocean. When the soil was washed away and the land became less green, it absorbed more sunlight, there were fewer clouds, the rains failed, and parts of the land turned to desert. By the sixth century before the Christian era these ecological problems led to a collapse in agricultural production, as recorded by Jeremiah.

From exile in Babylon, Jeremiah reflects on the nature and causes of the downfall of the House of David and concludes that at root the Israelites were vanquished because they had neglected to worship Yahweh. Instead they had idolised wealth and power and enslaved one another and the land in the process.[25] And so Jeremiah reads both exile and ecological collapse as consequences of idolatry and sin. Exile is their punishment for failing to care for the land, and their refusal of the terms that Yahweh had set in gifting the Promised Land to Israel.

The moral principles that guided the economic and agricultural activities of the Israelites were derived from the law of the Sabbath. The commandment to respect the Sabbath enjoined rest on human beings in order that they would remember that Yahweh had made the land, and not they. The Sabbath also reminded them that their work on the land could only be just and holy when it was in service to the God who made it. The word 'till' in Hebrew – *leawod* – indicates work on the soil, but it also indicates service.[26] The implication was that the tilling of the soil would only bring forth fruit when the earth was treated with respect. The practice of Sabbath celebrated the spiritual relation of the world to God by sanctifying time, and the work of the agrarian year.[27] It trained the Israelites to remember that the land was gift and not property, for 'the earth is the Lord's and all that is in it' (Psalm 24:1). In this way the Sabbath laws sanctified time and space, and indicated that work and making are moral as well as material activities, governed by transcendent principles as well as by biological laws and relationships. And when the Hebrews neglected sacred time, their work and making in material space come into conflict with these principles, laws and relationships.

From the Sabbath law the Hebrews adumbrated a range of related laws in the Deuteronomic code which moralised the material activities of agrarianism and making. Domestic animals as well as farmers were to rest on the Sabbath, and sufficient land was to remain unused to provide space for wild animals. Similarly, land was not to be deep ploughed since this implied lack of respect for the land. And the land was to lie fallow, to be given its own Sabbath, every seventh year. The Sabbath laws also restrained the economy of ancient Israel. Every seventh year in Israel those who had become debt slaves through ill luck or bad work were to be freed from their debts and sent on their way with provisions and animals to restart their lives. And every fiftieth or Jubilee year accumulated debt bonds in the form of land were to be returned to the families who had originally owned them. These laws enshrined the principle of Sabbath – that human success in working the land should be set within moral limits – in the economy of the land.[28]

Jeremiah identifies the neglect of divine law, and in particular the Sabbath law, as the cause of the downfall of Jerusalem: 'You must not carry any loads in through the gates of Jerusalem on the Sabbath day. If you disobey, I will set the gates of Jerusalem on fire. It will burn down all the fortified dwellings in Jerusalem and no one will be able to put it out' (Jeremiah 17:27). Judgment would not only befall the cities. The wild places of Israel and Judah would also be polluted by the covenantal disobedience of the people. The land would turn to desert, the rivers would run dry, the crops and vineyards would fail and the animals would not be able to bear young. The Book of Jeremiah records then not just a geopolitical event – the conquest and sacking of Jerusalem and the vanquishing of the House of David. It sets this event in a larger biopolitical and moral nexus: the land of Israel was conquered and polluted because of excessive ecological demands on the land by the late Israelite monarchy and the merchant class it spawned. We might even say that

Jeremiah is the first ecological prophet in literary and religious history. In neglecting divine justice and righteousness, he argues that Israel had upset the order of Creation.

For the ancient Hebrews, justice was not a human invention but a divine attribute set into the character and structure of creation.[29] When the people of God allowed their work on the land to displace the worship of God, and when they enslaved one another and the land to serve the greed of the rich, the land lost its fertility and rich and poor alike were exiled from the land. The belief that justice is set into the created order reflects a widespread assumption in the ancient world that biological and human communities are caught up in a nexus of relationships which also include heavenly bodies such as the sun, moon and stars, and the gods.[30] The rites and practices of temple and agriculture were supposed to mirror and order these earth–heaven and divine–human relationships. And when they did not, the forces of chaos that the gods restrained would overwhelm the land. In the ancient world the ocean was seen as the primeval source of chaos, and this perhaps represents a cultural memory of prehistory in which the oceans had covered far more of the land than they have since the end of the last ice age. For Jeremiah, the covenant that Yahweh made with the Israelites after their release from slavery in Egypt was therefore a cosmic covenant in which human work on creation ought to recognise the sustaining creative powers of the Lord of the earth in keeping the forces of chaos at bay:

> Do you not fear me? says the LORD;
> Do you not tremble before me?
> I placed the sand as a boundary for the sea,
> a perpetual barrier which it cannot pass;
> though the waves toss, they cannot prevail,
> though they roar, they cannot pass over it.
> But this people has a stubborn and rebellious heart;
> they have turned aside and gone away.
> They do not say in their hearts,
> 'Let us fear the LORD our God,
> who gives the rain in its season,
> the autumn rain and the spring rain,
> and keeps for us the weeks appointed for the harvest.'
> Your iniquities have turned these away,
> and your sins have deprived you of good.
> For scoundrels are found among my people;
> they take over the goods of others.

Like fowlers they set a trap;
they catch human beings.
Like a cage full of birds,
their houses are full of treachery;
therefore they have become great and rich,
they have grown fat and sleek.
They know no limits in deeds of wickedness;
they do not judge with justice
the cause of the orphan, to make it prosper,
and they do not defend the rights of the needy.

Jeremiah 5:22–28

In this passage Jeremiah links ecological disaster and exile with unfaithfulness to the laws and worship of Yahweh. When the Hebrews worshipped Yahweh they worshipped the spiritual source of created and moral order. They honoured this order when they followed the moral guidelines in their revealed law for care of the land, respect for their fellow citizens and compassion and justice for the poor. When they abandoned the worship of Yahweh, they began to worship the objects of material power which they had made from created things, and so idolised the creature instead of the creator. And when idolatry reigned the poor and needy were undefended and 'ill fared the land'.[31] As inequality and oppression grew in their society so domination and destruction came to characterise the human relation to the land and other species. Ecological collapse, local climate change and post-exilic political disenfranchisement are all read by Jeremiah as the consequences of Israel's neglect of the moral order upheld by true worship.

A Moral Climate?

The writings of the Hebrews narrate the relationships between God, humans and the earth as relationships which are intrinsically moral, in which the climate of the earth responds to human idolatry and immorality. For the Hebrews, as for the classical Greeks and the early Christians, morality and reason are vitally implicated in the character and structure of creation. If the earth shows signs of stress, if the land loses its fertility, if the oceans flood the land, the prophets and sages of these ancient cultures read such events as indicative of something awry in the relations between creatures and their creator, and between rich and poor. By contrast, most modern interpretations of the evidence that the earth's climate is changing focus exclusively on the material origins of global warming in the history and imperious spread of industrial civilisation.

The last two hundred years has seen the rise of the new gods of secular reason and technological power, and more recently the displacement of political sovereignty by the autonomous procedures of economic markets. These new gods have displaced faith in a divine Creator and sacred respect for the order of creation. They have displaced small farmers and agrarian communities and practices with corporate monocrops and industrial tillage. And they have displaced craft workshops and traditional skills with machine-driven manufacture. These displacements have ruptured traditional bioregional relationships between the cultures of food and farming, production and consumption. In consequence, the customer becomes an abstract consumer while the maker becomes merely a worker, compensated in monetary terms for the loss of meaning in industrial making.[32] And the core spaces of this rupture – airports, automobile highways, container ports, corporate offices, factories, retail sheds, shopping malls – spread across ancient cities, harbours, and rural hinterlands from London to Sydney, and from Quebec to Santiago.

Jeremiah's theological reading of the collapse of the House of David and the ecological collapse of the land of Israel offers a powerful narrative with which to frame the current ecological crisis. Rising sea levels, strengthening storms and prolonged drought and heatwaves threaten the collapse of the complex distribution systems of industrialism. The excess greenhouse gases produced by industrial capitalism are the fruits of the modern devotion to the gods of secular reason, technological power and monetary accumulation, and the sidelining of traditional understandings of community, justice and the sacred. Above all, these excess gases are indicative of the attempts of the great corporations and technocratic agencies which direct the global market empire to accumulate wealth so that the future of humanity is secured against all contingencies. But this attempt to store up nature's wealth in banks of assets, far from promoting welfare, actually generates deep insecurity and moral harms because it involves the systematic destruction of what Robert Putnam calls 'social capital', and Paul Hawken calls 'natural capital'.[33] The rate of change and the scale of value accumulation in the market empire draws power from local communities and ecosystems, breaking down established social habits of cooperation and exchange, destroying local knowledge of ecological and cultural conditions, and rapaciously trawling forests, savannah, oceans and subsoils for minerals, fuels and fibres to turn them into marketable objects. This process of concentrating material power sucks the life out of local and sustainable forms of exchange and destroys both human and species communities. Those who doubt this need only consider Putnam's observation of a 58 per cent decline in local associations in the United States in the last thirty years, the inexorable rise of the single-person household in the United Kingdom which has contributed to the nation's present housing crisis, or the enforced move from rural areas of millions of small and peasant farmers all over the world in the last sixty years.[34]

Judgment and Hope

When read in the light of Jeremiah's theological reading of the imperial geopolitics of ancient Mesopotamia, global warming, like the exile of ancient Israel, represents both the threat of judgment and the promise of a better way of living on God's earth than the neoliberal vision of a global market empire. Once the roots of ecological crisis in the destruction of small farms, local communities and local ecosystems are exposed it is possible to acknowledge in global warming the earth's judgment, and hence the judgment of the Creator, on the infection of trade and the perversion of making which has spread across the surface of the earth in the last fifty years. And yet paradoxically – and this is the source of hope in the prophetic tradition of the Hebrews – judgment is also and always redemptive. The divine word of judgment in Old and New Testaments is again and again the occasion of moral and spiritual conversion. Jeremiah prophesied that Israel's exile would come to an end; God would vindicate the people of God and all created order. But to *know* redemption the people would need to learn repentance, change their practices, and recover fidelity to the spiritual foundations of their society and of the earth.

In what follows I suggest that redressing the global and intergenerational injustices of global warming will also require new practices and new politics. If collapse in the earth system is to be prevented the global economy will have to be brought back into scale with the ecology of the earth, and the political ecology of particular places on earth. People will need to recover a sense of the spiritual significance of treasuring and guarding their own local ecosystems. Food, fibre and fuel will need to be grown and utilised locally and in ways which respect the regenerative cycles of life on earth. Human making and the money supply will need to be remade through local exchange and trading schemes. And where distant exchanges continue these will need to be re-moralised so that fossil fuel use, and the full ecological and social impacts of all traded commodities – such as the computer on which I write – are known and resolved at all stages in the production and consumption process.

If and when the economy of human exchange is brought back into harmony with the biological and geochemical flows and structures of this living planet it will be an economy which, instead of constantly destroying social and natural capital, builds up the relational nexus of friendships and partnerships in which true peace between peoples, between humans and the earth, and between creatures and Creator, is sustained. Such an economy will promote genuine human flourishing as well as the welfare of the earth.

At the heart of the pathology of ecological crisis is the refusal of modern humans to see themselves as creatures, contingently embedded in networks of relationships with other creatures, and with the Creator. This refusal is the quintessential root of what theologians call sin. And like the sin of Adam, it has moral and spiritual as well as ecological consequences. Resituating the human economy in the enfolding and relational ecology of the earth is therefore not a dispiriting task of merely constraining or limiting human making and creativity. On the contrary, it is joyous and spiritual work. And the root of this work in those who have already begun it is a renewed appreciation of the abundance of diversity that the relational networks of the earth – from ocean and forest floor to the upper atmosphere – have sustained through the evolving history of life and of the rich history of more beneficent forms of human interaction with life in all its diversity.

1
Message from the Planet

Thus says the Lord:
See, waters are rising out of the north
and shall become an overflowing torrent;
they shall overflow the land and all that fills it,
the city and those who live in it.
People shall cry out,
and all the inhabitants of the land shall wail.
At the noise of the stamping of the hoofs of his stallions,
at the clatter of his chariots, at the rumbling of their wheels,
parents do not turn back for children, so feeble are their hands.
Jeremiah 47:2–3

On 29 August 2005 Hurricane Katrina slammed into the United States coastline and caused more death and destruction than any previous hurricane in American history. Despite the hasty departure of more than a million people from the worst affected states of Louisiana, Mississippi and Alabama, thousands were left behind. Most lacked the resources to flee, and at least 1,570 people died in Louisiana, and around 500 in other states.[1] Most perished in New Orleans as the levees along the Mississippi River which protected the low-lying city from flooding were breached, and this large historic city was rendered uninhabitable. As with extreme climate events in other parts of the world, it was not the mobile and affluent who suffered from the Katrina event but the poor who could not drive out of the city, and even lacked the resources for a long-distance bus ticket.[2] Post-hurricane aid failed to materialise from the government of the richest country on the planet, and the world watched as the poor drowned on the roofs of their houses, or ran out of food and water, milk and diapers for their babies in the large football stadium which became their temporary home. Forced to forage in local stores, the poor found that the failing local state could still field police to arrest them for stealing food even though it could not prevent them from going hungry. Ironically Hurricane Katrina also caused the shutdown of oil production in the Gulf of Mexico and of on-shore refining on the Gulf coast. As gas prices rose to $3 a gallon many Americans were beginning to ask whether there was a connection between the exceptional number and ferocity of hurricanes in 2005 and the scientific claim, then fiercely resisted by the Bush/Cheney administration, and by influential American media such as the *Wall Street Journal*, that the human use of fossil fuels is changing the climate.[3]

Similar questions are asked in Bangladesh where, with far less Western media exposure, millions are displaced and thousands drown every year in flooding caused by rising sea levels and the growing ferocity of cyclones in the Indian Ocean.[4] Bangladesh is a crowded country of around 140 million people, more than 120 million of whom live and farm on lands located in the labyrinth of waterways and swamps which constitute the largest river delta on earth. The glaciers and snows of the Himalayan massif feed the three great rivers, the Ganges, the Meghna and the Brahmaputra, that create the delta. The Asian monsoon, which affects 60 per cent of the world's people, has strengthened in the last four decades and growing quantities of rain and snow in the high Himalayas strengthen seasonal river flow and increase the depth, extent and regularity of flooding of farmlands and village houses, and the erosion of river-banks.[5] Millions of people are regularly displaced by seasonal flooding in the river delta, and the number of flood victims has grown dramatically in recent years. In 2004, 750 people were killed and 30 million displaced by floods which were exacerbated by exceptionally strong cyclones in the Indian Ocean.[6] Rising sea levels are also causing the soil and the water table to become salty. Ground water can in many places no longer be drunk and instead of rice many farmers are harvesting shrimps which now live in the salinised soils.

The burden of suffering entailed by extreme weather in Bangladesh, as in New Orleans, falls on the poor. Most of the population of Bangladesh live on land less than 10m above sea level, and 35 million live on the coast. Around one million are permanently displaced every year by flooding, and the government of Bangladesh is too poor to provide anything but the most modest direct aid: most 'have nowhere to go and end up living on relatives' land or by the roadside or on embankments'.[7] Global warming will make many more permanently homeless. With a sea-level rise of 40 centimetres, which the Intergovernmental Panel on Climate Change (IPCC) predict will likely occur in this century, the numbers of people whose lands will be annually flooded in coastal areas and megadeltas will increase from 13 million to 94 million people, 60 million of these in South Asia.[8] Around 1.3 billion people live in areas affected by glacial retreat and they will also face destructive flooding as melt water increases, while many of these same people will eventually suffer water shortages as rivers like the Ganges receive half their water flow from glacial melt water.[9]

It is possible that the occurrence of more violent storms in the Atlantic and Indian Oceans is part of a natural cycle and unrelated to anthropogenic climate change. But recent scientific papers indicate that climate change is almost certainly responsible. The number of cyclones in the Atlantic, Pacific and Indian Oceans varies in a natural thirty-year cycle. This is sufficient to explain the increased number of tropical storms hitting the United States, Japan and Bangladesh in the last fifteen years. But the intensity of these storms has grown significantly: wind speeds are faster, they last longer and are sustained at a higher level of activity. Hurricanes are consequently more likely to reach land and to do extensive damage.[10] The operative cause is the role of ocean surface temperatures in generating hurricanes. The ocean has warmed faster than any other component of the earth system and represents 90 per cent of the increase in the heat content of the earth system between 1961 and 2003.[11] Tropical ocean temperatures show a warming of 0.5°C since 1970 and this marked warming creates greater updraft of moisture into the clouds and hence sustains more intense and violent storms.[12]

In 2005, meteorologists recorded the highest mean ocean surface temperature since instrumental recordings began in 1880, and consequently that year saw many other life-threatening extreme weather events across the globe.[13] China was hit by a succession of violent storms which destroyed 900,000 homes and displaced millions of people; forest fires and drought plagued Spain and Portugal while farmers in the Languedoc in Southern France were for the first time attempting to defend their vines and olives trees from an invasion of locusts from North Africa; in the North Indian State of Orissa temperatures reached 47°C in a life-threatening heatwave in June 2005, while extreme heat in Mumbai was broken by a monsoon which dumped 37 inches of rain on the city in a single day, and thousands drowned in flash floods; in sub-Saharan Africa prolonged drought and locust invasion in 2004 and 2005 affected more than 30 countries and forced millions to rely on food aid as crops failed on a considerable scale.[14]

The Scientific Assessment of Climate Change

The Intergovernmental Panel on Climate Change (IPCC) was established by the United Nations Environment Programme and the World Metereological Organization (WMO) in 1988 under the United Nations Framework Convention on Climate Change (UNFCCC) to report on the science of global warming. The consultation procedures of the IPCC are reminiscent of the ecumenical councils of the Christian Church in the era of the Roman Empire, drawing as it does on the peer-reviewed work of thousands of scientists from across the globe, and from a range of disciplinary perspectives, in a quest for consensus in its quinquennial reports on the state of the planet and the possibilities of saving it from meltdown. Its reports have moreover been remarkably accurate in their predictions of global warming, and of related weather and ecosystem events. In its *Fourth Assessment Report* the IPCC observes that, although the average surface temperature of the earth in the last hundred years has risen by 0.8°C, most of the warming has taken place in just the last three decades, with a rise of 0.2°C per decade.[15] At this rate global temperatures will rise by a further 2°C by the end of the present century. This *rate* of temperature rise is unprecedented in at least the last 650,000 years of planetary history, according to paleoclimatic indicators such as ice cores. In the IPCC's view this rise in temperature is driven by the growing emission of greenhouse gases, principally from the industrial use of fossil fuels. These gases enhance or 'force' the non-anthropogenic global warming effect which sustains the planet at a temperature warm enough for organic life. The most important of these gases is carbon dioxide. Using historical evidence from long-term climate indicators including ice cores, tree rings and historic weather records, climatologists have traced a close relationship between the quantity of carbon dioxide (CO_2) in the atmosphere and the temperature of the earth. Carbon is present in the earth's system in great quantities, but when too much is present in the atmosphere it acts like double glazing in a greenhouse, trapping in the atmosphere more of the heat from the sun's rays as it is emitted from the earth's surface.[16]

The science on which the IPCC draws for its reports suggests that even if industrial humans radically curtail their present production of greenhouse gases the planet is already committed to at least a further 1°C of warming by the end of the present century, in addition to the 0.8°C rise, which has already occurred over pre-industrial temperatures. This is because of inertia in the climate system. Gases emitted now will cause atmospheric changes up to fifty years hence. Consequently the full heating effects of greenhouse gases emitted in the last fifty years have yet to manifest in the earth system. If industrial humans continue to increase their production of greenhouse gases at the present rate, it is likely that the earth will warm between 3°C and 6°C in the *present* century.[17] This humanly-generated global warming effect or forcing of the earth's climate is likely to be magnified by a number of feedback mechanisms in earth systems, such as the release of billions of tonnes of methane, which has a global warming potential twenty times that of carbon dioxide, from frozen tundra in Siberia and Northern Canada. Events like this may move the planet towards a 'tipping point' that would make dangerous and irreversible climate change, beyond even 6°C, more likely by the end of this century.

A small minority of scientists, and many political commentators, still argue that the apparently slight average warming of the earth's surface of 0.8°C in the last one hundred and fifty years is explicable on the basis of natural climate variability, or even variations in sunspot activity, though there is no correlation between recent temperature rises and solar activity.[18] Some suggest that a warmer world will be greener, and will in any case provoke other feedback mechanisms, such as thicker and more reflective cirrus clouds, which will prevent the kind of extreme scenarios of dangerous climate change most predict.[19] Most scientists are however now in agreement that rising average global temperatures are caused by increases in humanly-produced greenhouse gas emissions since the industrial revolution, and more especially since 1970 as greenhouse gas emissions, driven by unprecedented economic growth and global trade, have risen exponentially.

Prehistoric reductions in atmospheric carbon which made life on earth possible in all its myriad complexity were in part a consequence of life's growing success over geological time in utilising atmospheric carbon in the formation of bone, plant, shell and other organic matter. As carbon was drawn down from the atmosphere into the earth's living systems, the planet cooled from earlier extreme warm periods, so producing the conditions which made mammalian life, and hence human life, possible. There are those who suggest that since Greenland was once green and not covered in ice, the present warming is not unprecedented, and is therefore not caused by industrial humans.[20] But when dinosaurs walked in Greenland most of the earth was so hot as to be uninhabitable by mammals. The earth in evolutionary history, and, Christians would add, through divine providence, has become cool enough to be inhabited across most of its surface by humans and other animals. If present humans are making it uninhabitable for future generations and species, and for some presently existing peoples, by their lack of prudence in burning excessive amounts of fossil fuels, then they have a clear and urgent moral duty to moderate such activities.

The close relationship between atmospheric CO_2 and the planetary temperature that prehistoric climatological data reveals is mirrored in recent changes in the climate. From 1000–1750 the proportion of carbon dioxide in the atmosphere was at 280 parts per million (ppm), a figure that has remained relatively stable since the end of the last ice age 15,000 years ago. Since 1750 the proportion of CO_2 in the atmosphere has risen by one third to 380ppm. And this rise correlates with a rise in the earth's temperature, although its full warming effects have been reduced by the extent of atmospheric pollution from soot and other industrially-produced particles which have acted as a sun shield and so reduced the full potential warming of the additional greenhouse gases already in the atmosphere. By 2030 pollution controls on particulate emissions, which are very harmful to human health and cause millions of deaths every year, will reduce their presence in the atmosphere; subsequently the full warming effects of greenhouse gas forcing will become evident.[21] As atmospheric chemist Paul Crutzen, who first posited the sun shield effect of particulates, observes, for the first time in geological time the behaviour of one species is forcing climatic change on a global scale; humans are in effect in charge of the climate of the planet.[22] Or as the IPCC more cautiously put it in their *Fourth Assessment Report*, 'it is very likely that greenhouse gas forcing has been the dominant cause of the observed warming of globally averaged temperatures in the last fifty years'.[23]

A Brief History of Global Warming

The greenhouse effect is essential to life on earth. Without the global warming effect of CO_2 the earth's surface temperature would be –18°C instead of the present average of 15°C.[24] The action of greenhouse gases in retaining around 20 per cent of the long-wave infrared energy of the sun as it is emitted from the earth's surface was first described by French scientist Jean Fourier in 1824. Fourier hypothesised that it is this global warming action that creates a climate unique in the known universe in being warm enough to sustain life, while being cool enough to prevent the planet from burning up. By contrast the planet Mars, which has a much thinner atmosphere, reaches an average of 37°C in the daytime but the temperature drops to –120°C when the sun goes down as almost no warmth is retained.

The possibility that industrially-generated greenhouse gases might contribute to an enhancement of naturally-occurring global warming was first suggested by a nineteenth-century Swedish scientist, Svente Arrhenius, in 1896.[25] But it has taken more than a hundred years for Arrhenius' thesis to become the consensus view of climate scientists.[26] Lay perceptions of a change in the climate preceded acknowledgement by climate scientists of this eventuality. People first began to notice a change in the weather in the Northern hemisphere in the inter-war period of the twentieth century. Rivers unfroze earlier, or hardly froze at all, winters were warmer with fewer frosts, and fish were found to be migrating further north than usual. First reports of the observations of local fishermen and hunters of changes in the strength and duration of ice and snow cover in subarctic regions appeared in Canadian newspapers in the 1920s, while the *New York Times* first reported possible climatic changes consequent on greenhouse gas emissions in 1936.[27]

It was, appropriately enough, a steam technologist, Guy Stewart Callendar, in a paper to the Royal Meteorological Association in London in 1938, who first linked these lay reports of warmer weather with Arrhenius' theory that the increase in carbon in the atmosphere, consequent on human burning of coal, oil and gas, could warm the earth.[28] But other scientists suggested that since the ocean contained so much more carbon than the atmosphere it would take up any additional carbon emitted by humans. And the oceans are the principal repository or 'carbon sink' on the planet. The North Sea and North Atlantic are particularly important in this respect, having absorbed half of humanly-produced carbon in the last fifty years.[29] But the oceans are only able to absorb around one third of the present increased output of greenhouse gas emissions.[30] The excess emissions are deposited in the atmosphere, enhancing the warming effect of the preindustrial carbon and methane layer.

Wider scientific acceptance of the theory of anthropogenic global warming was impeded by the lack of evidence that fossil-fuel burning had actually resulted in more CO_2 being present in the atmosphere. So long as no such data existed, the idea that there was no limit to the carbon the oceans could absorb could not be dismissed. Data that would convince scientists of the reliability of the hypothesis that the weather was changing because of a rise in levels of atmospheric carbon dioxide awaited the development of instruments to accurately measure the presence of this invisible gas in the atmosphere. With such instruments Charles Keeling at the Scripps Oceanographic Institute in California was able to demonstrate in 1960 clear year-on-year rises in carbon dioxide levels over the United States.[31] These increases were correlated by Michael Mann with rising land and ocean temperatures over the last millennium in a famous paper which represented the existing and potential temperature rises in the Northern Hemisphere in the form of a graph which looked like a hockey stick because it showed such a sharp rise in temperatures in the present century as compared to the preceding relative temperature equilibrium of the last one thousand years.[32] Though some still contest the hockey stick graph because it seems to under-represent variability in the cooler period of the late Middle Ages and the sixteenth-century Little Ice Age, it is now widely replicated in other scientific studies, including the great array of graphs and data presented in the IPCC's *Fourth Assessment Report.*[33] What the hockey stick graphically represents is the unprecedented extent to which industrial civilisation through its mobilisation of the earth's prehistoric store of carbon has inadvertently begun to re-engineer the climate.

The principal greenhouse gases are carbon dioxide, methane and nitrous oxide. The increase in the atmospheric presence of carbon dioxide since 1750 is 100ppm over pre-industrial levels. Atmospheric methane has increased by 1,000ppm since 1750, while its rate of change over the previous 11,500 years was less than 200ppm. Nitrous oxide, the third most significant greenhouse gas, is rising at 0.8ppm per year, and again had been stable for the previous 11,500 years. The increase in CO_2 is responsible for approximately 75 per cent of the human forcing of the climate since 1750, methane for approximately 18 per cent, and nitrous oxide for 7 per cent.[34] It is now customary to talk of the atmospheric presence of greenhouse gases in terms of ppm equivalents of CO_2 (CO_2e), since methane and nitrous oxide make up a small part of the total. Presently the quantity of CO_2e in the atmosphere is 430ppm.

Since the last major ice age ended around 12,000 years ago the earth has been in what is known as the Holocene period. In this period it has enjoyed a period of relative climatic stability, compared to previous eras.[35] The proportion of carbon dioxide in the atmosphere remained pretty constant at between 260 and 280ppm throughout this period, and methane, the other most significant greenhouse gas, was similarly stable. This stability in the atmospheric quantities of these gases only begins to change after 1750 with the industrial mobilisation of the earth's carbon and methane stores in the form of coal, oil and natural gas. In the first two hundred years of the industrial revolution from 1750 to 1950 the annual global output of carbon dioxide, the most significant greenhouse gas and the most long-lived, went from 1 to 20 billion tonnes. In the last forty years the annual output has increased to 38 billion tonnes and it is still rising. Humans have released approximately five times the quantity of CO_2 in the last forty years that they released in the previous two hundred. Output of other greenhouse gases has also increased dramatically. Between 1970 and 2004 global emissions of CO_2 equivalents went from 28 to 49 billion tonnes, the largest proportion of which was from the energy supply sector.[36]

Climate change sceptics are fond of pointing out that humans have been burning fossil fuels for hundreds of years and yet the extent of global warming is less than 1°C.[37] But this ignores the growth in greenhouse gas emissions in the last fifty years. As the industrial economy has grown rapidly in its use of fossil fuels, and in its global reach into forests, oceans and other carbon sinks, so emissions have grown dramatically. It is this sudden and rapid increase in greenhouse gas emissions that is responsible for the anthropogenic influence rising to a point that it is now the biggest influence on changes in the earth's climate, larger even than the sun or the ocean.[38]

Few scientists now contest that the quantity of carbon dioxide in the atmosphere is rising because of industrial activities, though some still claim that the raised levels of CO_2 are not responsible for the recent warming. The Mauna Loa observatory in Hawaii shows an annual increase in atmospheric carbon of 1.5ppm, with some years, such as 2002, showing much faster growth.[39] It is likely that recent spikes in the annual rate of $C0_2$ take up in the atmosphere indicate that forests, oceans and soil, the principal carbon sinks, are declining in their absorbative capacities because of systemic changes prompted by the rise of land-based temperatures.[40] Another possible factor is the considerable quantities of carbon dioxide and methane emitted by extensive humanly-set forest fires since the late 1990s in Borneo, Java and the Amazon.

Uncertainty and the Climate System

In the last ten years significant changes in ecosystems are being observed in many earth regions. Perhaps the most significant of these biological climate change 'footprints' is the evidence of a large-scale, globally diffuse and systemic movement of species in every earth region. As mountains, lowlands and oceans warm, animals, insects and plants are forced to spatially relocate, and most species are moving 6km towards the poles every decade. But not all species are able to relocate and some are consequently threatened with extinction by global warming.[41] Temporal changes in species behaviour are also taking place; spring events – bulbs flowering, shrubs and trees in bud, bird migration and nesting – are occurring 2.3 days earlier each decade.[42]

The biological footprint of climate change, combined with the warming of land and oceans reveal a significant change in the earth system. That this change is caused by human activity is indicated by the clear correlation between increased greenhouse gas emissions and marked biological and climatic changes. Triangulation is a well-established method in the social sciences and is based on the idea that data derived from three different kinds of interrogation of empirical reality are more reliable than data from a single source. The triple footprint of rising temperatures, extreme weather and ecosystem change is therefore much more reliable than earlier accounts of global warming based solely on temperature rise. Consequently great confidence can be placed in the scientific claim that climate change is happening and that it is caused by industrial humans.

But real uncertainties do remain. How long will the carbon dioxide remain in the atmosphere? Might more of it eventually be taken up by biomass and ocean? Does its presence in the atmosphere warm the climate permanently or is the present warming just a blip in the larger climate cycle, which could still end with a glaciation event in the next few thousand years? The biggest uncertainty concerns climate sensitivity, or how responsive the climate is to raised atmospheric levels of greenhouse gases. While the IPCC are confident that their ability to predict change is confirmed by the good fit between their first predictions in the 1980s and present observations and measurements, nonetheless the issue of climate sensitivity remains unresolved. The consensus view that the likely range of temperature change is between 2°C and 4.5°C has provided a point of agreement in the midst of a rapidly-developing field of scientific discovery and policy-making.[43] But it is increasingly open to challenge. The IPCC's latest report indicates that the figure of 4.5°C may be exceeded in the present century if greenhouse gas emissions are not moderated.

The central device for the generation of estimates about the responsiveness of the climate system to greenhouse gas emissions are Atmospheric and Oceanic General Circulation Models. General Circulation Models (GCMs) originated as weather forecasting computer programmes for modelling the behaviour of clouds, oceans, wind and water vapour. They have undergone considerable adaptation with efforts to factor in landmasses subjected to human influences such as agriculture, deforestation and urbanisation, and atmospheric gases.[44] GCMs are devices for interpreting data; they are not independent sources of wisdom. They rely on the quality of data inputted, the interpretations of data, and the methods and procedures involved in its generation. Perhaps the most controversial assumption on which these models rely is that change will be gradual since there have been times in planetary history – though not in the last 15,000 years – when climate change has been extremely abrupt.[45] Another controversy arises from the procedure for modifying information from land-based temperature measurements because of the influence from urbanisation, deforestation and other local or regional, rather than global, variables. Michael Crichton, in a widely-read novelistic debunking of the evidence for anthropogenic forcing of the climate, suggests that most of the rise in land temperatures in the models comes from local human activities rather than global temperature shifts, since most land temperature measuring stations are located near built-up areas; current climate models, he suggests, do not sufficiently discount this effect in their aggregation of land temperatures.[46] This view is roundly denounced in the IPCC's *Fourth Assessment Report.* That it needs denouncing reflects the scepticism that climate forecasting and computer modelling still provoke in some quarters, and particularly in those countries with the heaviest CO_2 emissions.[47]

Controversies and uncertainties remain about how soon the planet will warm and what effects this warming will have. But these uncertainties are areas for investigation around a central core of consensus which enabled meteorologists to issue the first major public warning of the likelihood of climate change resulting from the burning of fossil fuels at an international conference in Villach in 1985.[48] It was in response to this report that the UNFCCC was agreed and the IPCC established. The IPCC has issued four major consensus reports, with the fourth being formally published in stages through 2007. Each has been more detailed than the last and successively more alarming. In 2001 the third IPCC report indicated that the earth was warming faster than earlier anticipated, and at a faster rate than it had ever warmed before.[49] The same report also indicated that it is possible to say definitively that human activity in warming the climate has led to reductions in the extent of polar ice, the global retreat of almost all land-based glaciers and consequent sea-level rise, increased growing seasons at temperate latitudes and increasing extreme weather events, attributable in part to the increasing frequency and severity of El Niño ocean warming.[50] El Niño – which means literally 'the Christ child' because the phenomenon occurs around Christmas – is the name given to a periodic warming of the equatorial Pacific Ocean. This warming has a global effect on weather through oceanic and atmospheric perturbations. A stronger El Niño strengthens the Asian monsoon, which creates more flooding in some parts of Asia, and more extreme heat, and forest fires, in other parts. The *Fourth Assessment Report* indicates that the IPCC's earlier short-term projections of temperature change in the period 1990–2005 of between 0.15°C and 0.29°C equated to a realised change of 0.33°C in the period, giving some confidence in the reliability of IPCC predictive scenarios, although the realised warming was at the upper end of the prediction.[51]

One study in 2005 was less cautious than the IPCC. Using data from a new kind of networked computer modelling involving collated results from thousands of small GCMs run on personal computers all over the world, David Stainforth and his colleagues put the possible upper limit of average warming much higher at between 6°C and 11°C.[52] This higher range was generated by modelling what would happen if the earth system reaches a tipping point where global warming goes into overdrive as a result of feedback mechanisms which amplify humanly-generated warming.[53] In this scenario warming of land and oceans turns the earth system from a carbon sink into a carbon emitter. In the oceans it is likely this will happen as the water becomes more acidic as a consequence of increased carbon content and increased heat. This reduces the quantity of life and the rate of carbon absorption that the oceans can sustain; for example oysters and other shellfish will not be able to form shell if sea temperatures and acidity from absorbed carbon continue to rise.[54] This would mean that major oceanic mechanisms for drawing down carbon from the atmosphere would be disrupted and the role of the ocean as a carbon sink will therefore diminish or possibly cease altogether. It is also likely that land-based carbon sinks such as tropical forests will die off because of raised temperatures. But the bigger threat is from increased decomposition of subterranean peat, the release of more carbon from soils, and the release of frozen methane from land and ocean. The Hadley Centre predict that by 2050 raised land temperatures will turn the soil of the planet from its current role as a carbon sink, that draws down carbon from the atmosphere, into a net carbon source which releases more carbon into the atmosphere than it absorbs. The process, called soil respiration, is caused by microbes in the soil responding to raised temperatures by becoming more active, so turning more of the fixed carbon in the soil into atmospheric CO_2.[55] A further likely contributor to a possible tipping point is the changed albedo – or reflectivity – of the earth's surface, resulting primarily from the darker and less reflective character of land and ocean at the poles when ice and snow melt.

Climatological Threats to Life

Most of the effects of climate change so far charted by scientists are in the climate system and in the responses of glaciers, ice caps and species to temperature change. But the growing number of extreme weather events triggered by global warming is also beginning to have an impact on human beings, as we saw in the events of 2005. The year 2007, in which I finished this book, was shaping up to be the warmest since modern weather records began. In January of that year Alpine slopes were bare of snow below 700m, bears refused to hibernate in Moscow zoos as temperatures remained above zero, small flowering trees blossomed in London squares, and New Yorkers lay on sun-loungers in temperatures of 21°C.

The effects of global warming in the Northern hemisphere have been far less life-threatening than those in the South. Climate in parts of the Southern hemisphere is becoming so unpredictable that drought, flood or fire increasingly threaten the poorest and most populous regions of the planet. This is not just because people in the South are less insulated from the effects of climate change by wealth and technology, but also because many of the worst effects for humans of global warming occur in the tropical and sub-tropical regions. Droughts and floods have been particularly concentrated in the tropics as climate change magnifies the El Niño events, causing changes in tropical monsoons. The IPCC estimates that most of the severe negative impacts on human populations will occur in developing nations in the Southern hemisphere. These effects will include the spread of mosquitoes and water-borne pathogens to new areas, reductions in crop yields from drought, and reductions in air quality and ground water.[56] Sea-level rise will also have terrible effects in the South. As the 2004 Indonesian tsunami showed, developing world populations are very vulnerable to changes in sea level and storm surges.[57]

In its *Fourth Assessment Report* the IPCC predicts a sea-level rise of between 0.4 and 0.7m by 2100.[58] Presently average global sea-level is rising at 3.3mm a year. The principal causes are thermal expansion as more heat in the oceans increases the volume of water, and the melting of glaciers and polar ice caps. Jim Hansen, the chief NASA climate scientist, is more pessimistic and believes that there could be a sea-level rise of up to 25m by the end of the present century as feedback mechanisms send global warming into overdrive and melt the ice caps.[59] His reasoning is reflected in the IPCC's *Fourth Assessment Report* which notes that paleoclimatic studies indicate that when the earth's atmosphere last contained CO_2 at present levels – between 360 and 400ppm – air temperatures were 2–3°C higher than pre-industrial temperatures, and the oceans were 25m higher.[60]

There are signs that ice melt at both poles is already taking place much faster than scientists had expected. In Greenland scientists have found that the velocity of glacial flow has dramatically increased as Arctic warming in the last ten years has caused a rise of 3°C in air temperatures.[61] Warmer summer temperatures cause lakes of melt water to pond on the surface of glaciers. These lakes drain into crevasses in great cascades. The water then reaches land below the glaciers and lubricates the interface between ice and rock.[62] Consequently glaciers are moving faster towards the sea and drawing down ice from the heart of the continent towards the oceans at a much faster rate. In addition ocean warming is melting the base of glaciers close to the sea and as ice shelves collapse this contributes to a faster rate of glacial flow.[63] This effect was also noted in Antarctica after the collapse of the Larsen B ice shelf in 2002.[64] Its tributary glaciers sped up three to four times in the following three years. In Greenland speed increases of glaciers increase ice mass loss from 90km^2 in 1996 to 220km^2 in 2005, contributing .6 mm to the present annual sea-level rise of 1.5 mm.[65] Greenland's complete melting would contribute 7m to sea-level rise while Western Antarctica contains 6m of sea-level rise.[66]

Even the conservative estimate of 40–70cm of sea-level rise will be enough to inundate small island nations such as Tuvalu, Kiribati, the Maldives, Mauritius and the Seychelles. Before their lands are inundated by seasonal tides the water sources of these island nations will have already turned to salt, forcing their inhabitants to become environmental refugees.[67] There is a peculiar irony to this in the case of the Maldives and the Seychelles, which rely almost exclusively on air-flown tourists for income when air travel is such a climate-polluting mode of transport. People are already leaving Kiribati and Tuvalu under government resettlement plans, the majority moving to New Zealand.[68] But their governments' requests for compensation from the industrialised nations, or for sovereign territory on other continents, have so far met with no response.

Sixty per cent of the human population lives within 100km of the ocean, with the majority in small and medium settlements on land no more than 5m above sea level.[69] More than 1 billion live in tidal cities which include such major population centres as Bangkok, Hong Kong, Kuala Lumpur, London, New York, Shanghai, Singapore and Sydney.[70] A sea level rise of 0.7m will therefore affect a very large number of people who will experience increased risk of seasonal flooding, and a significant proportion of these will permanently lose their homes.[71] All of these effects will be magnified in countries where reliance on agriculture remains the primary source of livelihood, and where living conditions are less insulated from climatic events than in developed countries.

The most immediate threat to human life from global warming comes not from the oceans directly but from changing precipitation patterns caused by ocean temperature rises. Whereas in the Northern hemisphere climate change is likely to lead to more rainfall in some areas, in sub-tropical and tropical regions in the Southern hemisphere it is predicted there will be significant reductions in rainfall, probably leading to drought and famine on a large scale. And these effects are already in evidence. Prolonged droughts and associated famines have occurred in sub-Saharan Africa for thousands of years and the recent increased duration and frequency of droughts is partly attributable to widespread deforestation across much of the region. But there is growing evidence that the increased frequency and severity of drought in sub-Saharan and South Africa in the last ten years is a consequence not just of local climate change but of anthropogenic global warming.[72] Scientists report a general drying of the African continent from the Mediterranean coast to the Cape of Good Hope, and they note the spread of sand dunes from the Sahara and Kalahari deserts across Northern and Central Africa.[73] Desertification south of the Sahara in the Sahel may also be a consequence of rising temperatures in the Indian Ocean and not just of overgrazing.[74] Extensive famine in the region in the last twenty years is quite possibly the first major human catastrophe consequent on human-induced climate change. It is predicted that land temperatures in Africa will rise by 1.6°C by 2050, with a consequent further increase in drought and crop failure, threatening the very survival of millions of people.[75]

The Climate of Consumption

In *Collapse: How Societies Choose to Fail or Survive*, Jared Diamond demonstrates how throughout human history empires and civilisations have frequently collapsed because they expanded beyond the carrying capacity of the regions in which they were situated.[76] Easter Island is one of the most devastating examples of this tendency. It appears that its indigenous inhabitants gradually denuded the island of trees while at the same time enslaving one another to a building project which left behind the giant stone idols which survive to this day. But once the trees were gone the ecology of this fertile island collapsed and when the island was 'discovered' by Spanish explorers in the eighteenth century the few remaining human inhabitants had been reduced to the most straightened circumstances on an island which was near desert. It is a strange irony that the island was named Easter Island since the island's history speaks less of a renewed creation than it does of ecocide.

The fall of Rome, an empire whose legacy, culturally and legally, still shapes the Western world, is also instructive for our present planetary predicament. Recent studies of the energy and resource requirements of the late Roman Empire indicate that the pressures the empire placed on the lands around the Mediterranean eventually led to the collapse of the ecology of these regions. The demand for agricultural surplus to feed Rome's growing appetite for meat and its large mobile armies led to problems of soil erosion in North Africa and Southern Spain and Italy as topsoil was washed into the rivers from too much ploughing or, in the case of the region which is now the Saharan Desert, blown into the ocean. The consequent ecological problems on the borders of the empire spurred the barbarian invasions which ultimately caused Rome's downfall.[77] There were also cultural and moral factors in the collapse of Rome. As luxury and surfeit grew in the imperial cities of Rome, so their privileged populace increasingly turned to ever more lewd and violent forms of public entertainment and religion, and the classical virtues of prudence and temperance which had aided Rome's rise were lost. The fall of Rome may have begun in the ruin of the breadbasket of North Africa but its fate was sealed with the moral and spiritual decay of its citizens' lives.

The last forty years of industrial consumerism has conferred on a minority of the world's people levels of luxury and surfeit unimagined by the Romans. The consumer cornucopia of material goods is driven by advertising and fashion and by the entertainment industry, and it has fostered a culture of superficial hedonism and waste. Citizens of the industrial empire increasingly regard it as their birthright that they should continually buy new clothes, own cars, enjoy foreign holidays and fill their lives with the latest electronic entertainment devices while living in super-heated or cooled homes sparkling with every kind of lighting device. The constant turnover of consumer objects, and the waste of the precious metals, minerals and fossil fuels used in their making, fosters growing instability in ecosystems and now in the earth system. Industrial consumerism is a form of material culture which is entirely at odds with the regenerative and recycling patterns of natural systems. As the throwaway society mines precious metals and fossil fuels from beneath the earth's surface and later buries them in holes in the ground, or emits them to the atmosphere, it comes into conflict with the earth system and threatens its continuing vitality.

The energy-hungry systems of industrialism not only subvert the health of forest, ocean and soil, and the stability of the climate; they also corrode human moral character by removing from daily life the constitutive and embodied activities of human making, and the virtues that careful making and repair entail.[78] As Wendell Berry argues, farmers reliant on fossil-fuelled machines and fossil fuel-derived fertilisers and pesticides destroy the regenerative capacities of the soil, replacing microbes and worms with industrially-derived fertility and eroding the topsoil.[79] Analogously, industrial workers build machines which are designed not to be repaired and so householders lose the material skills of repair and maintenance, and the associated virtues, and are drawn ineluctably into the throwaway prodigality of consumerism. This prodigality produces irresponsibility as people are increasingly ignorant of the materials and the work required to maintain their daily lives. As Berry suggests, when people cease to be involved in the kinds of work which sustain human dwelling and the fertility of the earth they also lose touch with the 'perennial and substantial world in which we really do live' and whose necessities and mysteries guided the wisdom of predecessor cultures and traditions.[80]

Global Warming and Neoliberalism

The extent of social and ecological destruction involved in modern consumerism is obscured by the anonymous and remote supply chains which distance consumption from making. And this tendency has been accelerated by the neoliberal promotion of a deregulated and corporately-controlled global economy in the last forty years.

Trade and consumption of goods and services has grown exponentially, and it is estimated that a greater quantity of goods and services has been traded in this period than in the whole history of human civilisation.[81] This growth in trade has seen carbon emissions rise ten-fold. But the message that growth is not ecologically sustainable is fiercely resisted, and especially by the powerful institutions charged with proposing strategies to respond to global warming. National governments and supranational institutions like the European Commission and the World Bank continue to push through more highway and airport construction, because they believe that economic growth and international trade are the keys to human welfare. And while climate-change rhetoric in Britain and elsewhere tends to focus on the carbon footprints of domestic consumers and households, the global greenhouse gas emissions of British multinational economic corporations exceed all emissions within Britain many times over.[82]

The dramatic increase in consumption and economic activity in the last forty years has meant that industrial humanity has inadvertently become the principal driver of the earth's climate. This expansion of economic activity began with the United States' ending of the Gold Standard in 1969, and it has been legitimated by the economic ideology of neoliberalism. The Gold Standard limited the quantity of money in circulation – including liquid money and banking debt – to a politically determined ratio between the money supply and the physical quantity of gold held by national banks. The abandonment of the Gold Standard freed the American dollar from its material roots in the biophysical world, disconnecting the quantity and supply of money from the limited biophysical resource of gold. It also freed the dollar from social regulation. This 'freeing' or depoliticisation of money spread to other currencies and is now worldwide in extent.[83] It has turned money – which is the principal device for connecting consumers and producers – from a publicly-regulated common resource into a private commodity that is primarily created and controlled by bankers and corporations, and regulated by supply and demand curves in the new abstract value markets of derivatives dealers.[84] Consequently the money supply has grown exponentially in the last forty years and this has driven exponential growth in carbon emissions and ecological destruction.[85]

The political deregulation of money is a central plank of the neoliberal economic project. Its advocates suggest that the welfare of individuals and nations is advanced when economic actors – of whom multinational corporations and investment banks are the most powerful – are freed from legally-enforceable responsibilities such as payment of a living wage, payment of public taxes, provision of a healthy working environment, and respect for the environment. Money values have consequently accumulated at remarkable speed in the corporations and banks of the world's financial centres. This process has fostered levels of corporate, household and national debt and of resource use and ecological destruction unprecedented even in the history of industrial capitalism. It has also seen the radical demoralisation of making. The neoliberal phase of modern capitalism has seen multinational corporations move many of their operations to places where they can achieve the lowest wage rates, where they pay the lowest taxes, and where environmental regulations are minimal or non-existent.[86] Most products that are sold in Europe and the United States are no longer made there. Instead they are made thousands of miles away in often miserable working conditions in countries which are prepared to sacrifice air quality, forests, rivers and oceans to toxic pollution in the quest for rapid economic growth. But ironically Western politicians often point to the increases in CO_2 emissions which fuel the factories that now make the products on behalf of Western corporations in countries like China and Brazil as a reason for refusing to reduce their own CO_2 emissions.

The leaders of the rich nations are elected on the basis of a collective lie in which they collude with their corporate funders, advertisers and the media; that they will make the lives of voters better, and their communities fairer as well as more prosperous, by allowing economic corporations to continue to rip through the ecosystems of the earth and the social fabric of human communities in North and South as they accumulate monetary wealth at others' expense. Politicians will not tell the truth about climate change and ecological collapse, which is that addressing these long-run problems will actually require *reductions* in economic growth, a reigning-in of corporate greed and the re-regulation of the money supply. This is because the truth runs counter to the core assumption of neoliberalism that unrestrained economic growth, deregulated trade in goods and deregulated money markets are redemptive devices that make the world a better place.[87]

In reality, however, the rise of neoliberalism has accompanied a decline in collective welfare in those societies which have most avidly embraced it. States of reported happiness have not increased in Western developed countries in the last thirty years, and in some cases have actually markedly reduced.[88] A recent report on the welfare of children revealed that in Britain and the United States child welfare had significantly declined in the last twenty years in response to commercial and government pressures on both parents to work outside the home, and to purchase the many material goods that advertisers and corporations foist on their children through the mass media.[89] At the same time the destruction of working-class jobs, as these have been exported to overseas low-wage and low-regulation economies, has devastated whole industrial regions in America and Europe, and left millions without secure employment or adequate incomes from work. This is the principal driver behind the surprising growth of child poverty in Britain and America in the last three decades, also recorded by UNICEF. Wellbeing has also declined in many of the countries whose ecosystems and cheap labour support the material base of the neoliberal economy. Ecological collapse, increasing poverty and violent conflict have all followed in the wake of the rapacious demands of the neoliberal economy for increased quantities of precious raw materials such as coltan (for mobile phones), copper, diamonds, oil, tropical timber and uranium.

And neoliberalism has not only advanced child poverty and the destruction of ecological and human communities. It turns out that the mobility as well as the monetary accumulation of the deregulated global economy is in direct conflict with a stable climate. The global movements of goods and peoples that distantly connect consumption and production procedures on different continents are sustained through energy-hungry transportation. Although much of the focus in relation to global warming has been on air transport, shipping is responsible for at least as many greenhouse gas emissions.[90] And with the present trend of economic globalisation planes and ships between them are forecast to more than double their CO_2 output by 2020. Land transportation of goods and people is an even more crucial component of economic globalisation and presently produces three times the CO_2 emissions of planes and ships. And growth trends for land transportation emissions are even greater than those for ships and planes. All this mobility of people and goods advances a 'politics of speed' which is in conflict with both people and planet. It enables the rapid accumulation of monetary value by rich corporations and individuals without regard for the destruction and disruptions such accumulation rests upon. And it fosters a collective forgetting of the roots of real wealth and wellbeing in stable communities and ecosystems where governance is local and not remote.

Remote Control and Unaccountable Power

The remote command and control systems of the global economy enable the coercive concentration of economic, ecological and hence political power in the hands of rich corporations and individuals. The result is what Sheldon Wolin calls 'inverted totalitarianism', where, instead of the state wielding unaccountable and tyrannical power over the people, it is corporations which acquire this power through the technological devices and procedures which sustain the neoliberal project.[91] The irony is that the neoliberal economy is advanced in the name of progress and liberty. Corporations and consumers released from legal responsibilities to respect those who produce their goods or the environments in which they are produced are said to be 'free'. Conversely, societies that still insist on a moral relationship between producers and consumers are said to be hidebound by regulation and stuck in the past. But the resulting split between consumption and production leads to the moral perversion of making, which is at the heart of the human vocation on earth. This perversion undermines personal agency, replacing it with a sense of powerlessness before the large forces and powers which direct the global economy. Far from advancing freedom and democratic governance, as its advocates state, the global economy is tyrannical. Its growing power over the planet involves systematic erosion of local sources of power and freedom as the well-being of human communities and ecosystems are sacrificed to sustain the 'freedoms' of global corporations and wealthy consumers to accumulate regardless of the costs to others and the earth system.

In a critique of the United States government's response to the terrorist attacks on America in 2001 the Kentucky farmer and essayist Wendell Berry suggests that the violence of the terrorists is a reflection of the violent nature of the neoliberal regime of international trade which the United States has promoted in the last thirty years. The use of destructive weapons against the United States is a byproduct of a global corporate economy which accepts 'universal pollution and global warming as normal costs of doing business'.[92] It is also a reflection of the inequality implicit in the scale of American consumption of the earth's resources. American companies advance their wealth and the culture of consumption by advancing ecological and social collapse in other earth regions by their coercive harvesting of cheap labour, foods, fuel, metals, minerals and timber. The claim that American 'security' is advanced by the rhetoric of hate, the caricature of enemies, and violent wars is at odds, Berry suggests, with the denials of freedom and the growth in political fear that have accompanied this claim. If peace and security are the true aims of government policy, then the means to realising them is the recovery of meaningful local economic livelihood, political participation, and the education of the people of the United States in the virtue of peaceableness:

> The key to peace is not violence but peaceableness, which is not passivity, but an alert, informed, practiced, and active state of being. We should recognize that while we have extravagantly subsidized the means of war, we have almost totally neglected the ways of peaceableness.[93]

The reason the United States neglects peaceableness and advances violence is not only that war is profitable, which it is, but because its government supposes that 'it is possible to exploit and impoverish the poorer countries, while arming and instructing them in the newest means of war'. Berry suggests that since 2001 Americans have faced a choice between an economic system dependent on worldwide sourcing of goods and services which will require a hugely expensive worldwide police force to secure it and restraints on freedom and civil rights at home, or a decentralised world economy whose aim would be to assure 'to every nation and region a *local* self-sufficiency in life-supporting goods'.[94] The virtues which underlie the recovery of a self-sufficient economy are those of thrift and care, saving and conserving whereas 'an economy based on waste is inherently and hopelessly violent, and waste is its inevitable byproduct. We need a peaceable economy.'[95]

Berry's comments recall the claim of Ulrich Beck that globalisation involves a condition of risk which is in many ways novel, linking people and societies through systems and technologies which involve the multiplication of risks between people and organisations who are unknown to each another and places that are far distant from one another.[96] A globalised world is an increasingly borderless world where viruses, or greenhouse gases, cross national boundaries as easily as packets of drugs or plastic explosives. The side-effect of globalisation is the emergence of an international order in which security in one country can no longer be assured when other countries with which goods or personnel or pollutants are regularly exchanged are in states of insecurity or anarchy.[97] Fear of violence from non-state actors is advanced by globalisation as dependence on globally-traded foods, fuels and fibres is combined with trade in weapons. This trade sustains authoritarian elites in Third World countries, while their populations are immiserated by the larceny of their lands, forests and oceans by corrupt government officials in league with private corporations. For a superpower such as the United States, this new global condition produces a situation in which no territory can any longer be said to be distinct from the project of global economic dominance if it contains actors or resources on which the United States has become economically dependent, or if it threatens the United States or its economic interests with violence.[98] But behind this threat of global imperialist invasions is a refusal to recognise that it is precisely the imperialist nature of the global economy which is driving the ecological and social destruction which foments terrorism and violence. Neoliberalism rests on a collective lie of momentous proportions, which is that when economic growth is pursued without let or hindrance, and when material power is concentrated in the hands of economic corporations, the peoples of the earth will enjoy progress and peace. And it is therefore no coincidence that the prime sponsors of global warming denial are the United States government, multinational corporations and corporately-owned mass media.

The Character of Witness

The ninth commandment, 'thou shalt not bear false witness', is taken by many modern philosophers to be the supreme moral duty. Thus Immanuel Kant suggests that lying is the most heinous of all actions since lying subverts the reasonableness of human life and without reason humans are given up to instinct.[99] The problem of deception is at the heart of the presently entirely inadequate response of rich nations, consumers and corporations to the problem of global warming. The paradoxical contradiction between the extensive scientific record of what industrialism is doing to the planet and the continuing irresponsibility of modern consumerism is an instance of what Thomas Aquinas called 'affected ignorance', where individuals – or in the present case whole societies – 'choose not to know what can and should be known.'[100] Truthful witness, on the other hand, is the paradigmatic form of resistance to the deceits which collude with the power–knowledge nexus of technocratic society. Such witness constitutes what Aristotle called *parrhesia* – fearless speech – and Jews and Christians call prophecy.[101] In the art of *parrhesia* witness is born to the truthfulness of speech by the actions and character of the prophet. *Parrhesia* addresses the speaker as much as the hearer because it 'exonerates, redeems, and purifies him; it unburdens him of his wrongs, liberates him, and promises him salvation'.[102]

Truthful speech is also a kind of confession: it owns a share in the burden of wrong and therefore resolves the problem of self-deception or affected ignorance. In *parrhesia* the one who speaks the truth is also the one who is prepared to witness to the truth by living in service of truth and in solidarity with those who suffer at the hands of the powerful. As Berry suggests, this kind of truth-telling involves a preparedness to 'stand by words' and to match deeds to words.[103] Such truth-telling is foundational to ethics, for humans are only capable of responsible action when they acknowledge and respond to the authority of the perennial and substantial world from which all life derives, and which Christians call Creation. The glaring gap between public speech and responsible action, or 'the dangerous confusion between public responsibility and public relations',[104] is linked with the presumption that 'the human prerogative is unlimited, that we *must* do whatever we have the power to do. Specifically what is lacking is the idea that humans have a place in Creation, and that this place is limited by responsibility on the one hand and by humility on the other.'[105]

Stanley Hauerwas narrates the character of truthful speech, and of the community required for its utterance, by engaging with Richard Adams' novel about rabbits, *Watership Down*.[106] Hazel is the courageous young rabbit whose youngest brother, Fiver, perceives in an estate agent's noticeboard a coming cataclysm which will destroy the warren of Sandleford. Hazel approaches the leader of the warren, but is dismissed as a doom-monger and so Hazel and a small band of followers embark on a long and arduous journey at the end of which they eventually found a new warren. Along the way they make mistakes but through it all they remain true to their original intuition. They are heroic not because they are strong but because they have the courage to obey the truth. Like Frodo in the *Lord of the Rings*, Hazel is an unlikely hero, for he has no intrinsic strength except his preparedness to rely upon his friends, and even on strangers, in bearing the burden of truth, and his preparedness to give up the known for the unknown, to leave a comfortable life and to suffer on the road in order that others might find life again.

Climate change threatens to disturb our planetary home, just as much as the property development threatened the warren of Sandleford. But we should not imagine that the international treaties and arrangements for mitigating it and adapting to its effects, hammered out by self-interested and powerful corporations and state technocrats, will produce the kinds of truthful and prophetic speech which Christians call witness. For this to occur, these institutions would have to acknowledge the extent to which their shared goals of profit maximisation and growth in global trade involve and require systematic and irreversible destruction of the earth, and subvert local forms of knowledge and politics and local care for places and ecosystems. Against the putatively impartial discourses of modern economics and natural science, there is a need for 'partial witness' which names the sufferings of those people and species which are threatened by climate change for what they are: vicious sacrifices demanded by the destructive gods of the global economy.[107]

Like the judgment that befell Israel at the time of the Babylonian captivity, climate change is indiscriminate; the poor suffer more than the rich. Jeremiah was a partial witness who told the rich and powerful that they had the blood of the poor on their clothes, and were responsible for what had befallen Israel. He prophesied that when the people had learned again to worship God and respect the Sabbath, then they would return to the land. In the mean time he called on them to pray for the welfare of the cities to which they have been exiled, for in their welfare they would find their own welfare (Jeremiah 29:7). This call to pray for the welfare of the foreign peoples amongst whom they found themselves was a call to faithfulness. Instead of railing against God for excluding them from the land, as they had the poor and other species, Jeremiah suggests they find new moral purpose in their relationships with those they had previously regarded as alien invaders.

The governments and corporations who manage the fossil-fuelled global economy claim to direct the course of history in the interests of citizens, customers and investors. But global warming witnesses to the same truth that Jeremiah uttered; economic relations that neglect justice and the health of the land ultimately bring ruin to all. It also reveals that the human tenure of this earth is one of temporary guardianship and not control. Far from controlling history and the fate of the earth, industrial civilisation, like all previous empires, is still subject to radical contingency. But by distancing consumer from producer, industrialism obscures the contingent ecological sources of material wealth. It also obscures the extent to which the rich exclude the poor from the natural wealth of their own lands to maintain their present levels of consumption. Global warming represents an intensification of this exclusion as many in the South are increasingly exiled from their own lands by persistent drought or flood.

According to Jeremiah the exile would train the Jews to give up their claims to privilege and recognise that divine providence is directed at the welfare of all the peoples of the earth and not just the 'chosen' people of God.[108] The rich are not yet the ones who are experiencing exile as a result of global warming. But acknowledging the suffering of those who are already being exiled from their lands involves the recognition, which Jeremiah also embraced, that the suffering, and not the powerful, hold the key to history. In their radical dependence on God, on one another and on natural systems, the poor and the exiled know, as the rich do not, that history is not in human hands.

The discourse of suffering is a discourse 'from below', for it is the boldness of those who are weak and yet speak truth to power which validates their testimony. Testimony to the suffering caused by the energy-hungry economy arises like a tide in those regions which are rich in fossil fuel and which consequently pay a heavy price for the extravagant industrial use of fossil fuel. Coal- and oil-rich regions suffer from extensive pollution of ground water, rivers, soil and ocean, the destruction of ancient forests, and the removal of the tops of mountains or the bottoms of valleys in the case of the strip mining of coal in some earth regions. Truthful action to redress these kinds of systemic abuses of earth and humans originates not in the discourses and strategies of the powerful but in what liberation theologians call an 'option for the poor'. This option involves identification with the suffering and pain of those who are persecuted and oppressed, and it invites the powerful to confess their role in the systemic and structural evils which sustain this suffering.[109]

For the truth to be heard, and to produce a moral response, requires that the suffering are permitted to speak.[110] And yet polar bears and tropical trees threatened with extinction by the climatic extremes which global warming is provoking cannot speak. As Bruno Latour points out, they have no voice in the science-informed deliberations of modern parliaments and corporate boardrooms, though their fates are determined by such deliberations.[111] Similarly those whose islands and coastlands are threatened with inundation are too poor for their influence to be felt in the centres of financial and political power.

The World Council of Churches (WCC), the global ecumenical association of churches from North and South, has for more than twenty years prophetically witnessed to the threat presented by global warming on behalf of those who are already suffering from global warming. The core assumption of WCC reports on climate change is that it represents a problem between nations which is not resolvable by conventional economic assumptions and procedures because these involve neglecting the moral limits put on human use of the land by the biblical command to respect the Sabbath and to do justice to all the inhabitants of the land. The WCC has consistently called on member churches to practise solidarity with the victims of climate change, and to resist the values of wealth and power which drive the global economy in visiting ecological harms on the powerless.[112] Many churches have already begun to look at ways in which they can reduce their carbon footprint and many are installing renewable power systems. There is no guarantee that such symbolic efforts will be successful in turning the world around. But this partial witness to the moral dilemma of climate change is sustained by the Christian claim that the future that God ordains for the world is known not in fear but in faith, hope and love. And 'love does not depend on the assurance of success', for living in truth is not about effectiveness.[113]

Partial witness in word and deed finds authenticity not in its success in modifying the nefarious bargains of the powerful, but in living true to the character of creation by bringing human making back into service of God's creation, and of local ecological and human communities. By contrast, much of the political effort to match deeds to words in relation to global warming has focused on technological fixes and the construction of new economic instruments such as the new market in carbon, of which more below. This focus mirrors the global move towards market instruments over moral and political deliberation and the perversion of human making that the neoliberal project has already advanced. Against these kinds of economistic responses to global warming, and against the various forms of global warming denial which continue to make waves in the media and in 'think tanks' sponsored by the oil companies, I suggest in what follows that truthful response to global warming requires a proper accounting and confession of the intrinsic connection between global warming, modern imperialism and neoliberal global capitalism. Only when the rich confess the ecological harms, or 'ecological debts', with which they burden the poor and other species through the deregulated global economy will it be possible for the poor to gain justice and for climate debts to be justly redeemed.[114]

But this is not to imply that the poor have no agency in the present global crisis. Far from it. The end of the fossil-fuelled tyranny of corporate power and global markets can only be enacted by the recovery of radical democracy in local communities where the disempowered recover agency, and overturn the sovereignty of monetary values over their lives and over the ecosystems they inhabit. As Berry suggests, citizens and local communities need to regain control over food, fuel and fibre. They need to reclaim personal responsibility in their own economic exchanges, and recover the situatedness of such exchanges in ecological and social relationships. In this way it will be possible to retreat from forms of making which alienate consumers from producers and economy from ecology, and which seem to require a war of all against all. Dwelling in buildings which conserve heat and light and draw on locally-generated power, travelling in ways which respect the earth, growing and buying food locally – these are ways in which local communities and households can repair human making and exchange so that they build up rather than break down the relational networks which are so central to human and ecological flourishing.[115] And in this way they will recover a sense of political responsibility and of being in place. As Swiss theologian Lukas Vischer, who has long been a prophet of global warming in the European churches, observed to me last summer in his Alpine home, 'there is something democratic about solar panels'.[116]

2
When Prophecy Fails

Does the snow of Lebanon leave the crags of Sirion?
Do the mountain waters run dry, the cold flowing streams?
But my people have forgotten me,
they burn offerings to a delusion;
they have stumbled in their ways, in the ancient roads,
and have gone into bypaths, not the highway,
making their land a horror, a thing to be hissed at for ever.
All who pass by it are horrified and shake their heads.
Like the wind from the east,
I will scatter them before the enemy.
I will show them my back, not my face,
on the day of their calamity.

Jeremiah 18:14–17

The snows of Mount Kilimanjaro in Tanzania are projected to disappear completely by 2020.[1] Kilimanjaro is the highest mountain in Africa and the world's tallest freestanding mountain, reaching 5,000m above a plain of 1,000m. It has three single peaks called Kibo, Mawenzi and Shira, of which only Kibo, the highest, still retains its glaciers. The Swahili name Kilimanjaro means 'shining mountain', which is a reference to its white ice cap. The disappearance of the glaciers on this famous African mountain is a powerful metaphor of the likely magnitude of the impacts of global warming on the African continent. The causes of the disappearance of the ice cap on Kilimanjaro are multiple and include local deforestation, but few doubt that the single most significant cause is anthropogenic global warming. The snows have survived since the last ice age, and for the millions of Tanzanians who can see them on the horizon their loss will be symbolically significant.[2] The decline of the ice cap will further undermine the hydrological cycle of the mountain, which is already at risk because of deforestation on its lower slopes. Deforestation raised temperatures, causing the decline of the fog and rains which sustain the extensive cloud forests of the mountain and an increase in forest fires. These factors will eventually threaten the mountain's most significant water catchments, and this will have deleterious consequences both for the many species which live in the national park on the mountain slopes, and for the Chagga people who have dwelt there for millennia.

Chagga smallholders produce 30 per cent of Tanzania's high quality Arabica coffee in the Kilimanjaro region, as well as other cash crops such as sisal, sugar cane and cotton, and food crops which include bananas, beans, rice and millet.[3] The decline in the clouds and fogs which sustain the forests and the rise in forest fires from higher temperatures both present major threats to the survival of the Chagga people. They are adept at managing the precious water of the forests, using small irrigation channels, small dams and other devices to cultivate a rich range of crops. But with declining rains and river levels women and children now spend large parts of each day in the dry season fetching and carrying water for household use. Around 130,000 people presently live in the forests of Kilimanjaro. The waters of the mountain's rivers also supply three hydroelectric projects in the rivers downstream and these provide 20 per cent of Tanzania's electricity. The hydroelectric dams also yield 4,000 tonnes of fish annually. These rivers feed the rice growing area of South-East Moshi, and they sustain the wetlands of Ol Tukai and Kimana, which are home to thousands of Masai pastoralists and many wetland bird and mammal species.

The mountain's slopes and forests are home to around 900 unique tropical species and 25 per cent of Tanzania's biodiversity. The forests are particularly rich in epiphytic plants which grow on the branches and roots of the trees, and to a great variety of ferns. There are 405 bird species, including many unique highland species, and 140 species of mammal, including many varieties of antelope and bat. The forests are also home to the largest surviving population of the globally threatened Abbott's Duiker antelope, and to 418 reptile species. Kilimanjaro, like all tropical forests, is also very rich in insect fauna, with hundreds of beetles, butterflies and other insect species, many of which are unique to Kilimanjaro.

Global Warming in Tanzania

The warming of Kilimanjaro is indicative of the general drying of the African continent noted above. Extensive drought in the last ten years has been accompanied by significant temperature rise across Africa, which in Tanzania has occurred at 0.275°C *per annum* in the last quarter century, and which is ten times the rate of globally-averaged land-temperature warming.[4] Tanzania is among the poorest countries on earth, with an annual income per head of around $270 and an average life expectancy of just 43 years. Less than one quarter of Tanzanians can read, just 0.6 per cent of its population attends college or university, and only 5 per cent of its roads are paved. More than half of the country's income is earned from agriculture and more than 80 per cent of its population still live in rural areas and rely on agriculture as their primary source of income. Both drought and flood cause frequent problems in Tanzania. In the late 1990s two years of persistent drought significantly reduced crop output while a very strong flood in 1998 washed away many roads and agriculture-related infrastructure, exacerbating problems of food supply and malnutrition. Since 2000 Central Tanzania has experienced a persistent drought, which has hardly abated in six years. The length of the drought, and its closeness to previous periods of extreme stress, has meant that 'traditional coping strategies are breaking down' and people are 'increasingly unable to meet their basic needs.'[5]

Donald Mtelemela, the Anglican Archbishop of Tanzania spoke of the terrible effects of climate change on the people of his land at a climate change conference in Edinburgh in 2005. He recounted how the agrarian people of his country still see rain as the blessing of God on the land, which alone gives them the means to life when it waters their crops. But because of climate change, caused by the burdens of the rich North's industrial emissions on planetary systems, the rains come less regularly and droughts come every three or four years instead of every ten years as in the recent past. This recent change has produced social breakdown in rural areas as hunger and an absence of local sources of food has forced many parents to leave their villages for the cities in a desperate search for food. Sometimes older relatives are able to care for the children while their parents go in search of food but many are left at home unsupervised, malnourished and unable to attend school.[6]

Climate change in Tanzania is projected to get much worse. Under a doubling of CO_2 in the atmosphere over pre-industrial levels – around 550ppm – GCMs indicate that Tanzania can expect an average temperature rise of 3–5°C.[7] At these temperatures the growing season will shorten, precipitation will decline further, and yields of maize – the country's staple food crop – will decline by at least a third over the entire country and by up to 85 per cent in the central regions of Dodoma and Tabora. More than 40 per cent of Tanzania is still covered by tropical and sub-tropical forests and these are also expected to be affected by climate change. As the fogs needed by the highland cloud forests fade and Kilimanjaro's last glacier disappears, rains will further diminish and water levels in the country's main rivers will decline by at least 10 per cent. Tanzania also has an 800km coastline and many millions of Tanzanians live on the coast and are consequently at risk of flooding with a projected sea-level rise of 0.7m. The country's coastal capital, Dar es Salaam, would also be at risk.

Because so much of Tanzania's population still relies on agriculture it is among the most vulnerable countries in the world to the effects of global warming. Like many developing countries Tanzania is attempting to develop a two-pronged strategy in relation to climate change. On the one hand it is seeking to mitigate its own contribution to global warming by reducing its reliance on fossil fuels and by developing renewable and lower carbon energy sources. On the other hand it is planning for significant adaptation to climate change including efforts to limit desertification through new water management strategies, and proposals to develop new crops which can be grown with less water than the staple crop of maize, as well as researching more drought-resistant strains of food crops. That Tanzania is already thinking about both mitigation and adaptation strategies in relation to global warming is indicative of the extent to which the effects of global warming fall unequally between rich and poor countries. The peoples who inhabit tropical regions have for some decades already been witnessing the life-changing and now life-threatening effects of global warming, while the industrialised nations are only just realising that global warming will affect them as well, though not in life-threatening ways, or at least not in the next fifty years. Furthermore for the industrialised nations the costs of adaptation to climate change – improved flood defences, changes in crop husbandry, new building techniques and regulations – will represent a relatively small proportion of their income. But for low-income countries such as Tanzania adaptation to climate change will take up a rising proportion of national income, while for many individuals and communities the costs will be so high as to force them to abandon their crops and homes or die of hunger.

While many nonhuman species on highland slopes in Africa have nowhere else to go when these slopes lose their forests, humans have long been capable of extensive migration by land and sea, and human migration from Africa to Europe is already very significant. Every year tens of thousands of Africans risk their lives, and many die, in precarious fishing barks and fragile rafts in their efforts to cross the Atlantic Ocean to the Spanish Canary Islands, or on the passage from the coasts of North Africa to the Spanish and Italian mainlands and the Mediterranean islands of Sardinia and Sicily. This present flow of migrants, many of whom are summarily shipped back to Africa by wealthy European countries who resist international migration, is nothing compared to the tide of people who will leave Africa if global warming continues unabated. Tens of millions will be forced to flee their homelands in a long march north to the only remaining inhabitable lands in the hemisphere, which are presently the abode of wealthy Northern Europeans who are themselves the principal cause of global warming in Africa. Just one British power station, Drax B in Yorkshire, emits more CO_2 into the atmosphere than the combined carbon emissions of Kenya, Malawi, Mozambique, Tanzania, Uganda and Zambia.[8] All of these countries have suffered drought and in some cases near famine conditions in the last ten years. The latest IPCC emission figures indicate that the whole continent of Africa puts out one sixteenth of the CO_2 annually that is emitted by the United States. The inequity is far more serious even than the inequity in basic incomes between Africans and industrial citizens. As Grace Akumu of Climate Network Africa puts it, 'Africa's hopes and aspirations are being dashed by the blind pursuit of economic development in the industrialised countries.'[9]

Inequity in Emissions and Effects

In many parts of Africa, Asia and Latin America annual per person emissions of carbon dioxide average at just 0.2 of a tonne, compared to between 12 and 20 tonnes in industrialised countries.[10] This is in part because more than one and a half billion people in developing countries live off grid and so have no access to electricity. Instead they rely on wood and charcoal for cooking and heating. Smoke and interior pollution from fires causes cardiovascular and respiratory disease in millions of African households.[11] And dependence on timber for cooking and heating also places an increasing strain on the land.[12] Trees prevent soil erosion and their root systems help to retain water in the soil. As forests are thinned for fuel, soils thin as well, and the water table drops, so making conditions even harder for food growing and reducing the availability of potable water. As the global Millennium Ecosystems Assessment Report indicates, extreme poverty puts extra pressure on 'ecosystem services' and leads to a downward spiral which damages the capacity of ecosystems to deliver services.[13]

Despite the extreme poverty of tens of millions of Africans,

European and North American corporations continue to extract surplus value from African lands through agribusiness which uses large areas of prime agricultural land and water sources to rear exotic vegetables and flowers, and even 'organic' vegetables, which are then air-flown to Western supermarkets. Whereas Africans still experience shortened lives and the diseases of poverty, in part because of the continuing exploitation of their lands for the benefit of others, rich industrial citizens who mostly live in the Northern hemisphere enjoy increasing longevity with the aid of their fossil-fuelled lifestyles. North Americans produce a claimed average of 19.7 tonnes per head of population a year. Although the United States has only 4 per cent of the world's population, it produces 18 per cent of world greenhouse gas emissions from its internal activities. Britons, with a claimed national output of 542 million tonnes of carbon dioxide per annum, produce an average of 9.2 tonnes per head, which is close to the European average.[14] Europe and America together claim responsibility for 30 per cent of present world CO_2 output, while representing only 10 per cent of world population. But the way in which these internal carbon budgets are estimated masks the true extent of their global warming contribution. Europe and America between them are responsible for more than 90 per cent of the historic emissions which are already in the atmosphere.[15]

The figures are misleading in another way since they are based only on greenhouse gases produced inside the industrialised countries. In a global economy large parts of the economies of these countries, and hence their climate and ecological footprints, are located overseas in the form of farmlands, factories, oil wells, mines, plantations and quarries. In the case of Britain, if all the external emissions of the banks and businesses quoted on the London Stock Exchange are factored in the real carbon footprint of the United Kingdom is not 2 per cent, as the government claims, but 15 per cent of global greenhouse gas emissions.[16] For the United States the same calculation would indicate that its consumers, corporations and bankers are responsible for more than half of all global greenhouse gas emissions. It is commonly argued in the United States that because India and China are fast increasing their domestic greenhouse gas outputs, no international agreement on limiting greenhouse gases can be agreed between industrialised nations which does not also require India and China to reduce their emissions. But the great majority of the greenhouse gases already present in the earth system were emitted by the older industrialised nations and this is why there is a strong moral argument that they should act first. The global injustice of climate change will remain unresolved until the rich industrialised nations both radically reduce and curtail their emissions and compensate other parts of the world financially, and so provide resources to enable people to adapt to their globally-warmed environments.

Preventing Dangerous Climate Change

The IPCC is charged with advising the UN and member nations on how to prevent 'dangerous interference with the climate system'. The IPCC accepts the view of a predecessor UN body that a temperature rise of 1°C is likely to cause serious damage to ecosystems and that a 2°C rise represents an upper limit beyond which such damage is likely to be grave and to increase rapidly. The European Union has taken the same view in adopting the policy position that emissions should be limited in order that planetary heating should not exceed 2°C. This target can also be described in terms of CO_2 in the atmosphere; presently there are 380ppm of CO_2 and 430ppm of CO_2e. The threshold of atmospheric CO_2 which it is thought likely equates to 2°C of planetary heating is around 450ppm of CO_2e, a threshold the planet will reach by 2015 on the present emissions trajectory, though some estimates suggest that this threshold has already been passed.[17] To prevent the planet warming beyond 2°C requires that greenhouse gas emissions are reduced to the point where the planetary carbon sinks can again absorb them, as they did before the industrial era. This is the strategy known as global warming *mitigation*. The other response to global warming, as we have seen, is that of adaptation, and adaptation is already a feature of the planning and management of climate change by local and city authorities as they seek to deal with increased storm and flood events and raised temperatures.

Global greenhouse gas emissions are still rising. Efforts at decarbonising economic growth have increased economic output relative to greenhouse gas emissions but have not staved off the continuing global growth in emissions. On present trends, which include efforts at more sustainable development, annual emissions will increase by between 10 and 37 billion tonnes between now and 2030. The IPCC's *Fourth Assessment Report*, published in Bangkok in May 2007, indicates that strenuous reductions are needed by around 2015 if the earth system is not to be driven into an irreversible warming trend, and by 2050 emissions would need to be around 70 per cent lower than they are now.[18] The below-2°C emissions scenario requires developed countries to reduce emissions rapidly by 80 per cent, which will mean an emission target of around 2 tonnes of CO_2 per person. Some suggest that the political and economic cost of this kind of radical change is beyond the capability of politicians to deliver. Lord Stern, in a report for the British government, suggests that the United Kingdom should commit to a medium-term goal of energy cuts by 2050 which will result in an atmospheric concentration of 550ppm of CO_2 equivalents, and the certainty of average temperature changes above 2°C.[19] However, the IPCC believe that there are many positive economic outcomes from radical emissions reductions and that if the 2°C target is pursued vigorously and emissions peak by 2012 this could be achieved at the cost of a reduction of only 3 per cent of global Gross Domestic Product (GDP).[20]

Against those who suggest that it is feasible to delay action to reduce emissions, some believe that even the existing quantity of emissions now present in the atmosphere and oceans threaten fundamental changes in planetary systems, given that these emissions have yet to translate into climatic effects because of the inertia in the climate system. Ice core records show that the present Holocene era of a relatively stable and warm climate is unique in planetary history. This climatic stability coincides with the rise of human civilisations. Climate stability, and the related development of agriculture, are also central to the growth in human numbers which is in part responsible for *Homo sapiens* becoming the most influential creature in evolutionary history. The present number of humans, and the extent of industrial civilisation, sees humans using 40 per cent of the total biomass of the planet. This greatly reduces the chances that humans can simply adapt and migrate, as the very small numbers of their pre-historic forbears did in response to prehistoric climate change.[21] With over 6 billion at present, and a possible maximum of 9 billion by mid-century, it will be hard indeed for all these people to find water, grow food, and keep roofs over their heads in the context of the increased extreme weather events which global warming will generate.

The precariousness of the human situation in the face of global warming is exacerbated by the length of many industrial supply chains. The global nature of these chains requires large greenhouse gas emissions to sustain them. But at the same time the more distant the places from which populations find their food, fuel and fibre, the more climate change events put this novel form of civilisation at risk. And yet so dedicated are Western governments and international organisations such as the World Trade Organization (WTO) to the neoliberal idea that global free trade will automatically lift everyone out of poverty and advance human welfare, that the world remains set on a course where more and more goods and services will be bought on one continent and sold on another. This is the reason why since 1990 the principal driver of increased emissions has been the transport sector, and why it is projected still to be the principal driver for decades to come. The global and local movement of goods and people continues to rise as governments remain committed to mobilising the factors of production across geographical regions and state borders.[22]

Negotiating International Emissions Reductions

The long delays in implementing genuine emission reduction strategies by industrialised nations reflect tremendous political pressure exerted on international efforts to establish a legal framework for global reductions in greenhouse gas emissions reductions. At the Rio Earth Summit in 1992 a process was established which led to the UNFCCC being signed into international law in 1994. The UNFCCC commits member nations to a process which is designed 'to achieve stabilization of greenhouse gas concentrations in the atmosphere at a low enough level to prevent dangerous anthropogenic interference with the climate system'. The meetings of representatives in the UNFCCC are known as Conferences of the Parties (COPs), and they are divided into two groups, Annexes I and II (developed and developing countries respectively). At the third meeting of the COP in Kyoto in December 1997 a Protocol was agreed which formally committed developed nations to reducing their greenhouse gas emissions by an average of 7 per cent of 1990 levels by 2012. The European Union went into the Kyoto process seeking a goal of 15 per cent reductions on 1990 emissions levels by 2012, but under pressure from North America, Australia and Saudi Arabia this original goal was reduced to the much weaker 7 per cent target, and even that was watered down by carbon trading procedures.

The *realpolitik* of climate change negotiations reflects the influence of global power relations between the United States, the other industrialised nations and the developing world. The biggest greenhouse gas producer, the United States, is a signatory of the UNFCCC and is hence in principle committed to global mitigation of global warming. However, the Bush/Cheney administration has refused to ratify the Kyoto Protocol and in 2004 domestic US greenhouse gas emissions rose by 1.7 per cent to 7.1 billion tonnes.[23] The story in the European Union is only a little better. CO_2 emissions in member nations are said to be on track to fall to 2.5 per cent below 1990 levels by 2010.[24] But this is still below the agreed reduction target under the Kyoto Protocol and indicates that Europe's primary commitment to expanding trade within and beyond Europe is incompatible with its laudatory attempts to lead the international response to climate change.

While citizen awareness of the problem of global warming is growing in Europe, and a growing number of people are aware of their 'carbon footprint' and are beginning to explore ways of reducing it, such as driving or flying less or using low-energy light bulbs in their homes, on present trends European countries are not on target to reduce their emissions to the intended average reduction of 7 per cent on 1990 levels by 2012. The British government has adopted among the most ambitious targets of any European country, committing itself to a 60 per cent reduction by 2050. But this would not meet the reduction strategy needed to keep the earth from warming beyond 2°C, as indicated in the latest IPCC report. Despite publishing a Climate Change Bill in 2007 which commits future governments to publishing biannual greenhouse gas budgets, the British government continues to resist legal caps on emissions other than those in the cap and trade scheme inaugurated by the European Union, of which more below. Instead it is still commissioning new airports and roads to facilitate more movement of products and people.[25] It also resists including the overseas emissions of British corporations from air transport, shipping, and manufacturing in the British greenhouse gas inventory.

In the United States there is even more resistance to real reductions in greenhouse gas emissions. The majority of Americans still do not believe that fossil-fuel burning is driving global warming.[26] And Congress and the Bush administration have refused to commit even to the minimal greenhouse gas emission reductions mandated by the Kyoto Protocol. Here too though, there are signs of a hopeful shift. In 2006 the megastore chain Walmart announced an intention to begin to reduce its energy consumption and the ecological impacts of its operations. And many city authorities and state governments have begun to explore ways of reducing energy consumption, even although the Bush/Cheney administration has forbidden the Environmental Protection Agency from assisting in such efforts.[27] But even the Bush/Cheney administration publicly acknowledged in 2007 that global warming is happening and that it may be linked to fossil-fuel use. Nonetheless Bush and Cheney remain committed to strong economic growth, and to continuing growth in domestic fossil-fuel use, and refuse to embrace legally binding greenhouse gas emission reduction targets of the kind mandated by the Kyoto Protocol.[28]

Since 1970 total greenhouse gas emissions have increased by 50 per cent. The World Energy Council's 2004 World Energy Assessment suggests that the likely scenario is that fossil-fuel use will continue to grow at around 1.7 per cent per year for the next quarter century and fossil fuels will still make up the bulk of the world's energy supply in 2030.[29] The IPCC's 'business as usual' scenario indicates that if emissions continue on the present trajectory there will be growth of 62 per cent in greenhouse gas emissions by 2030 over 2002 levels. Two thirds of the growth in demand for energy in this period will come from the developing world. If these rates of fossil fuel use continue throughout the twenty-first century, by 2100 there will be at least 650 ppm of CO_2 in the atmosphere, and some estimates suggest even 1,000 ppm is not impossible. At the higher level of 1,000 ppm a runaway climate event becomes highly likely, leading to extreme warming, the consequences of which would be so catastrophic as to render most of the earth's surface outside of the polar regions uninhabitable, a scenario to which Gaian scientist James Lovelock believes industrial humans have already committed the planet.[30]

The Immorality of Global Warming

The predicted extent of climate change that industrial civilisation is in the course of provoking is not only unprecedented in human history but it is also a novel moral problem. In the 15,000-year history of civilisation humans have altered *local* climates by such actions as deforestation, over-tilling, poor irrigation and urban sprawl. But never before have they been responsible for planetary climate change. Equally the range of causation is more widespread, and reaches more fully into the everyday lives of individuals than any previous moral dilemma in the growth of industrial civilisation in the past three hundred years, including slavery. Most people in the British Empire did not own slaves, even though an important proportion of the wealth of empire was built on profits from slave labour plantations. But every individual who has driven a car or flown in an aeroplane, lived in an energy-hungry modern house, bought clothes or computers made 10,000 miles away, or bought shares in a large corporation, is fractionally involved in anthropogenic global warming.

The strongest moral case for mitigating global warming is that it is already life-threatening to those who are least able to defend themselves, and have no responsibility for its causation. Although they suffer the worst consequences of climate change, and they will be responsible for two-thirds of the projected *increase* in production of greenhouse gases in the present century, the inhabitants of the Southern hemisphere, whose growing numbers are often decried by environmentalists and scientists, are not those who place the greatest burden on the planet's atmosphere, and its other life-support systems. Instead it is the far smaller numbers of Europeans and Americans who produce carbon dioxide and other greenhouse gases at rates so much higher than their Southern planetary neighbours.[31] As Henry Shue argues, the moral problem is not emissions in general: all humans require moderate levels of greenhouse gas consumption to grow and cook food, keep warm, build shelters and make festival. These are livelihood emissions and typically amount to around 0.2 of a tonne of carbon per person per year.[32] These livelihood emissions of people in less developed countries contrast with the 'luxury emissions' of the rich, who live in almost every nation, though in smaller numbers in the Southern hemisphere, to sustain their unparalleled levels of ownership and consumption of everything from property and cars to foreign holidays and entertainment devices.[33]

The Christian tradition has always been critical of luxury. Christ frequently warns the rich against the spiritual dangers of wealth. Perhaps the sharpest of all his parables on the subject is the story of Dives and Lazarus (Luke 16:19–31). Dives (his name is the Latin word for a rich man) dresses in purple, lives in luxury and his table is laden with fine foods. At his gate is a beggar, Lazarus, who is covered in skin lesions which indicate the disease of leprosy, and he longs for leftover food and drink from Dives' table. Whereas on earth Dives lived in luxury while Lazarus lived in extreme poverty, in the afterlife the parable describes their situations as reversed. Dives languishes in the fires of Hades and longs for Lazarus, who is in paradise, to dip his finger in water and relieve his thirst. The implication is that the rich may deny the enjoyment of their wealth carries moral obligations to the poor, and that the accumulation of wealth may even have been gained at the expense of the poor. This is why in the Middle Ages Thomas Aquinas describes excess wealth accumulation as theft in his reflection on the eighth commandment, 'thou shalt not steal'.[34] He suggests that if a rich man takes possession of something which was common property for his own use, and in such a way as to exclude others from using it, he steals common goods from others and therefore sins. And Aquinas quotes Ambrose with approval when he says that 'he who spends too much is a robber.'[35] The basis for this claim is that God has sovereign will over all things and confers on all persons a right of use of them for bodily sustenance. Where wealth accrues to some such that this sustenance is denied to others then this is theft.

On this account the purloining of the commons of the atmosphere as a sink for the excessive emissions of the rich is a form of theft because it directly impedes the bodily sustenance of those in drought or flood-prone regions. That the poor are the rightful claimants to excess wealth makes the injustice of global warming all the more immoral, given that the cause of the problem is not the subsistence emissions of the poor but the luxury emissions of the wealthy. There is also evidence that the richer individuals are, the more carbon they tend to consume. Even within Western societies there are considerable differences between the carbon emissions of the wealthy and of the relatively poor.[36] Luxury emissions represent moral malfeasance by the rich against the poor both nationally and internationally, since they give rise both to local and global forms of pollution and other kinds of moral harms. As Shue suggests, beyond a certain level of emissions the wealthy should properly be held morally responsible for the damaging effects of their emissions on the poor.[37] Once they know of the damage their emissions are doing, the rich have no excuse for not reining in their emissions, and for not compensating the poor for the damage their emissions are doing.

It is of course also important to recognise that wealthy individuals, and industrial consumers more generally, do not act alone. Nation-states and corporations are together responsible for most of the energy infrastructure, marketing and regulation with which individuals interact in their uses of energy. Some developed countries – most notably the United States, Australia and the United Arab Emirates – have exceptionally high energy use levels, while others at an equivalent level of economic activity – Sweden and Switzerland for example – use energy much more carefully. And as we have seen, British corporations in their overseas operations are responsible for greenhouse gas emissions at least three times those of all domestic householders in the United Kingdom, and though data has not yet been collated there is no reason to believe a similar ratio does not exist in other countries.[38] A number of individual multinational corporations produce greenhouse gas emissions greater than many nation-states. Corporations also seek every opportunity to minimise their costs and to locate their polluting activities where regulations are less onerous. If the lion's share of the greenhouse gas emissions of corporations with headquarters in Britain and the United States is produced on other continents then caps on greenhouse gas emissions would therefore not resolve the tendency of corporations to reap where they do not sow. This means that a legal and regulatory framework designed to physically limit and reduce greenhouse gas emissions cannot be exclusively based on national accounts. It will also need to involve procedures which apply to the inter-state activities of organisations such as multinational corporations, and to the mobile super-rich who often reside in more than one country.

Climate change is not unlike the set of moral problems presented by the growth in the last one hundred and fifty years of a global economy characterised by global flows of goods, people, and resources within and between nations. As these flows make nations increasingly economically interdependent, they produce structural injustices and forms of economic coercion resulting in extreme poverty and even famine in some regions, which require amelioration between nations. And yet there is no adequate political mechanism for such amelioration. Nations and corporations mostly act in their own economic interests in the international arena and commit few of their resources to the amelioration of the ills their economic activities may visit on peoples from other nations. The scale of economic aid between the richest and poorest nations of the earth is still only a fraction of 1 per cent of the GDP of the richest nations, and most of this aid is organised in such a way as to generate business opportunities in donor nations. Far outweighing foreign aid are the net transfers of wealth from developing countries to developed countries that the neoliberal policies imposed on developing countries by the International Monetary Fund, the World Bank and the WTO have produced in the last twenty years. Monetary estimates of the value of these transfers show they have been rising dramatically under the present neoliberal global trade regime. In 2001 they stood at $155 billion and in 2002 at $193 billion.[39] Despite the issue being highlighted by the UN Secretary General, and at various intergovernmental summits, by 2006 net transfers from developing to developed nations had grown to an annual total of $650 billion.[40] By contrast net foreign aid in 2002 was only $57 billion.

These existing transfers of wealth from poor to rich are exacerbated by greenhouse gas emissions. Through the medium of the climate, nations, corporations and wealthy individuals are effectively physically harming and even *invading* other nations with their greenhouse gases; this produces a situation which is analogous to a state of war between the territories and agents concerned. The coercive and unequal nature of climate change is reflected in the original UNFCCC, which emphasises the inequity of the problem and therefore proposes that the developed countries 'should take the lead in combating climate change and the adverse effects thereof'. As Michael Grubb observes, the nature of the inequity between nations in causes and effects of global warming raises unprecedented questions about 'the nature and ethical basis of international and economic relationships'.[41]

Extinction and Climate Change

International relationships between human beings are not the only form of grave injustice arising from anthropogenic global warming. Industrial development is already provoking what biologists are calling the sixth great extinction, where one species – *Homo industrialis* – has become responsible for extinguishing many hundreds of others, and hence for intervening dramatically in the evolution of species.[42] Climate change threatens an even more catastrophic rate of species extinction because of the speed at which global warming is altering local climates. In evolutionary history species were able to adapt to major climate change because they happened more slowly than the present humanly-caused one. The prehistoric climate event in which the most number of species were extinguished, including the dinosaurs, was the one that occurred most rapidly; a massive asteroid hit the earth causing an explosion so large that particulates from it in the atmosphere cooled the planet and triggered a sudden ice age. In the present climate change event, while more mobile species including some amphibians, birds, insects and mammals will be able to migrate to cooler climate zones in response to local warming, others, particularly trees and shrubs, but also mammals such as polar bears and birds such as puffins or penguins, will be unable to adapt, and their numbers are already being affected.[43] At a warming of 2°C it is estimated that between 15 and 37 per cent of species will become extinct.[44] At a warming above 2°C an even greater extinction is likely, in the range of 37–52 per cent.[45] Extinctions to the upper end of these numbers depends on what happens to species-rich tropical forests in a warmer world. If, as scientists predict, raised ocean temperatures reduce precipitation over the tropical forests this will likely cause a die-back which will extinguish the most species-rich habitat on the planet.

The moral problem of species extinction has been extensively discussed by scientists in recent years and is often presented in a human-interest perspective. According to this approach, humans are the most conscious species and therefore the most capable of experiencing pleasure or happiness and so when the interests of humans are set against the interests of other animals, human interests invariably over-ride those of other animals. But even human interests are not well served by too much extinction. Variety in plant life can help humans respond to changing climatic conditions. Some wild species of rice or maize may be more drought-tolerant, others more resistant to certain pests. If all wild varieties are extinguished then this genetic bank of variability is not available for present or future humans. Similarly, variability in plants provides a major resource for pharmacological treatment of human disease. Extinctions can also be seen as posing foundational problems to the ecosystems on which humans as well as other life forms depend. For example if human fossil fuel-burning continues to burden the atmosphere and oceans with an excess of CO_2, then the oceans will become so acidic that calcification in shellfish becomes impossible, thus dramatically altering the ocean food chain and reducing further oceanic effectiveness as a carbon sink.[46]

The human-interest argument for species conservation is encapsulated in Paul Ehrlich's famous rivet-popping metaphor, in which the earth is represented as a spaceship and the earth's species are represented as the rivets which hold the spaceship together.[47] The more rivets that come off, the more likely the spaceship is to fall apart and to endanger the lives of those who navigate and pilot it. Climate change threatens to bring this endangerment about more rapidly than even Ehrlich imagined.

The interest argument for species conservation has considerable limitations, not least because it implies that value in human life and in nature is not intrinsic, but a human cultural construct. In an interest perspective value is said to reside only in the mind of the one species that is capable of rationally valuing itself, and hence of imputing value to other beings. Such a location of value implies that the material world does not itself already present a source of value to the person who encounters it, as classical and Christian philosophers have traditionally maintained.[48] It reflects instead the modern attempt to locate moral values exclusively in individual expressions of preference and interest, an attempt that turns ethics into a calculus of consequences. From a Christian perspective, the value of nonhuman species arises from their having been made by the divine

Creator who made them in their myriad diversity as a reflection of the divine nature.

The Global Household and the Illiberal Economy

The collective nature of the modern economy and the reach of its polluting impacts from the private household to the global commons of ocean and atmosphere indicate the peculiar nature of modern civilisation. In premodern societies households were responsible for meeting the necessities of life through the tilling of the soil. Almost everything that was consumed or used in the household – building materials, clothes, drinking water, food, fuel, furniture, lights, pots, pans and utensils – was created from local resources. Consumption and production were located in the same place, and so the human economy was governed by the local needs of households and in relation to the carrying capacities of their local ecosystems.

In ancient societies public life often involved agonistic conflict between princes and burghers. But these conflicts only spilled over into the household in times of great wars or when tyrannical rulers over-reached the ancient limits of the political and disturbed local patterns of growing, making and provisioning. The roots of the modern age may be traced to one such occasion when Henry VIII took to raiding the monasteries of his kingdom to fund his foreign wars and the luxurious lifestyle of his imperious court. The dissolution of the monasteries was the beginning of a two-hundred-year period in which the village economy of England was gradually broken up and peasants lost rights to land and the use of ancient commons to the growing demands of an aristocratic, and then an industrial economy. Instead of food and fibre for peasant smallholders, land was increasingly devoted to sheep, to the creation of large surpluses of cereals and other crops to fund large estates and luxurious aristocratic lifestyles, and to providing surplus cash for investment in the new mechanised textile mills which were the beginning of the industrial revolution.[49] Meanwhile, a mobile army of labourers was created as newly disenfranchised and homeless labourers were forced to take up employment as wage slaves in the industrial factories or on increasingly mechanised farms.[50]

The division of labour is the collectivising device which more than any other marks the industrial from previous eras. Its rise explains how the majority of humans are no longer involved in the growing or making of the necessities of daily life. As a result the private household has lost many of its traditional functions to external economic and political agencies and is turned into a site of consumption but not of making. The nation-state and economic corporations form the social medium in which the material necessities of life are produced and distributed. As Hannah Arendt suggests, the social form of industrial civilisation is one in which all households are gathered together into a great household and genuine private property and self-sufficiency disappear. In this new 'social as household' individuals become job-holders, their lives are governed collectively through the new science of economics and their estates are mortgaged to banks.[51] This new social structure, and its government by statistics, is the device by which the family and the private household are drawn into the distinctively modern form of government as bureaucracy which is the 'rule of no-man'. The social as household erodes private property and production in self-sufficient households. It also erodes the distinctively political sphere of public life in which individuals compete in the display of excellence in art, rhetoric, sport and virtue beyond the sphere of the household. Instead politics becomes the 'art of necessity' in which management of economic resources becomes the primary business of the political as economic management subverts the truly political.[52]

As Arendt suggests, the modern economy manifests an irresistible tendency to grow in its influence so that it eventually corrodes both the sphere of the truly private – the self-sufficient household – and the sphere of the truly public – the arts, education, politics and play. The result is the emergence of a distinctively modern form of tyranny. The 'rule of no-man' produces a new kind of collective coercion in which both private property and politics are subjected to the material necessities of collectively governed making. *All* of human life – and not just household management – is subjected to the necessities of labouring to provide the necessities for living. And this devotion of all to the survival of all produces a new and tragic condition in which the extinction of the human species seems now to be threatened in the future by ecological collapse and runaway global warming as every ecosystem, and the earth system itself, is drawn into the increasingly global form of this 'economy as household'.[53]

The invention of the social as household, and its management by economic science, involves a new form of political economy in which all human life becomes subject to the imperative of the collective accumulation and distribution of surplus value. The governor of this new form of society is identified by Adam Smith as the 'invisible hand'. And the driving force of the invisible hand is individual selfishness. The classical economist assumes that individuals and firms are driven by self-interest above all else and that they therefore only engage in exchange activity in order to further their own interests. Smith shares with Thomas Hobbes the novel assumption that individuals and households only act together in order to mediate their original selfishness in ways which produce a larger social realm governed by contract. In this way both promote a unique form of modern individualism in which the self-sufficient household and the traditional realms of political and public life are submerged in a coercive economic and social contract which binds individuals *as* individuals to something called society. And this new coercive social contract is increasingly represented as best governed not by law or virtue but by economics and statistics. Through the medium of statistics, production and consumption activity, having once been alienated, are reconnected into the new amoral realm of the national, and later the modern imperial, economy as greater household.[54]

This situation produces a strange bifurcation in the life of the individual in which meaning and moral purpose are increasingly reserved for the realms of intimacy and family, while a new collective moral calculus governs individuals in their relations in the economy as household.[55] The name of this calculus is utilitarianism. The novel claim of utilitarian philosophers such as John Stuart Mill is that it is possible to advance collective welfare in the new mass society by estimating the benefits and costs of particular courses of collective action, and choosing those actions which are calculated to advance the greater good. It is precisely this instrumental approach to value that we see advanced in relation to the extinction of species. But the inability to affirm any measure of the human and more than human good other than a hedonist calculus of welfare reduces all human activity to an agonistic calculation of costs and benefits, and an agonistic competition for natural resources and space whose ultimate conclusion is the present global contest over the atmospheric sinks of the planet.

It is under this calculus that governments and corporations continue to advance the politics of speed over the prior claims of private property and the beauty and fecundity of the earth on which the household traditionally relied for its preservation and self-sufficiency. Thus when a decision is made to locate a motorway near a group of existing houses, or through a beautiful landscape, despite the detrimental effect on householders and species, it is justified on the basis of a utilitarian calculus; decreased journey times are said to represent an efficiency gain which benefits a larger group than those who are deleteriously affected by traffic noise and air pollution. Instrumental economic reasoning justifies such behaviour as rational because it promotes 'efficiency' and hence is said to promote the interests of all rather than the partial and particular goods of certain individuals, even when the welfare of the planet is at increasing risk because of this preference for mobility over place.[56] This is why the distinctive modern form of the social as household is so much in contest with the health of the planet. The alienation of making from consuming and the subjection of each individual and particular households to the national and increasingly global household economy require that the connections are constantly remade by the promotion of speed of movement. The costs of this mobility on the welfare of particular places and peoples and species are discounted against the statistical calculus of the efficient promotion of the greater good.

The modern division of consumption from production produces a situation in which economic management and contractual relationships become the over-riding means of social interaction. The division of labour and the mechanisation of production and trade are promoted in the name of reducing human subservience to the necessities of tilling the soil and of advancing scientific control over nature. And yet they advance a social condition in which households increasingly lose their own property rights, and every individual is increasingly dependent at every moment in their lives on the necessity of their connection with the labour market, and with an economy in which natural wealth is transformed into stored money values. The values that govern this collective economy are not the preservation of life and property but their continual destruction and 'efficient' transformation into monetary wealth.

Philosophical attempts to rescue the realm of ethics from this collectivisation of social life take two characteristic forms. For David Hume, the truly moral is pre-social and is contained in the capacity of each individual for feeling, whereas for Immanuel Kant morality is located in the capacity of each individual for rational thought. But moral values beyond the realm of private feeling and thought become increasingly hard to identify or articulate. And this is why the moral climate fostered by modern political economy is peculiarly maladjusted for responding to natural signs of the kind presented to modern societies by the increasingly strange physical manifestations of global warming.

Utilitarian (Mill), emotivist (Hume) and deontological (Kant) forms of modern moral reasoning refuse the perennial and traditional claim that there is a deep moral structure in the biophysical ordering of life on earth precisely because they all collude in the invention of society *as* household, and of mass society as a collection of individuals.[57] As Arendt points out, the modern description of morality as a realm of values indicates just how strongly the economic paradigm has reshaped the moral and social world. What philosophers now call moral values are represented as human constructs, or distinctive forms of language analogous to exchange values in the contracts which increasingly govern all human and human–nature interactions. On this account the good, the true and the beautiful are no longer seen as embedded in the intrinsic nature of creatures, recognisable *through* their relations with other creatures and with the divine Creator. Instead capitalism turns all that human beings count worthy into exchange values. And hence modern societies suffer a foundational loss of intrinsic worth; 'nothing any longer ever possesses an "objective" value independent of the ever-changing estimations of supply and demand' which are 'inherent in the very concept of value itself'.[58] And just as all forms of goodness, truth and beauty are de-objectified and rendered exchangeable by the 'economy as household', all forms of human making, including even the family and human natality, along with all forms of species being, are drawn into the process of the accumulation of monetary values, and increasingly subjected to the governance of the mechanistic procedures of market economics.

Cosmos as Mechanism

The economic governance of the society as household through the construction of mechanistic market procedures – and especially the laws of supply and demand – finds significant analogy with the new construction of nature as mechanism in early modern science. Isaac Newton represented the attempt to describe fixed and universal laws of cause and effect, such as those that govern time and motion, space and matter, as an attempt to reveal the mind of God, who had laid out the universe according to these eternal rules. The use of experimental science to validate such rules is what underlies the ability of modern science to now predict the sensitivity of the climate system to human use of fossil fuels. But this approach soon leads to the adoption of a cosmological model of the earth, and the universe, as a secular autonomous mechanism governed by laws that operate independently of divine or human volition.[59] The resultant disenchanted description of the universe as a cause–effect mechanism is implicated in the inability of modern people to imagine that their actions collectively have an effect on the earth system. It is also implicated in the increasing relations of domination between humans and the earth which, as Keith Thomas shows, come to characterise discourse about human–species relations in early modern Europe.[60]

Premodern Europeans regarded creation as a nurturing mother and this attitude was not only expressed in art and literature but also in a sense of constraints on what humans could do to 'mother earth'. Each household learned of these constraints through its tending of ancestral land and its efforts in conserving soil and managing animals, trees and water sources so as to preserve life rather than destroy it, while still providing for the household. By contrast, the modern economy becomes increasingly reliant on forms of production which involve tearing up the fabric of the earth, mining substances like coal from beneath the earth's surface, clear-cutting forests or 'harvesting' oceans. As Carolyn Merchant suggests, the demise of the conception of earth as mother and the rise of the idea of earth as an impersonal mechanism fosters a new attitude of domination towards the non-human world in relation to other species and natural systems. This also reduces the sense of limits on human interference with the earth that the older more feminine cosmology involved.[61] The new mechanistic conception of the cosmos also desensitised those who took it up to cruelty to animals and to destruction of ecological systems. Hence it was imaginable for early modern philosophers such as René Descartes to think that animals were simply machines – since they lacked rational minds – and did not feel pain. And it becomes imaginable for modern economists to resist the idea that wealth accumulation has biophysical or moral limits while embracing the idea that it can be governed mechanistically by the invisible hand of the market.

The early modern mechanistic cosmology also influences human discourse about natural disasters and disturbances in the normal rhythm of the seasons. The ancients viewed such events as indicators of divine anger at human behaviour. But the cosmology of mechanism trains modern people to read the earth as governed by arbitrary mechanical laws. Human greed can no more be punished by the earth than the Danish and English King Canute, in the famous story, could prevent the tide advancing by his own will; and hence the radical difference in the interpretation of natural signs between the modern household *as* economy and the premodern household. This also helps explain a key conundrum in the ethics of global warming, which is the assumption among natural scientists that provided the problem of climate change can be described rationally and with the minimum of uncertainty – and the science of climate change has moved in this direction – then corrective actions by individuals, corporations and nation-states will follow.[62] That this has not happened is indicative that the assumption is false. The cosmological story of mechanism, and the mechanistic social practices it seems to legitimate, train modern people to refuse the moral or spiritual significance of natural signs.

And there is a related problem here. Economic corporations and the consumers whom they hold in their thrall continue to argue that so long as there is *uncertainty* about the meaning and interpretation of the climatic footprint of global carbon then there is no need to stem the tide of emissions.[63] But the quest for complete certainty, with which science to an extent colludes, involves an unreasonable standard of proof – *unreasonable* because it obviates the need for reasoned debate about probabilities and particular cases in moral deliberation. The quest for certainty takes the characteristically modern form of dogmatism which, as John Dewey argued in his 1929 Gifford Lectures, is the ironic outcome of the influence of the cosmology of mechanism on the world of ideas.[64]

The turn to certainty has been tremendously influential in the culture of modernity because it privileges proof over argument, logic over rhetoric, reasons over causes, and universals over particulars.[65] According to Aristotle the Good has no universal form that can be specified apart from the particular action and agent. Each agent aims at a good for it, but the only universal quality of this good is that its fulfilment leads to *eudaimonia* or flourishing for the agent which pursues those virtues which enable the realisation of its own particular good.[66] By contrast modern moral rationality reduces the good to reductionist rules which are said to operate independently of context, role, story and tradition. The utilitarian formula 'the greatest good for the greatest number' is just such a rule, collectively applied without regard to its effects on *particular* individuals, bodies, or ecosystems.

This move in Western philosophy and culture is characterised by Stephen Toulmin as the 'hidden agenda of modernity', which privileges formal logic over practical wisdom, and replaces prudential judgment on particular cases with abstract and universal rules.[67] This decontextualises and distances moral judgments from cosmological stories or historical narratives. The sixteenth-century metaphysical poet John Donne argued that the result of this new thought world was a loss of coherence as the worldview of the early modern age embraced atomism instead of unity:

(and) new Philosophy cals all in doubt.
The Element of fire is quite put out:
The Sunne is lost, and th' earth, and no mans wit
Can well direct him where to looke for it.
And freely men confesse that this world's spent,
When in the Planets, and the Firmament
They seeke so many new; they see that this
Is crumbled out againe to his Atomis.
'Tis all in peeces, all coherence gone;
All iust supply, and all Relation:
Prince, Subject, Father, Sonne, are things forgot,
(F)or euery man alone thinkes he hath got
To be a Phoenix, and that there can be
None of that kinde, of which he is, but he.[68]

It is intriguing for our purposes here how Donne specifically links atomistic physics and rationalist individualism with the inability to understand the cosmic significance of fire and sun, or to discern the sacred Trinity. What more powerful metaphor for the crisis of global warming could there be than that every individual imagines herself as her own unique phoenix, her imagined uniqueness displacing the gods from the cosmos so the individual becomes her own world of meaning and disregards the impacts of the inner world on the larger biophysical order? This kind of solipsism trains modern individuals to refuse any relationship between human morality and natural systems, and hence they find it hard to believe that there is a relationship between changes in the physical climate and individual and collective behaviours.

From John Donne to William Blake and John Ruskin there is a notable literary tradition of dissent at the ecological and moral consequences of cosmological mechanism, utilitarian economism and instrumental rationalism. Blake and Ruskin both viewed coal-fuelled industrialism as the illegitimate progeny of these earlier mistaken traditions, and in particular of their progenitor Newton.[69] And both criticised the loss of community and tradition that the cornucopia of industrial consumerism, the inhuman disciplines of industrial labour and the alienation of consuming from making brought in their train. Donne is an authentic conservative voice, and he stands with Milton in *Paradise Lost* and with Yeats in his poem 'The Second Coming' in lamenting a culture in which 'things fall apart' and 'mere anarchy is loosed upon the world'.[70]

For Donne and Yeats, as for the Gospel writers and the Hebrew prophets, a storm on a lake, the movements of a star, melting snow, are all figural exemplars of a larger unitive order which links *cosmos* and *polis*.[71] Christ's stilling of the storm on Lake Gennesaret is not for the Evangelists mere happenstance, any more than the star that guides the Magi to Bethlehem. The storm which threatens to engulf the disciples on the lake before Christ calms wind and wave symbolises the attempts of the imperial and religious authorities to resist the divine power and wisdom of the Son of God come to earth (Luke 8:22–4). The storm, like the star, is both cosmic and political.

The Relational Science of Gaia

The early modern ruptures of *cosmos* and *polis*, and of household and making, are called into question by the scientific description of global warming. Scientists who find a biophysical response to the modern industrial economy in the behaviour of the planet are involved in the recovery of a relational conception of physics and the biosphere. And this new cosmology recalls the claims of Jeremiah and Aristotle, Aquinas and Donne that household, politics and morality are inextricably linked with the beauty, order and purposefulness which the Creator has set into Creation. This recovery takes the form of James Lovelock's Gaia hypothesis which, like sacred cosmology, describes the world not as a mechanism but as a finely-balanced relational nexus of life forms.[72] In the Gaia hypothesis Lovelock was the first to propose what climate scientists now take for granted, which is that all of life is connected by the carbon cycle and involved in maintaining proportions of gases and chemicals in the earth's rocks, soils, waters and atmosphere which constitute a planet capable of sustaining life.

Biogeochemistry is the more formal scientific name given to Gaian science.[73] But it is a young discipline and its cultural influence remains weak. Gaian science trains those who understand it that there are moral and biophysical limits to what humans can do on planet earth, just as the ancients held. However, the cosmology of mechanism remains the reigning paradigm for human–nature relations, and nowhere does this paradigm hold more cultural power than in the pseudo-science of modern utilitarian economics, whose protagonists hold, after Adam Smith, that the choices of individual consumers or firms are intrinsically rational when they each act severally and in their own interests and without regard for the common good. The principle of rational choice takes flesh in the procedures of Cost Benefit Analysis, where exchange values override any other goods. The principle and the procedures between them sustain the illusory claim that human welfare is advanced by infinite growth in the money economy, because such growth is said to advance the possible range of choices through which individuals and firms may maximise their preferences. Trained by the quasi-scientific logic of these procedures in an increasingly global market place, consumers, corporations and civil servants find it hard to imagine that the biophysical world represents any kind of limit to preference expression. The only theoretical limits to such expressions are the enactments of contracts, and their enforcement by law. In practice, governments do regulate beyond the strict requirements of the defence of private property, but even these arrangements are conceived in the form of a social contract without any ecological or metaphysical roots. In liberal political theory since Hobbes and Smith, society is said to be constructed by contract rather than by the interactions of households governed by moral relationships of care, nurture and respect.

The modern moral climate is then a construct of three assumptions: that the human moral agent is an autonomous reasoning sovereign, that human economic exchange is a realm of contractual mediation between autonomous agents, and that the social contract is relationally independent of the cause–effect mechanism of the cosmos. All three elements of this conceptual architectonic manifest seriously inadequate descriptions of the cosmos, of the good life, and of human and species being. These arise from Newton's failure to describe correctly the relational character of space and time in its original dependence on the being of God as Creator, and the relational inter-connections, subsequently observed by Einstein, between the different dimensions of space and time; from Mill's failure to ascribe intrinsic value to creaturely being, and hence the failure of the utilitarian calculus to value individuals and households as well as society; and from the classical economists' failure to conceive of society as more than a contractual aggregate of sovereign individuals. These mis-descriptions produce and shape a moral climate in liberal societies which is increasingly demoralised and disenchanted, and which stands in urgent need of reformation if the planet is not to be committed to irreversible ecological breakdown.[74]

These connections between pre-Einsteinian physics and classical economics are not just ones of similitude. Philip Merowski shows how nineteenth-century economists modelled their mechanistic accounts of human exchange relations and the valuing procedures of the 'laws' of supply and demand on nineteenth-century mechanistic physics.[75] And under the influence of neoliberalism, this mechanistic modelling of human behaviour by economists has actually deepened. Neoliberals in the last forty years have sought to redescribe human behaviour in purely mathematical terms after Newtonian-style laws of object relations, time and motion, and also by analogy with binary computer codes and mathematical models of genetics. These mathematical models have acquired near-mystical power in British and American government circles, as well as among bankers and corporate managers. Consequently the United States and British governments have both sought to confer on markets powers over more and more areas of human life while shrinking political mediation in economic and social policy. Even government agencies have adopted mathematical mantras as markets are introduced into education, healthcare and, most recently, into climate-change mitigation with the inauguration of carbon trading as the principal social mechanism for responding to global warming. The effect of all this has been to produce what Walter Wriston calls the 'twilight of sovereignty', in which the powers of local human communities and even national governments to order their affairs according to shared deliberation on moral ends is given up to autonomous market instruments based on the movement of bits of mathematical information between computers.[76] But of course these instruments are not truly autonomous, but humanly made. They further exalt exchange values over intrinsic worth. And they exalt the corporate holders and accountants of money wealth as sovereigns, or even gods, in the neoliberal global economy.

The Climate Narratives of Noah and Joseph

Modern descriptions of the cosmos, the good and human and species being are blind to the enchanted relational nexus in which all life is lived, because they are ecologically as well as morally reductionist. And yet these inadequate descriptions have acquired considerable cultural power since descriptions, even erroneous ones, shape character. This is because humans and communities are 'story-formed'. By listening to stories humans learn of the kinds of behaviours that are to be admired and mirrored, and the kinds that are to be proscribed and avoided.[77] This is why reading scriptural stories is central to Jewish and Christian worship. In the public retelling of biblical narratives worshippers find resources for moral discernment and wisdom, and they learn about the moral character of God, and of God's creation. There are two foundational biblical stories where individuals act prudentially to stave off climatic disasters, these being the stories of Noah and Joseph. Both these stories are in the Book of Genesis, the Bible's most cosmological book. They offer a range of practices and tactics in respect of anthropogenic global warming.

In the first story God tells Noah that he intends to make it rain for months and to flood the earth because 'the wickedness of humankind on earth was great' (Genesis 6:5). The precise nature of this wickedness is not clear but its marks included corruption and violence, and an apparent desire of these ancient humans to become godlike. Their pride and violence corrupted human community and polluted the earth.[78] God therefore announces an intention to 'bring a flood of waters, to destroy from under heaven all flesh in which is the breath of life' (Genesis 6:17). And God calls on his faithful servant Noah to save himself and his household and two of every species on earth by building an ark, of whose design and proportions the text provides precise detail. Noah begins his great ark project according to the divine design, and his neighbours subject him to ridicule. They are content in their lives, they do not suspect that disaster will befall them, and they think Noah and his family are off their heads. When the flood comes the only ones who survive are those, animal and human, whom Noah has seen safely aboard his ark as members of his extended floating household.

The biblical story of the flood echoes similar accounts of primordial inundation in the ancient Near East, such as the Gilgamesh Epic, which likely have their roots in an actual geological event.[79] Around 5,600 BCE the rising waters of the Mediterranean overflowed at the Bosphorus and inundated the deep basin that contained the freshwater Lake Euxine, forming what we today call the Black Sea.[80] This great flood, no doubt precipitated by heavy rain, was a consequence of the combined effect of the flow of fresh water from the retreating European ice sheet, and of rising sea levels from melting Polar ice, both processes emanating from the gradual warming which ended the last great ice age.

The ancient saga of Noah is not only a powerful story of human survival in the face of a climate-related catastrophe but also a moral tale in which the flood is seen as divine punishment for a generation of humans whose wickedness was such that it affected the life of all flesh on earth, and endangered even the earth which was as a result 'full of violence' (Genesis 6:13). The narrator here picks up a theme from the opening chapters of Genesis in which the sin of Adam and Eve is said to have effects not only on their children and children's children but on the earth itself.[81] Noah is a redemptive figure who in many ways anticipates the redeeming work of Christ. God has Noah build the ark to train him in living *with* creation, and to discover his fellow creatures as companions. Noah responds to the impending flood by detailed attention to the materials he will need to build the ark, and then to the accommodation and food requirements of the species he finds he is called to save. The narrative indicates that Noah spends twelve months on the ark, which he had carefully built and provisioned with appropriate food for his fellow creatures. Rabbinic commentators suggest that this careful attention to the needs of animals and other creatures was the key lesson of the Noah saga. It is as if God has Noah build the ark to train him in caring for creation.[82] In contrast to the prideful destruction visited on the earth by his violent neighbours, Noah submits patiently to the humble demands of caring for the many animals he intends to save from destruction.[83]

The Noahic narrative reflects a cosmological belief, shared by the ancient Hebrews and other Mesopotamians and by the early Christians, that the forces of creation, the natural elements, are given order and restrained by divine power. And it reflects the moral and spiritual insight that only when humans acknowledge the sacred character of creation and recognise divine limits on human action will their reproduction and work within creation not result in its destruction and pollution.[84] Hence when wickedness threatens creation, and humans substitute themselves for God, the waters return and reinvade the land. The sea in these texts, as in climate and evolutionary science, is seen as the original source of life. Life emerges first in the oceans and only later on land (Genesis 1:20–21). Similarly the Creator had 'founded the earth upon the seas' (Psalm 24:2). Divine restraint of the sea, as Jeremiah suggests (as we noted above), is what makes life on land possible: when humans dishonour the poor and neglect the moral order, then the land is again at risk from the ocean and all of life is threatened.

Recent scientific descriptions of climate change have some striking analogies with these early creation stories. Scientists have discovered that ocean currents, the salinity and temperature of the oceans, and the quantity of phytoplankton and other life forms they sustain, are critical determinants of atmospheric gases and hence of the earth's climate. And ice core records show that at those times when the sea regularly encroached over far more of the earth's surface than it does now, levels of greenhouse gases in the atmosphere were very unstable and hence the earth did not provide the kind of stable climate that humanity has known for the past 15,000 years.[85] Before this period, when mammoths roamed the earth alongside small groups of neolithic humans, relatively rapid shifts of climate to extremes of heat and cold were commonplace. Data from fossilised tree rings provide a reliable record of these extreme changes, and alongside trapped gas bubbles from prehistoric atmospheres contained in ice core samples, reveals a clear correlation between fluctuations in atmospheric CO_2 levels and in land and ocean temperatures. These fluctuations were caused by prehistoric wobbles in the earth's orbit around the sun and changes in solar output. As we have seen, it only became possible for the earth to sustain human agriculture and thence urban civilisation across much of its surface when this pattern of extreme fluctuations changed and the earth settled into a more stable climate pattern at the end of the last major ice age.[86]

In the biblical narrative, the dominion of humanity and the peace and order of creation which make possible the first development of agrarian civilisation in human history are represented as a result of the providential ordering of God in constraining and directing the tremendous powers of the ocean and so producing a stable climate and a fruitful earth. Contemporary science also finds that the ocean is the most significant terrestrial influence on the earth's climate. The ocean covers two-thirds of the earth's surface. It is the principal heat store and cold store of the planet. Its currents are the primary means by which heat and cold are moved around from the tropics to the poles, warming some shores and cooling others. The reason that human actions are beginning to affect the climate more noticeably is because of delayed reactions occurring in the oceans to the build-up in the last century of industrial pollution in the earth's atmosphere, reactions evidenced in rising ocean temperatures and changing thermohaline currents such as the Gulf Stream, which cycles heat from the warm tropics northwards in the North Atlantic Circulation. These changes in turn precipitate the more extreme climate events – such as the greater strength of El Niños, increased strength of hurricanes and changing monsoon and rainfall patterns – which are already having such deleterious effects on developing regions in the South.

The covenant which God made with Noah and his descendants after the Flood promised that the earth would not again be threatened by the bursting forth of the chaotic waters. It was a covenant which included not only humans but 'living things of every kind'.[87] It was, as Robert Murray suggests, a 'cosmic covenant' in which confrontation between the chaotic and elemental powers of ocean and climate and the ordering and sustaining power of God play a central role.[88] The Promised Land 'flowing with milk and honey' is in its fertility and order the more particular gift of this covenant, being a land in which beneficent climate and soils would allow Israel to develop a culture whose religious and moral arrangements are deeply intertwined with agrarianism. For the Psalmist, the regularity of the rains which water the crops are evidence of the sustaining action of the Creator, but they are also seen as a sign of good government; the good king oversees a land in which the rains come regularly (Psalm 72. 1–6). And as we have seen the Hebrew Prophets suggested that the burdens which greedy kings and merchants placed on Israelite households and lands in the latter years of Israel's tenancy of Canaan caused the droughts and spreading deserts which afflicted not just Canaan but the whole region of Mesopotamia from the eighth century BCE. Pollution is seen as a consequence of the failure to follow divine law: as Isaiah has it, 'the earth lies polluted under its inhabitants; for they have transgressed laws, violated the statutes, broken the everlasting covenant' (Isaiah 24:5).

After Noah and before Isaiah, the Old Testament speaks of another patriarch who not only correctly foretold an impending climate disaster, but helped an alien nation, through them his own family, and ultimately the chosen people of God, to survive it. This man was Joseph, whose story offers a more hopeful scenario for our current climate change predicament than that of Noah: Joseph's warnings of imminent climate change were heeded by the Egyptians and prudent preparations were made which saved them, and subsequently Joseph's own family, from calamity. Like the story of the great flood, the story of Joseph may also have historical roots in the memory of a great drought in the ancient Near East, which studies of Greenland ice cores indicate took place in a three-hundred-year period beginning around 2200 BCE. Although the drought was not caused by Egyptian agricultural or imperial practices, it nonetheless would have shown up any weaknesses in those practices. Above all, the claim of their rulers to be gods was shaken by the devastating famines and, as Fagan suggests, the famine would have taught them to act less like gods and more like shepherds of the people.[89] The lesson they learned was the strategy recorded in the Joseph saga – investment in agriculture and social sharing of stored surpluses was a more effective means for managing the vagaries of the climate for the people of the Nile region than the imperial claim that the rulers were gods and could control the forces of nature.

The Joseph saga is not only a powerful narrative of the role climate plays in shaping the destinies of civilisations: it is also a description of divine providence. Through his betrayal by his brothers, who were jealous at the favouritism shown by Jacob towards Joseph in the gift of the multicoloured coat, through his patient perseverance during servitude and imprisonment in Egypt, and through his prophetic wisdom in interpreting dreams, Joseph became vice-regent in Egypt. In that capacity he established a system of tithes and a centralised system of agricultural storage, and he was in a position during the famine to save his own kin from the effects of the drought. When his family visited the court of Pharaoh to plead for help they were in awe at the power and wealth of the court, and doubly so when they later realised that Joseph, the brother they betrayed and nearly killed, was now a ruler in this great empire. As Joseph put it, the evil they performed in selling him into slavery was taken account of in the divine plan: 'God sent me before you to preserve life' (Genesis 45:5). Divine providence works even through the evil his brothers did to him, and the tragedy of undeserved suffering serves to 'preserve for you a remnant on earth, and to keep alive for you many survivors' (Genesis 45:7). The story shows in a paradigmatic way how Yahweh works through history, both human and planetary, and through undeserved suffering, to bring about the salvation of the people of God. From the selling of Joseph into slavery by his jealous brothers to the years of drought, the divine hand works to save the sons and daughters of Jacob whose descendants, through the Exodus, eventually become the heirs of the land of promise.

The fathers of the early church read the Joseph saga as an analogy for the Christ events. Just as God preserved his people through the undeserved suffering of Joseph, so God redeems the new people of God through the undeserved suffering and patient endurance of his Son Jesus Christ. The grace of God uses the human temptation of jealousy, which drives Joseph's brothers to assault their brother and sell him into slavery, and turns it towards God's own purposes, just as the betrayal of Jesus by his friend and disciple Judas presages the saving death and resurrection of the Lord. As a friend of a prisoner, Joseph is ultimately brought before Pharaoh to interpret a dream. And just as Christ ascribes his wisdom and power to God the Father, so Joseph suggests that any facility he may have in interpreting Pharaoh's dream comes from God and not from himself (Genesis 41:16). There were no doubt many in Egypt who scoffed at Joseph's prediction of the years of drought and who argued, like climate-change deniers today, that the prophecy of coming cataclysm was mistaken and that, even if it were true, it would be better to deal with it when it arrived than make expensive and elaborate preparations in advance. But against the naysayers Joseph read the dream as a message from God to the King of Egypt. In a very important sense the science of global warming finds a powerful analogy in this story.

There is however a dark side to the Joseph saga. Ultimately all of Egypt, and then Israel, is ensnared by the centralised system of food storage and feudal tithes that Joseph inaugurates. Joseph responds to the coming climate disaster with a scheme of control and of management in which he stores up the good gifts of the creation in such a way as virtually to monopolise the product of the land. Noah, on the other hand, shows a deep respect for God's creatures while not seeking to coerce them into a grand scheme. In caring for the creatures in the ark Noah becomes not only their servant but their companion; and he develops a love for them.[90] When the dove leaves the ark in quest of dry land Noah is sad to see it go and the rabbinic commentators suggest that he was glad when it returned the first time. Joseph, on the other hand, seems to have little love for those whom he ensnares, and ultimately enslaves, in his grand agricultural scheme, as witness his highly manipulative encounters with his brothers in their journeys to his court. In this respect Joseph is like those who respond to climate change with technological schemes to engineer the climate – for example by putting mirrors into space to reduce solar heating or seeding the oceans with iron filings, and so attempt to stave off global warming without reducing industrial civilisation's reliance on the wasteful burning of fossil fuel. And yet, as Lovelock suggests, if industrial humans are forced to manage the climate of the earth by extreme climate change they will in effect enslave themselves and future generations to a task of such momentous proportions that it may ultimately overpower even industrialism's seeming prowess in technical management.

It is then fitting that the Noah saga has a stronger hold on the Jewish and Christian imagination than the Joseph saga. Noah respects the contingency of creation and shows love for God's creatures, and it is this love, and not the managerial skills of the overseer, which is ultimately redemptive. There is also an affinity between the Noah saga and the Christian account of the origin of Creation, and in particular the doctrine of *creatio ex nihilo*. The idea of creation out of nothing emerges quite late in Jewish history but becomes a central tenet of Christian orthodoxy from the second Christian century. In this doctrine Christians affirm that God's relation to the world is not one of complete identity, nor one of controlling or instrumental dominion. Isaiah represents creation as pure summons: 'My hand laid the foundation of the earth, and my right hand spread out the heavens; when I call to them, they stand forth together'

(Isaiah 48:13) and this representation of creation as the 'free utterance of God' in summoning being from nothing means that the act of creation is not well expressed in the language of power or control or management, but is rather an act in which intrinsic worth is created by divine freedom and generosity.[91] As Rowan Williams puts it, 'creation is not power, because it is not exercised *on* anything': creation simply affirms that what is is so because God wants it so.[92] This confers a relative freedom on creation, creating biological and metaphysical space for the unfolding of life in the many convergences which have at a number of points in evolutionary history produced cellular organisms, and ultimately more complex beings with eyes and hands and knee joints and brains.[93]

The human vocation of finding agency by working on creation means attending to the world in its diversity and in its becoming, just as Noah does. In the biblical account, God makes space in creation for humans, in particular, to image divine freedom. But the difference between divine and human freedom is that humans are dependent on the material and embodied order of creation for their constitution as agents. The Orthodox put this in a powerful way when they argue that humans are appointed the priests of creation. As John Zizioulas suggests, this means that humans depend upon their right relationship with the material creation for their own redemption.[94] Only when humans treat the earth as gift, and offer the fruits of the earth to God in true worship, can they achieve a right relation with the created order.

By contrast modern humans identify their salvation with progress in economic development, material wealth and technological control. This belief in progress is linked to the liberal conception of the person as an autonomous self-creation, and the technological quest to secure for modern society material independence from the generative environment – both divine and material – which is the source of being. As Williams puts it, the 'illusion of being an individual self-regulating system' generates a quest for 'piecemeal securities' that refuse the relational dependencies and cyclical processes in which life truly flourishes.[95] Against this illusion, the doctrine of creation, like the Noah saga, affirms that both 'I' and the 'environment' exist because of the divine will, and that being creatures, and even priests of creation, means attending to a relationship of lived dependence on the other selves and other species which constitute the earth. This means that the explanation of every event in the world cannot simply be God, for if this were so there would be no freedom, no agency, no non-competitive space in creation where God allows the creation to be what is not God. But it is wrong to imagine human mastery of creation as an analogy to divine power in that non-competitive space. On the contrary the life, death and Resurrection of Jesus Christ, like the ark of Noah, reveal 'a creator who works in, not against, our limits, our mortality: the creator who, as the one who calls being forth from nothing, gives without dominating.'[96] Recovering the proper human relation to creation involves mirroring this relationship of giving without mastery, of conferral of being without control. Such a relation is characterised by humility, a preparedness to recognise human limits, both in knowledge and in the ability to control larger than human forces, and the willingness to offer all creation back to God as gift and not as human possession.

There are those who suggest that God will not again allow the earth to be subject to judgment as happened at the time of Noah. Conservative Christians, particularly in North America, are misled by the theological claim that humanity recovers its rightful dominion over creation after the coming of Christ, and cannot imagine that the planet could be in danger of running out of control as the global warming scenario suggests. Some dispensationalist evangelicals even maintain that fear of global warming is a sign of the end of time since Christ warned of 'men's hearts failing with fear for looking after those things which were coming on the earth: for the powers of the heavens shall be shaken'.[97] For the millennialist Hal Lindsay fear of global warming is a precursor of the apocalypse that he has long been predicting is just around the corner.

Others are trained by the technological powers released by scientific rationality to imagine that if the world does stand in such danger then technological power will give humans the ability either to wrest the climate back from impending disaster by geoengineering, or to adapt successfully to dangerous climate change. Against these kinds of theological and technological hubris, the Noah saga suggests that turning away from the ecologically destructive path on which humanity is headed requires humility and a preparedness to change direction in response to the clear signs of impending danger. It also suggests that the religious narrative of the sacred character of creation, and of the sacred charge on humans to care for creation through careful stewardship, are central to the vocation of humans in caring for and working on and in creation.

If the signs of disturbances in the climate truly signify a potential cataclysm from fossil-fuel burning, then the Genesis stories of Noah and Joseph, in which God acts in history to save humanity and other species from destruction, educate those who read them that there is a divine will to preserve the creation from cataclysm. They also indicate that God does not act alone to preserve the earth, but through the prophetic insight and fearless witness of individuals like Noah and Joseph. This means that there is a particular responsibility on those who discern the links between these ancient stories and the natural signs of modern climate change to seek to challenge and repair the instrumentalising forms of making and consumption which have advanced the intergenerational, international and inter-species injustices of global warming.

3
Energy and Empire

For thus says the Lord *concerning the house of the king of Judah*
'You are like Gilead to me, like the summit of Lebanon;
but I swear that I will make you a desert, an uninhabited city.
I will prepare destroyers against you, all with their weapons;
they shall cut down your choicest cedars and cast them into
the fire.

And many nations will pass by this city, and all of them will say to one another, 'Why has the Lord *dealt in this way with that great city?' And they will answer, 'Because they abandoned the covenant of the* Lord *their God, and worshipped other gods and served them.'*

Jeremiah 22:6–9

The end of the nineteenth century was a dark time in imperial history. Twenty million people died from hunger in Southeast Africa, Brazil, China, India, Java, Korea and the Philippines because of droughts triggered by a series of back-to-back El Niño events between 1896 and 1897 which affected a quarter of the earth's land area. But drought turned to famine as a consequence of the pressures of the colonial economy.[1] British imperial rule demanded so much tax from India that it effectively destroyed the peasant economy, while at the same time British tariffs against homespun cotton almost destroyed the indigenous cotton industry.[2] As a consequence, droughts which had in previous eras not precipitated famine resulted in the deaths of 4.5 million people. The British believed that their imperial 'improvements' to Indian agriculture and communications systems meant that the colony was well prepared for drought. But callous destruction of the peasant economy through excessive taxation, the use of famine relief funds to support imperial wars and the refusal to release stockpiled grain led to death on a large scale.[3] While human health and longevity improved dramatically in the imperial cities of Northern Europe and North America in the late nineteenth century, the colonial project saw an apocalyptic collapse of human and ecological health in the South. A Kenyan man is said to have commented to an English missionary, 'Europeans track famine like a sky full of vultures'.[4]

In twenty-first century India life prospects for many rural peoples have not greatly improved since colonial times. One of the obstacles to improvements in human health in India is fuel poverty. Millions of villagers in South Asia still gather firewood and animal dung by hand to fuel their stoves, and to heat and light their homes. In some districts in the State of Jharkand in Northeast India women who normally engage in the backbreaking work of gathering fuel have an additional constraint because the forests in the area have recently been protected by an environmental order that prevents the gathering of fallen branches. This means the women can only gather fallen leaves or prickly twigs from the *putus* thorn bush, and they must travel much further to gather sufficient fuel to cook for their families. Such women share the fate of one quarter of the world's people in having no access to electricity. According to biogeographer Philip Stott, such people can only be saved from their plight by gaining access to the hydrocarbons to which the rest of the world already has access. Stott dismisses the claim that global warming is caused by burning hydrocarbons as the 'new witchcraft' and suggests that the energy-poor can only be released from their poverty *by* hydrocarbons.[5]

Indian economist Bina Agarwal suggests that the principal cause of fuel poverty in South Asian villages is not so much the absence of electricity as the widespread destruction of forests by colonial authorities, who left much of the felled timber to rot on the ground. Peasants are very careful to conserve forest resources and mostly do not chop trees down; instead they prefer to gather fallen wood because they know that their future needs can only be met when the trees live.[6] With the gradual disappearance of common forests because of colonial and post-colonial appropriation of forests for paper pulp, building material and urban fuel-wood markets, the area of forest is drastically reduced. Fuel poverty is leading to malnutrition in many parts of South Asia since without fuel for cooking it is hard to achieve adequate nutrition from cold and uncooked food sources.

The problem is exacerbated by the commodification of land in postcolonial India; many common lands have been privatised by wealthy high caste families, leaving the landless with fewer fuel sources.[7] Agarwal argues that privatisation of common lands, inequality in land holdings and access to fuel sources are critical for understanding the differential access of wealthy and poor households in South Asia to adequate fuel for cooking, heating and lighting. Even where there is no electricity, the land-rich have no difficulty in gathering enough fuel to provide warmth and light for their households. The key to addressing these inequities is not hydrocarbons but political participation in the management of land and forest resources. Where people, and especially women, enjoy equitable access to such resources they flourish. Where such participation and equity is denied they suffer ecological exclusion and fuel and food poverty.[8] And this ecological exclusion now takes a new form as global warming exacerbates problems with declining water sources and excessive heat in poor rural states such as Rajasthan, Udapradesh and regions such as Jharkand.

Inequity in access to land is one of the principal bequests of European colonial history in South Asia and in other parts of the world. Accumulation of land and its product was the primary mechanism for transferring value from the colonised regions of the earth to the European regions, and this process disrupted the traditional land management procedures which had stewarded common resources before the coming of the Europeans. This is something that Garret Hardin's much lauded account of the 'tragedy of the commons' misses.[9] Hardin suggests that common resources are invariably abused because they do not belong to any one in particular and therefore no one has a vested interesting in preserving them. But his North American, private property-driven worldview misses entirely the role of customary law and traditional community in managing common resources in precolonial societies. The real tragedy of the commons ensued when new patterns of land ownership were imposed by the colonial state in which settlers claimed exclusive proprietorship to traditional lands and overused such shared resources as rivers and wells. Europeans appropriated these for the accumulation of personal wealth and surplus value. But traditionally they had been regarded as resources to be shared equitably by procedures which limited their abuse, and which passed them on to succeeding generations in good condition. Colonisation destroyed the complex social arrangements for preserving common resources and the rich fund of ecological knowledge these arrangements represented. It was this heedless colonial process of surplus value extraction which brought so much poverty and eventually famine to Africa, and to other colonial regions.[10] As J. K. Galbraith observes, 'the modern age opened with the accumulation of capital which began in the sixteenth century', and in particular the discovery of gold in Latin America.[11] Europe and North America could not have achieved their industrial prowess without the subjugation of large imperial terrains, from which natural resources and the power of human labour were extracted mercilessly over hundreds of years.

Even as many of the colonists have withdrawn to their homelands and granted their former colonies political, if not economic, independence, the peoples of the South are becoming aware that their skies are now overshadowed by a less tangible but still very real presence from the North. Scientists at the World Health Organization (WHO) estimate that around 150,000 people in the South are dying every year from the side-effects of climate change, including more frequent extreme weather events, monsoon instability, flooding or drought.[12] Malnutrition and increased exposure to malarial mosquitoes are the main specific health threats resulting from these climate-induced changes, and the WHO expects that the numbers of those dying will double by 2020. The IPCC suggest that up to 3 billion people, mostly in the tropics and the Southern hemisphere, will be displaced or suffer serious threats to their wellbeing if greenhouse gas emissions continue to rise unchecked in the twenty-first century.[13]

Argentinian economist Joan Martinez-Alier suggests that ecological conflict over atmospheric sinks is the latest in a long list of modern ecological conflicts between industrial producers and residents of communities of place.[14] Such conflicts occur in tropical forests between mining and plantation companies and indigenous peoples, between traditional fishing communities and shrimp-farming companies in mangrove swamps, and between indigenous peoples and petroleum companies who extract oil in ways which damage local lands and ecosystems. In Indonesia, for example, there have been thousands of conflicts arising from palm oil plantations where local people have been forcefully evicted from farmlands and forests.[15] Transboundary ecological conflicts also occur in respect of industrial pollutants such as sulphur dioxide, which causes acid rain. Similar problems arise from the export of industrial and toxic waste from rich developed to poor developing countries. Greenhouse gas pollutants are a form of transboundary ecological pollution where the rich are using the atmosphere of the poor to absorb their waste carbon.

The expropriation of the lands and atmospheres of people in poorer countries by rich people who live elsewhere is represented by Martinez Alier as 'ecological debt'.[16] Ecological debt is defined as 'the debt accumulated by Northern, industrial countries toward Third World countries on account of resource plundering, environmental damages, and the free occupation of environmental space to deposit wastes, such as greenhouse gases, from the industrial countries'.[17] The North has already benefited financially from deferring action to reduce greenhouse gas emissions while poorer Southern countries are already incurring increased costs from such deferrals as a result of increasing extreme weather events. At the same time the average use of carbon by poorer countries in the South is around 0.2 tonnes per head, compared to an average for Northern consumers of around 12 tonnes per head. Consequently atmospheric pollution is now the biggest single form of ecological debt. Putting hard figures on climate-change related ecological debt is not easy but one estimate has the debt of just the G7 at $13 trillion. This is a monetary estimate of the costs of adapting to present and future impacts of climate change caused by existing greenhouse gas emissions.[18] It seems an impossibly large figure, and yet it represents just 13 years of the total world spending on weapons of war, which has stood at over $1 trillion per annum for the last fifteen years.[19] Andrew Simms suggests that in the future poor nations will be able to sue the carbon polluters for the damage that climate change inflicts.[20] And indeed under the UNFCCC developed countries are legally mandated to help poorer countries to adapt to global warming. However, though the rich nations have committed to meeting the estimated £230 million a year costs of adaptation in the developing world, they have so far only committed £93 million in total to the relevant funds.[21]

Since the rich owe the poor an inestimable debt because of the temperature rises and climatic extremes which are already imminent as a consequence of the effects of the excess carbon dioxide and other greenhouse gases with which they have polluted the atmosphere, then, Simms suggests, the rich industrialised nations ought to begin financial restitution by forgiving outstanding financial debts owed by the poor nations to the bankers and governments of the North.[22] These debts represent a tragic burden on the indebted nations, which have paid the principal many times over but are still shackled to debt repayments because of the compound interest rates of creditor banks, governments and international financial institutions, including the publicly funded World Bank and International Monetary Fund (IMF). For more than a decade Christians have led international calls for debt cancellation, and this contribution has been deeply grounded in the biblical narratives of jubilee and debt forgiveness.[23] Continuing insistence that at least a proportion of these debts must be serviced, and that as condition of their forgiveness debtor nations must submit to the humanly and ecologically damaging Structural Adjustment Programmes imposed by the economists of the World Bank and the IMF, has meant that indebted nations are forced to enslave the labour of their people to this debt burden. They are also forced to devote their forests, hills, croplands and coastlines to the production of cash crops such as coffee, cut flowers, logged timber, monocrops such as bananas and mangetout and farmed prawns for export to the North in an imposed quest for foreign exchange with which to pay interest on the debt.

Neglect of the imperial legacy of ecological debt owing to the poor in the South reflects the timeless and placeless nature of modern moral rationalism and of the workings of the money economy. The focus in the reports of the IPCC and the International Energy Agency is on present and future effects of greenhouse gas emissions and the need for those who will be worst affected to adapt to changing climatic circumstances. And yet the looming climate catastrophe has roots in colonial history, and even further back before the dawn of historical time, when the sun first shone on the earth and warmed its surface, giving life, through the process of photosynthesis, to terrestrial plants, ocean corals and phytoplankton. Energy from the sun which hit the earth before humans walked the earth was absorbed and transformed by living creatures into a storage bank of carbon which until very recently lay deep in the bowels of the earth. The plants and the creatures that ate them lived and died in their billions. Through the powerful forces of erosion and sedimentation they were gradually buried in the earth's crust in the carboniferous era, and became what Sieferle calls the 'subterranean forest'.[24] This underground store of carbon played a key role in arbitrating the temperature of the planet, helping to lock carbon dioxide out of the atmosphere so that the earth became temperate enough to sustain the great array of biological species which have since dwelt on it. The process of photosynthesis on land and in the ocean is the single most powerful means by which the earth absorbs the heat of the sun on its surface instead of emitting it back out into the atmosphere.[25]

Describing the Carbon Cycle

Most of the earth's store of carbon is locked in geological substrata of rock, and can only be released by human extraction. The principal forms of extraction are fossil-fuel production and cement-making. Coal, oil and natural gas are all composed of fossils, from which carbon is released when they are burnt as fuel. Limestone is also composed of billions of fossilised shellfish, and when it is crushed and heated to make cement it similarly releases the stored carbon into the atmosphere. The ocean is the largest store or sink of moveable carbon by a factor of five; next in size is the soil, then the atmosphere, and finally land plants.[26] The ocean is like a giant carbon sponge and it is capable of taking up excess carbon in the atmosphere, even in the quantities that industrial humans have put there, but not at the rate we are putting them there.[27] So much carbon is present in the ocean that only 2 per cent of oceanic carbon would be enough to double the amount of atmospheric carbon.[28]

The oceans act as giant biological pumps that recycle nutrients and gases from the atmosphere to the deep earth every year. The peak of activity occurs in spring when the oceans in temperate and sub-arctic regions are replete with phytoplankton fostered by the upflow of nutrient-rich waters from the deep cold ocean and by spring sunlight.[29] Their presence produces changes in the colour of the ocean which can be seen in pictures from space. In winter phytoplankton begin to sink in the ocean, taking nutrients and carbon molecules with them. Some decompose as they sink, some are eaten by larger creatures such as shellfish, while others reach the ocean floor where they may remain in the deep ocean carbon pool.[30] The movement of carbon between ocean, land and atmosphere is known as the global carbon cycle. Though each molecule of carbon may only remain in the atmosphere for four years it may live for decades if it moves from the atmosphere to the surface of the earth, for hundreds of years if it moves to the deep ocean, and for millennia if it ends up as part of the geological store of bedrock beneath the ocean or in permanently frozen tundra.

As complex and multifaceted as this biological pump is in its many constituent parts, it nonetheless reached a point of equilibrium 15,000 years ago. This equilibrium is now being disturbed by industrial gases with troubling consequences. Recent data indicate that as the oceans warm they lose their capacity to act as 'carbon sinks'. The term carbon sink indicates the capacity of elements of the earth system to absorb and store carbon including atmosphere, forests and soils as well as oceans. Warmer oceans are also the origin of the growing strength of storms, as we have already seen. The Pacific Ocean has been far less pacific in recent years as rising water temperatures are driving the increasing frequency and severity of the El Niño effect.[31] And rising ocean temperatures are not confined to the tropics. Risen temperatures in the North Atlantic and Arctic Oceans are the most likely cause of a significant reduction in phytoplankton growth in the last ten years. This sudden decline in phytoplankton threatens the myriad species which feed off plankton.[32] It is the probable cause of the reduced ability of sea birds successfully to rear their young around Scotland's extensive coastline, for example. And it threatens a larger event, since phytoplankton play such a crucial role in the carbon cycle. It is likely that declining phytoplankton production indicates that in the last ten years the amount of warming that greenhouse gas emissions have now generated is actually reducing the effectiveness of the planet's largest living carbon sink.[33] What this means is that far more of present and future greenhouse gas emissions will remain in the atmosphere, as less are recycled into the oceans. This will lead to an acceleration in present warming trends, a possibility which throws out even the latest projections of the IPCC, whose panel of scientists assume for the purposes of their estimates of climate sensitivity – that is, how much temperature change a given amount of CO_2 emissions represents – that the carbon cycle maintains a constant capacity to absorb humanly-generated greenhouse gases.

All mammals, and not just industrialised humans, are constituent parts of the global carbon cycle. Mammals breathe oxygen and expel carbon dioxide from their lungs and methane from their stomachs, and the plants mammals eat take up available carbon in the atmosphere and turn it, through photosynthesis, into oxygen and biomass.[34] Interactions between humans, other animals, insects, oceans, soils and vegetation are thus the principal determinants of the climate of the earth, as Lovelock's Gaia hypothesis first revealed. Without photosynthesis on land and in the oceans there would be too much heat absorbed by CO_2 in the earth's atmosphere, and the planet would warm to a point where it would become uninhabitable; there would also be no oxygen in the earth's atmosphere.

Photosynthesis is not only a giant form of air conditioning which prevents the planet from overheating. It is also the source of all life on earth. Without plants capable of turning heat and light into matter, and carbon dioxide into oxygen, there would be no mammalian life on the planet and humans could not exist. As the first chapter of the Book of Genesis indicates, plants were created to make the earth both beautiful and habitable and hence they are to serve as the food source of animals and humans alike. There is beauty and not just biochemistry in the rich diversity and fecundity of the original forests, savannas, mountain slopes, oceans and rivers of this wondrous earth, though precious few of these habitats remain for us to admire today. It is conceivable that a planet which had far fewer plants to do the 'job' of photosynthesis could still have a stable enough climate to make it possible for warm-blooded creatures like ourselves to survive. But beauty and diversity, not just biochemical order and control, were clearly part of the divine plan for the earth from the beginning. This is just one of many reasons why Christian philosophers have traditionally believed that this is the 'best of all possible worlds'. But in just three hundred years the influence of industrial humanity has penetrated every region of the globe, and in so doing has turned this best of all worlds into a place far less beautiful, diverse and wondrous. Not only have humans driven 15 per cent of the earth's species to extinction in the last one hundred years,[35] but the manufactured and waste products of industrialism are now evident on every continent, and on every patch of ocean.

The primary cause of the recent increase and spread of human influence over the ecosystems of the earth is the facility with which industrial civilisation has mined, drilled, piped and burned the carbon store of the planet in the forms of coal, oil and natural gas. After three hundred years of industrialism 600,000 million tonnes of the carbon formerly locked in the earth's crust now resides in the moving carbon cycle. This increase in available carbon has resulted not only in a warmed climate but in a notably greener earth as plants have responded to increased atmospheric carbon with faster rates of growth. Some have even suggested that this greener earth is a good thing and that this upside to industrial carbon emissions is a reason not to reduce them. But this effect will not last for long as rising temperatures will eventually have a negative impact on rates of biomass production.

In the 1930s H. G. Wells suggested that the industrialisation of the earth had turned human history into ecology, as the power released by this great store of energy enabled human impact to spread over the whole globe. The fate of humans became quite literally the fate of the planet.[36] Wells partially misread the situation, for human history has always been ecological in the sense that humans, like all animals, are dependent on plant photosynthesis for their existence. The illusion of human control over the earth that the industrial revolution has conferred upon humanity enables modern humans to imagine that their history is *not* ecological. But Wells is right in one key respect: the difference that the industrial revolution made was that it enabled the expansion of human influence across the globe because of the limitless use of the earth's store of the light and heat of the sun which industrial humanity had begun to draw upon. Hence, on a purely materialist reading, the ecological crisis and the extreme form of this crisis that climate change represents begins with the dawn of the industrial era. The transformation of the earth's stored carbon into mechanical 'Horse Power' through steam and electricity, the internal combustion engine and the jet engine made possible the range and speed of industrial humanity's modern conquest of the planet. Humanity dominates the globe not because of its superior biomass – there are still far more beetles than humans on the planet – but because of the power which energy and the machine age confer upon the human species to clear land, cut forests, dam rivers, rear domestic animals, mine minerals, quarry mountains, and move materials, people and other animals at speed across land, ocean, and in the air.

The Promethean Age of Steam

The modern industrial revolution began with the invention of the first working steam engine.[37] Before the age of steam waterpower drove textile mills and a few other devices, but the steam engine made available a quantity of mechanical power that radically transformed the organisation of human work and its relation to the environment. Ironically enough, the first industrial engine was used in a coalmine to pump water and so facilitate more efficient fossil-fuel extraction. As timber supplies dwindled around Britain's growing cities in the seventeenth century, so demand for coal rose and by 1700 consumption was half a tonne per head per annum: to meet demand the mines had to be dug deeper and they consequently began to flood.[38] Thomas Newcomen was the first to come up with a steam engine capable of pumping water out of the mines.[39] An unsophisticated device which worked at atmospheric pressure and used a lot of coal, the Newcomen engine was nonetheless very simple to maintain and the original installation in a Cornish mine was patented and copied in mines all over England.[40]

It was the Scotsman James Watt who transformed Newcomen's simple but wasteful device into the efficient and powerful condensing steam engine which fuelled the industrial revolution. In so doing he did more than any other single individual to enable modern humans to turn the stored energy of the earth's fossils into heat and thence into mechanical energy, and so extend humanity's command over the forces of nature.[41] This invention set off the dramatic industrial revolution and the transition from an organic economy, dependent upon muscle power and renewable biomass from land and ocean, to an economy in which technology transformed the fossils and minerals of the earth's crust into mechanical power, and thence into 'iron, pottery, bricks, glass and inorganic chemicals'. The products made from these include machines, tools, buildings, vehicles, clothes and other consumer products.[42] Steam power released human production from traditional limitations of land area and animal and human muscle and was therefore the key device in making possible the unprecedented growth in economic production that characterised the industrial revolution in its successive phases.

The use of such phrases as 'the atomic age', the 'post-industrial era' and the 'knowledge economy' to describe recent stages of the industrial revolution sustains the illusion that the dependence of the modern economy on steam power and mineral extraction is now increasingly historic.[43] But most electricity is still produced by steam turbines, and between 40 and 60 per cent of the available energy from hydrocarbons is still wastefully emitted as heat into the atmosphere in the process, while no efficient method yet exists for storing electricity other than pumping water uphill into mountain reservoirs, as we do in Scotland. In material terms the 'knowledge economy' still firmly rests on fossils and minerals, or what Wrigley calls the 'mineral economy', and it still relies on energy derived from fossils to fuel steam-driven turbines. The laptop computer on which I write this book may be small, but its production and sale involved the industrial use of hundreds of gallons of water, considerable amounts of heat energy and mechanical power and a multitude of chemicals and metal alloys synthesised from minerals in the earth's crust, the processing of which produced toxic waste.[44]

At the same time as hundreds of millions of Europeans, Americans and Asians derive a living from mineral-based technologies, around three billion humans still meet most of their household needs directly from the organic economy. Many presume that the transition from an organic to a mineral economy is the necessary condition for the release of humanity from poverty and ill health.[45] However, the physical limitations of the earth's carbon cycle represent a fundamental challenge to this foundational assumption in modern political economy. If all 6 billion humans lived the industrial way the life-support systems of the planet would soon fail.[46] Mathis Wackernagel *et al.* estimate that in 1999 the global economy *as* household was using the equivalent of 1.2 earths, and that it was already overshooting the planet's biological capabilities.[47] And this overshoot is caused by industrial humans, not developing country farmers; for example, Scotland's overshoot is three times the size of the land area of Scotland, with each individual using a land area equivalent to five soccer pitches.[48]

The beginning of the transition from timber economy to coal economy was driven by deforestation in England, which was significantly advanced by the expensive wars of the late medieval and early modern periods. This transition represented an unprecedented move from an economy which relied on renewable biomass to an economy dependent upon the earth's store of carbon.[49] The other major change that took place was a dramatic increase in the quantity of energy used. Modern humans in industrialised societies use roughly 600,000kj per person per day, at least 100 times the energy that primitive humans do.[50] The quantity of energy humans require to consume as food remains basically unchanged at just 10,000kj. The difference is used for space heating and cooling, industrial food production and transportation, to create and drive machines and transportation devices, and to produce and distribute the myriad quantity of consumer goods around which modern industrial society is ordered.

The food economy produces around 25 per cent of all carbon emissions. Air transportation produces only 5 per cent, but at present rates of growth it will produce a much greater proportion in the next few decades. Simms describes the movement of the amount of oil needed for a transatlantic plane journey from the Lower Palaeocene sandstone layer in the North Sea via oil terminals, pipelines, ships, refineries and trucks to the atmosphere via the engines of a jumbo jet. It takes less than 10 days for the oil to travel from under the ocean floor in its 67-million-year-old sedimentary bed to the atmosphere. The speed of this process involves a 'mechanical collapse of time and space' which is 'driving the accumulation of carbon debt'.[51] An even larger displacement of oil from subterranean pools to atmospheric sinks is caused by the automobile, which has become the supreme icon of the consumer society. Motor vehicles are responsible for 20–22 per cent of greenhouse gas emissions in the United States and the United Kingdom.[52] But this figure only estimates the direct contribution of the fuels and oils utilised in engines.[53] A full accounting would need to consider the energy involved in manufacturing, maintaining, selling and scrapping vehicles, and the considerable energy expended in creating concrete and tarmac road spaces and garages for them to move around in. And this account would also need to include the quantities of carbon utilised in extracting, refining and transporting the oil to drive them. As Ivan Illich suggests, the car is the supreme object of the energy inequity that is at the heart of industrial civilisation.[54] It is also central to the ambiguous promise of liberty presented by liberal capitalism. Promising freedom to millions, the constant rise in personal car ownership and use has in reality brought congestion, traffic jams and pollution to much of the built space of the planet, while pedestrians, cyclists, children and wild animals are all denied quiet, unpolluted and safe space in which to move without aid of fossil fuels. The pursuit of motorised transport, like air transport, has also advanced the mechanical collapse of space and time and one of its most pervasive consequences, which is the loss of a sense of place, and of the local ordering of place by households where making as well as consumption occurs.

This loss of a sense of place, and of local politics, is deeply implicated in the cultural pattern of denial about the human and ecological consequences of industrial consumerism, of which denial about global warming is perhaps the strongest manifestation. Consumer culture promotes an obsession with the present, and with satisfaction derived from present experiences, and a correlative loss of a sense of the significance of both history and posterity. And as Paul Virilio suggests, there is a relationship between the modern obsessions with instantaneity and speed and the growing rate of accidents and natural disasters. As people move their bodies, consumer goods and money around the world ever faster, they neglect the ecological and moral implications of this ever-increasing speed of movement.[55] But the politics of speed advances the depoliticisation of space as local communities lose control over their local ecosystems and geographical terrains; the end result is an illusion of freedom as movement while true freedom – freedom from the necessity of wage labour, freedom for the preservation of self-sufficient farms and households and locally governed and sustainable communities – is lost.

The Geopolitics of Oil

The politics of speed is intricately connected with fossil fuels and with the geopolitics of oil. Global demand for fossil fuel, and for oil in particular, is at unprecedented levels and still growing rapidly. Oil discoveries have already peaked and some suspect that we are now approaching the peak of oil production from conventional oil wells.[56] To fill the gap between rising demand and static supply oil companies are turning to other forms of oil production, including the tar sands in Canada which contain approximately 2800 billion barrels or more than ten times Saudi Arabia's remaining reserves.[57] Extracting oil from tar sands is energy-intensive but it is 'economic' with the price of oil above $70 a barrel. No matter the cost to the climate, industrial governments are determined to extract every last barrel of oil from the earth. As a Canadian environment minister put it, 'there is no environment minister on earth that will stop this oil from being produced'.[58]

Despite the dangers that a warming world poses to human and other species, the industrialised nations which produce more than 80 per cent of greenhouse gases are still firmly set on the path of the colonial extraction economy, and less industrialised nations continue to bear most of the costs. The conservative argument against energy taxes and statutory regulations of the production of carbon and other greenhouse gas emissions finds its ideological core in the enhanced role that such actions give to the authority of the state over economic affairs and property regimes. Wentz argues that Americans resist centralised government-directed action to stem climate change because it runs counter to their political preference for dispersed democratic governance and small government. He also suggests that the record of centralising regimes on environmental regulation in the twentieth century is a poor one because the concentrations of power such regimes involve mean that 'people in charge who do not have to deal personally with environmental problems, and who can ignore others who care about or suffer from these problems, are unlikely consistently to act in environmentally responsible ways.'[59]

That there is much truth in this observation is demonstrated in the extent to which the modern nation-state is so deeply implicated in the procurement of natural resources, and especially fossil fuels, on which the industrial economy depends, although this is typically overlooked by conservative critics of the state. The prosperity and comfort delivered to citizens in developed countries by means of the modern industrial economy is narrated by conservatives as the outcome of the enterprise and imagination of individuals such as James Watt and Henry Ford, who between them are two of the key figures in the emergence of the mechanised division of labour which characterises modern factory production systems. However, without the actions of governments in creating the legal and social structures in which the modern market economy grew, and in garnering natural resources from across the globe to fuel industrial capitalism, these inventions would never have had the revolutionary consequences that they did.

Oil remains the key natural resource that enables the global movement of goods and services, without which the contemporary global economy of trade without borders would not be possible. Colonial governments played a central role in subjugating and annexing oil-rich regions, without which the oil industries of the North could never have laid claim to the large proportion of the earth's stored fossil fuel reserves that they currently do. And this history of government support for the oil industry is by no means over. The governments of the United States and the United Kingdom pump billions of dollars of public funds every year into subsidising the activities of enormously profitable oil companies. One of the commonest forms of subsidy is government support for the export of know-how and equipment to oil fields in developing countries. In the United States even more direct government subsidies passed from the Bush/Cheney administration to the oil companies, to the amount of $2.5 billion between 2001 and 2004. Subsidies also come in other forms. Government interventions in the Middle East and other oil-rich regions, including the recent and very costly war in Iraq, are designed to secure oil production for Western companies, and represent the largest publicly-funded contribution to the corporate wealth of the oil sector. A related subsidy is government underwriting of arms exports to countries like Saudi Arabia and Nigeria, where corrupt governments are sustained by revenues derived from foreign oil companies, and where pollution from oil extraction is a long-standing cause of grievance among local residents.[60]

It is only necessary to name some of the biggest oil-producing countries in the world, and to note the long-standing history of British and American intervention in the affairs of these countries, to indicate the extent to which oil procurement remains a priority for Western governments in a putatively postcolonial world. Angola, Columbia, Iran, Iraq, Kazakhstan, Kuwait, Nigeria, Sudan and Venezuela have all been the subject of covert or overt Western military interventions in the past fifty years. Western powers have intervened in Iran more than ten times in the last one hundred and fifty years, and as I write they are threatening yet another military intervention.

A report by Christian Aid into the effects of oil production shows how the governments on which national sovereignty was conferred in oil-rich regions by the former colonial powers have been corrupted by bribes and backhanders from foreign oil companies. The people of these countries not only suffer from the pollution of their lands and water sources as a result of oil production, while they themselves have no access to fossil fuels of any kind, or even to reliable clean water or hygienic drainage systems; they also live under corrupt and frequently dictatorial forms of government sustained by oil revenues.[61] The process of oil extraction often inflicts ecological harms on the people and other species who live in oil-producing regions. Oil revenues also inflict economic problems on oil-rich developing nations because sudden and large flows of oil revenues produce a 'wealth shock' which drains economic capacity from other forms of production and promotes political corruption. As a recent study indicates, oil wealth appears not only to slow economic growth but actually to create poverty and increase wealth inequality:

> Nigeria is perhaps the starkest example of oil fuelling poverty. It has Africa's largest oil reserves and began exploiting in the 1960s. Since then, human development has been at a virtual standstill, with poverty levels rising dramatically. According to the Nigerian office of the World Bank, the proportion of households living below the UN's US$1 per day absolute poverty line has grown from 27 per cent in 1980 to 66 per cent in 1996. While rural areas have been worst affected, the incidence of poverty in towns and cities has climbed from 17 per cent to about 58 per cent.[62]

Noting the greater numbers of poor people in developing countries rich in primary commodities such as oil and diamonds, Pauline Luong and Erika Weinthal call this the 'resource curse'.[63] There is also a correlation between oil discovery and exploitation and the absence of democratic representative government.[64] This effect, often noted in relation to Middle Eastern countries, is equally notable in Latin America, Africa and Asia. And this association is confirmed by the many recent court cases in which oil-company executives have been found guilty of paying large bribes to corrupt regimes or warlords in places such as Nigeria, Sudan, Kazakhstan and Angola.[65]

There is also a very clear association between oil extraction, violent conflict and war. From Columbia to Venezuela, from Iraq to Sudan, developing countries with large oil revenues are characterised by civil unrest, civil war, foreign military intervention, military rule, and a high incidence of death by violence.[66] This association between oil and war runs even deeper, for oil is the key fuel utilised by the military in fuelling aircraft, tanks, troop carriers and warships. Oil lies at the heart of the military industrial complex which dominated the warring history of the last century, just as wars over oil, and fuelled by oil, continue to dominate the first decade of the present one. The First World War marked the transition from muscle-based war to oil-based war. At the outbreak of war men and guns were moved more by horse and cart than by any other means.

But the military began to commandeer Parisian taxis, and then designed armoured oil-driven tanks and troop carriers to move men and large guns around the dreadful battlefields of France.[67]

By 1917 most warships were also converted from steam to diesel power and submarines ran on nothing else, while oil-driven aircraft were increasingly used for reconnaissance and then for warfare in and from the skies. Oil and access to oil eventually determined the outcome of the war as German submarines attempted to cut oil shipments from the United States to Europe and the allies responded by cutting Germany off from the Caspian oil fields,[68] thus disabling the German economy and forcing an eventual German surrender. Having learned the lesson of the First World War, Hitler's first objective in launching the Second was the oil fields of Poland and Russia. And the principal opening tactic of the war – *blitzkrieg* – involved oil-driven motorised land invasions. The war in the Pacific was similarly closely tied up with the availability of oil. One of the reasons Japan attacked the United States seventh fleet at Pearl Harbor was because the US had halted its oil exports to Japan. And the Japanese conquest of Southeast Asia was driven by its intent to capture the oil fields of the Malay Straits and the Indonesian archipelago.

The oil-fuelled wars of the twentieth century presaged a new era of technological warfare which seriously compromised traditional Christian just war teaching designed to limit the evils of war.[69] Indiscriminate killing of civilians from oil-fuelled machines deployed on land and sea and in the air were the unique features of this new era of total war; aerial bombers, armoured troop carriers, submarine warfare, rocket-propelled ballistic missiles, factories turning out millions of tonnes of ordnance – all of these depended upon the availability of large quantities of liquid fossil fuel. And the connection between oil and war does not end there. Throughout the second half of the twentieth century world geopolitics has been heavily determined by oil. From the overthrow by the United States of the democratically elected government of Prime Minister Musadeq in Iran in 1952, to the 9/11 attacks on the United States' East Coast by disaffected Saudi Arabians and Yemenis, international relations have been dominated by industrialism's appetite for oil. War and terrorism in the Middle East, and in Central Asia, are all crucially linked to the fact that these regions have been dubiously privileged in containing large underground reserves of oil.

Given this clear connection between the oil-based economy and war, the contention of Wendell Berry that the present industrial economy is not only unsustainable but immoral seems all the more percipient.[70] This is why, as Berry suggests, the industrial economy commits Americans to a state of war both with other nations and with the planet, of which the terror attacks were just one, albeit terrifying, instance of a backlash. Al Gore is therefore entirely right when he links the failure of the Bush administration to respond to immediate prior warnings concerning the terror attacks, the failure of the Bush administration to respond to Hurricane Katrina, and its failure to respond to the threat of global warming.[71]

In addition to the costs of oil extraction there are significant human and ecological negativities associated with oil transportation, refining and end-use. The most direct health hazard arising from end-use, apart from the fuelling of war, relates to the millions of people every year who die or are injured in road accidents. Death on the road is one of the biggest killers of people between the ages of 5 and 35 in the world today, and this holds true for developing countries in Africa just as much as for developed countries such as the United States and Britain. The burning of oil in power stations, furnaces and vehicles also has health effects because of the pollution of the atmosphere from vehicle emissions and particulates. The health of millions of people is detrimentally affected by particulate pollution which is linked with asthma, leukaemia, lung disease and other cancers.[72] In addition, on hot days in summer, levels of the toxic gas nitric oxide reach life-shortening proportions in cities such as Mexico City, Bangkok, Los Angeles, London and Beijing. Pollution from oil tankers and pipelines is another major problem. The Exxon Valdez disaster in Alaska was a particularly terrible example of the general tendency of the industry to degrade the ecosystems in which oil is extracted. Ten years after the disaster scientists are finding continuing ill effects from oil pollution on sea birds, ocean fish and even land mammals. Furthermore, ExxonMobil still refuses to financially compensate local communities, and it is still using single-hulled tankers, which are much more prone to oil spills, in this ecologically fragile region.[73] The dangers of shipping large quantities of toxic substances around the world are further enhanced by the irresponsibility of the oil companies in employing so many rusty and poorly-maintained oil tankers chartered from 'flag of convenience' countries such as Liberia, where regulations are weak and labour is cheap.

Oil is by no means the only fossil fuel to have negative effects on people, species and ecosystems where it is extracted. Coal preceded oil as the great fossil-fuel driver of the industrial economy and coal continues to be the largest single source of electricity in the United States, Australia, Germany, Poland and China, which all have very extensive coal reserves. Coal extraction is a notoriously dangerous and messy business. Coal mining has become increasingly unpopular in Western countries because of the risks involved and the consequently high levels of remuneration sought by miners. Coal companies in the United States have mostly abandoned deep mines and have turned instead to the deeply destructive procedure of mountain-top removal.[74] Coal-burning power stations also do considerable damage to human health and regional ecosystems. These plants emit mercury and sulphur in quantities high enough to raise the incidence of autism and cancer among those living downwind, and also create acid rain, which damages millions of acres of forest in affected areas around the world. Again, despite the heavy toll of coal extraction on local communities and ecosystems, and of coal-burning on the environment and human health, coal industries in the United States, Germany, and China receive extensive support from government. In the case of Germany, the industry receives $8 billion in public support per annum. Subsidies to coal companies have also been a major new item of expenditure of the Bush/Cheney administration in the United States.

Hydroelectric power is often represented as a clean and renewable resource without the associated human and ecological impacts of coal and oil and natural gas.[75] But hydroelectric dams are one of the most destructive of all modern engineering projects in terms of their impact on human communities and ecosystems where such dams are located. The 800,000 hydroelectric dams worldwide have uprooted between 40 and 80 million people, flooded thousands of precious ecosystems and created a host of other ecological and human health problems, including new waterborne diseases and the drying up of river beds used by fisherfolk and as local water sources.[76] Western engineering companies have built most of the world's gargantuan hydroelectric dams with backing from international financial institutions such as the World Bank and the International Monetary Fund because they putatively represent 'development' opportunities for the world's poor. However, the impact of the dams on the peoples and other species on whose lands and water sources they are imposed, is almost universally negative – they typically have no access to the electricity the dams generate, nor even access to clean water. Among the most notorious dam projects in recent years is the Sardar Sarovar dam project in India, which will flood thousands of square miles in Madhya Pradesh and other states in Central India and see the construction of over 3,000 dams in the Narmada River valley and its tributaries. According to its advocates in the Indian government and academia, the dam will provide secure drinking water, irrigation and hydroelectricity for millions of Indians. Special wildlife reserves will be established for threatened species and all displaced residents will be offered land title in other areas, so the environmental and social costs are far outweighed by the benefits.[77] But according to its more numerous opponents the dam project will displace 320,000 people and flood a million acres of forest, while irrigation and electricity production from the dams will only last two or three decades because of the silting up of the dams and the salinisation of the irrigation systems which afflict all such large-scale hydro projects.[78]

Vinod Raina suggests that the real issue is the contest in India between two models of development, the Gandhian model, which would revive the village economy, and the industrial model, which would turn India into a microcosm of the former colonial economy of Britain.[79] The reason the second model wins over the first in India is the caste system. Tribal and indigenous peoples and rural farmers, such as the inhabitants of Jharkand, are mostly from the lower castes. But the powerful elites who live in India's burgeoning cities, and enjoy electricity and piped water, are from the higher castes. The thousands of dam projects undertaken by the government of India since Independence, which take electricity and water from low caste communities and provide it for higher castes, are examples of the continuing role of caste and class in the political economy of postcolonial India.

The Narmada dam project, and similar ongoing hydro projects in China and Laos, are further instances of the myriad sacrifices made by human as well as other species for centralised power delivery systems and the fossil-fuelled industrial economy, which sustain the imperious power of the modern nation-state and the economic corporation. Those who suggest industrial societies cannot afford to reduce their energy consumption on the basis of unverifiable predictions of consequences for future generations neglect entirely the injustices and ecological damage inflicted by these industries on *present* generations and ecosystems.

In India, there are, however, some signs of hope, which may in part be driven by the recognition that global warming is a problem which affects everyone, rich and poor, and requires a range of solutions which includes the tribals and the Dalits. The Ministry of Non-Conventional Energy Sources (MNES) in New Delhi is proposing a series of power-generation initiatives that will bring renewable power technologies to tribal and low caste villagers, and so reduce their dependence on fuel-wood gathering. These technologies include a tidal wave project in West Bengal, a project using coconuts as biofuel in Tamil Nadu, and solar power and oil from the Jatropha plant in Jharkand and elsewhere.[80] In many parts of India the national government is struggling to build and maintain a reliable electric power grid. But in the more remote areas the MNES is pioneering a model of power delivery which may ultimately be both more sustainable and more reliable. Given its base in the local economy, and the small scale of the technologies involved, it is also likely to be more equitable, and as we have seen it is inequity, and not hydrocarbons, which is the principal source of fuel poverty in Jharkand.

Again here we see that there is a hopeful 'upside' to the global warming scenario. Industrial civilisation in its genesis and history has been imperialist in the way in which it has sucked control and power from local communities and ecosystems into centralised pools of wealth and reserves of energy. This process has seen the nation-state and large economic corporations draw ever greater powers to themselves, while households and small farmers are disempowered and lose their property, and control over local ecosystems, to distant bureaucracies and corporations. Addressing global warming with the new dispersed energy technologies which are already being pioneered in developing as well as developed countries will enable a reversal of this process of power accumulation; it will open up space for new forms of radical democracy in which respect for place and the health of ecosystems again become determinative goals, and households and small farms again become sites of careful making as well as consumption, and hence of care for the earth.

Geohistory and the Cedars of Lebanon

While the history of the modern Middle East has been marked by its wealth in hydrocarbons, in the lands known as the Fertile Crescent the most highly prized commodity was not oil – whose uses had yet to be discovered – but the cedar tree. The country of Lebanon was for thousands of years the greatest geopolitical prize of the Middle East, because its famous snow-capped mountains were once covered with great cedar trees.[81] These trees grew in dense forests at an altitude of between 1,400 and 1,800m on the slopes of Lebanon's mountains, and at similar altitudes in other parts of the Near East including Syria and Southern Turkey. Few stands of these trees remain today because of the immense value of the cedar to civilisations in the Near East from ancient times.[82] Valued for its great weight, density and resin, cedar was the supreme wood of the Near East and was used by Phoenicians, Israelites, Egyptians, Babylonians and Assyrians to build large powerful warships, fishing vessels, palaces and temples. The tree and its resin had many other uses; cedar resin was applied to ships' hulls to preserve them, it was used to waterproof earthenware, and in cough medicines and other medical treatments.[83] Cedar trees were particularly prized by the Egyptians. They used the timber to make sarcophagi and coffins as well as boats and they used cedar resin to preserve bodies in the practice of mummification.[84]

Because of its large cedar forests Lebanon was regarded as the greatest strategic prize in the region, for whoever had control of the hills and forests of Lebanon had the power to build great warships and weapons and so control the region. The second book of the 5,000-year-old *Epic of Gilgamesh* describes the cedar forests as the garden of the gods, and it goes on to recount how the hero Gilgamesh has the courage and temerity to clear-cut the forest where the gods lived. The vivid forest episode of the Gilgamesh Epic speaks of a 'green mountain' where tall cedars raised aloft their luxuriance and cast a 'delightful shade'.[85] A possible location of this great forest was Mount Hermon. Though bare of trees today, an ancient forest between Mount Hermon and Mount Lebanon is referred to in Egyptian cuneiform tablets. Three thousand years ago this valley would have been 'a scene of surpassing sylvan beauty, with the two great mountains, deep in forest, soaring on either side'.[86]

This throws new light on references to the beauty of Mount Hermon in the Old Testament. Other more overt references to cedars in the Old Testament see the tree as a metaphor not only of beauty but of the enduring righteousness of Yahweh and of those who worship him:

> The righteous flourish like the palm tree,
> and grow like a cedar in Lebanon.
> They are planted in the house of the LORD;
> they flourish in the courts of our God.
> In old age they still produce fruit;
> they are always green and full of sap,
> showing that the LORD is upright;
> he is my rock, and there is no unrighteousness in him.
>
> Psalm 92:12–15

Cedars are also seen as symbols of strength (Psalm 29:5), durability (Jeremiah 22:14), majesty (2 Kings 14:9; Zechariah 11:1–2) and stateliness (Isaiah 2:13; Ezekiel 17:22).

While these Old Testament texts clearly reflect the admiration of the Hebrews' neighbouring cultures for these great trees, Israel's involvement in trade in cedars is treated with more ambiguity in biblical narratives, and in particular its trade with Tyre. Tyre was a city-state in the region of the Promised Land of Canaan but, though formally granted to the tribe of Asher, it was never actually occupied by them. As coinage and other artefacts indicate, the Tyro-phoenicians were a powerful people because of their control of the heights of Lebanon, and the powerful warships they built with Lebanon's cedars. Their prowess in shipbuilding made Tyre a force to be reckoned with throughout the Levant, and it was not until the time of Alexander the Great that this powerful military base, with its fortified deep harbour, was eventually sacked.

Seeing the strength and splendour of their neighbours in Tyre, Egypt, Ur and Assyria, the kings of Israel apparently also aspired to be like them, as the Book of Samuel predicts in its account of the reluctance of Yahweh to give Israel a king. Solomon was the first who sought to trade with Tyre in order to gain access to the cedars of Lebanon for his great project, the Jerusalem temple. As 1 Kings 5 records, Solomon made a treaty with the King of Tyre, King Hiram, to exchange Hebrew produce for cedar trees (vv. 1–9). But the treaty was not between the people of Israel and Tyre but between two royal households. What is also notable is that the treaty involved conscripted labour from the tribes of Israel, a practice which ran counter to the terms of the covenant society established under Moses and described in Exodus, Judges, and Deuteronomy (1 Kings 5:10–14). Under Moses, Israel is said to have understood its vocation as a people set apart by its covenant with Yahweh, and a significant part of its sense of election was connected with the concept of covenantal justice in Israel, said to reflect the compassionate nature of Yahweh. Under the terms of the covenant practices such as slavery, forced labour, child sacrifice, which were commonplace in neighbouring countries such as Egypt and Assyria, were forbidden among the Israelites.

The institution of the monarchy in Israel is presented in the Books of Samuel and Kings as having the effect of turning the more egalitarian and participative social form of Israel under the federation of the Judges into a society whose people were gradually conscripted into great building and military projects in service of the kings of

Israel. Solomon was the first in a long line of Israelite kings who turned the project of Israel into an imperial project. And the seeds of this turn are clearly indicated by the Book of Kings in the sacrifices which the building of Solomon's great temple required of the people of Israel. With the building of the Temple begins that longer narrative of the Old Testament which laments the turning of the nomadic and pastoral ethic of the patriarchs and early settlers of Canaan into the more elitist urban social forms of the later Hebrew monarchy.

The Book of Ezekiel describes in great detail the military character of the political economy of Tyre, whose people built masts for her warships with the cedars of Lebanon (Ezekiel 27:5). The prophet laments the trading of men and the produce of the land for wares of silver and ivory and gold by nations such as Rhodes and Tarshish, and notes with particular shame that Judah and Israel had also indulged in such trade. Ezekiel also seems to recall the Gilgamesh Epic when he speaks of the beauty of the forest of cedars before it had been laid waste for the military and economic projects of Tyre and other nations. Consequently in the Book of Isaiah we find the cedar has become a metaphor for the pride which resists the sovereignty of Yahweh, a pride which had afflicted not only Tyre and Tarshish but the rulers of Israel.

> The haughty eyes of people shall be brought low,
> and the pride of everyone shall be humbled;
> and the LORD alone will be exalted in that day.
> For the LORD of hosts has a day
> against all that is proud and lofty,
> against all that is lifted up and high;
> against all the cedars of Lebanon,
> lofty and lifted up;
> and against all the oaks of Bashan;
> against all the high mountains,
> and against all the lofty hills;
> against every high tower,
> and against every fortified wall;
> against all the ships of Tarshish,
> and against all the beautiful craft.
> The haughtiness of people shall be humbled,
> and the pride of men shall be brought low;
> and the LORD alone will be exalted in that day.
>
> Isaiah 2:11–17

Just as the laying waste of the lush forested slopes of the mountains was a sign for Ezekiel of the effects of empire written onto the land, so Isaiah uses the destruction of the cedars of Lebanon and the oaks of Bashan as a metaphor for what God will do to those peoples who raise themselves up in pride and who would rule the earth instead of Yahweh. Isaiah is in effect declaring that Yahweh is against those who raise themselves up with the aid of cedars; like the cedars they cut down to achieve their power they too will one day be cut down and laid low. Amos and Zechariah reproduce similar judgments:

> Yet I destroyed the Amorite before them,
> whose height was like the height of cedars,
> and who was as strong as oaks;
> I destroyed his fruit above,
> and his roots beneath.
>
> Amos 2:9

> Open your doors, O Lebanon, so that fire may devour your cedars!
> Wail, O cypress, for the cedar has fallen,
> for the glorious trees are ruined!
> Wail, oaks of Bashan, for the thick forest has been felled!
>
> Zechariah 11:1–2

It is sometimes suggested that the Old Testament is weak in ecological reference and understanding. But it is clear from the prophetic denunciation of the trade in cedars, and their use in fuelling the wars and imperial designs of the peoples of the region, that the ancient Hebrews had a great love for these majestic exemplars of creation. And these great trees are also seen as moral and spiritual analogies for the covenant relations between Israel and Yahweh. When the cedars flourish in their majesty and strength they represent the worshippers of Israel, and the lush forest of cedars is reminiscent of the Garden of Eden in which Yahweh once walked with men. When cut down cedars represent the infection of unrighteous trade and militaristic empires in which the bodies of men are enslaved in the grand designs of idolatrous and proud kings, who have forgotten that Yahweh is the true Lord of all, and that he gifted the land to those who would treat it, and each other, with respect and dignity.

The fossil-fuel economy of our own day, like the ancient trade in cedars, requires a constant onslaught on the natural resource base of this region, and it has for more than one hundred years produced misery and blighted the lives of the ordinary people of the Middle East, even as their rulers, and the oil companies and investors who trade with them, have become enormously wealthy. An end to this trade seems highly unlikely, and yet, as we have seen, the end of oil is no longer a chimera; the rate of new discoveries has fallen in the last twenty years and present levels of production cannot be sustained for much longer. As the price rises and production declines the Middle East will eventually be freed from the burden of this trade. Perhaps then its people will at last be able to enjoy peace, provided that by the end of oil the world's climate is not so over-heated that the region is reduced to desert.[87]

Isaiah also dreamed of the end of empire and the cessation of its demands on the forests and peoples of the Levant. In a wonderful image, he suggests that when Yahweh lays low the proud empires which had destroyed the forests, the trees will at last breathe a sigh of relief and clap their hands in praise:

> The LORD has broken the staff of the wicked, the sceptre of rulers, that smote the peoples in wrath with unceasing blows, that ruled the nations in anger with unrelenting persecution. The whole earth is at rest and quiet; they break forth into singing. The cypresses rejoice at you, the cedars of Lebanon, saying, 'Since you were laid low, no hewer comes up against us.'
>
> Isaiah 14:5–8

For Isaiah, the collapse of the kingdoms of Israel and Judah has come upon them because they mimicked the greed, injustice and pride of their neighbours. Their fall at the hands of other nations will therefore see a great liberation for the land of Israel, and for those who had become poor and destitute under the burden of debt bondage and slavery. The analogy for this recovery of justice in Israel is that the land itself will be restored to its former moist and verdant glory, before the trees were cut down:

> Thus says the LORD who made you,
> who formed you in the womb and will help you:
> Do not fear, O Jacob my servant,
> Jeshurun whom I have chosen.
> For I will pour water on the thirsty land,
> and streams on the dry ground;
> I will pour my Spirit upon your descendants,
> and my blessing on your offspring.
> They shall spring up like a green tamarisk,

like willows by flowing streams.
Isaiah 44:2–4

In this text we see how the great biblical story of salvation is vitally tied up with the liberation of nature, and of trees. And this message is by no means absent from the history of Christianity, which emerged from the deep roots of the religion of Yahweh. Christians believed that the world was remade in the resurrected body of Christ, which is a new form of creaturely being. And for more than 1,500 years the monks of the monastery in the Kadishar Valley in Lebanon protected the last stands of cedar trees in Lebanon. The monks saw themselves as the guardians of the forest even as the Roman Empire collapsed around them. In 1998 these last remaining stands were declared a UN Natural Heritage site.

Power Politics

The Old Testament critique of the empires that the cedars of Lebanon sustained is fundamentally a critique of the great burdens which domineering empires and remote social structures visit on people and planet. At the heart of the greatest piece of political writing in the Christian tradition, Augustine's *City of God*, lies a related critique of the moral failings which beset human political arrangements. Augustine reflects at length on the insecurities and threats experienced by the general populace as a consequence of the violent politics of empire, and the sacking of the City of Rome by barbarians.[88] According to Augustine, social rivalry and competition for power and control lie at the heart of the problem of sin which infects the human social condition. He also comments on the foundational narratives of Roman society, which included the story of Romulus and Remus, and argues that the heroism of these accounts was at marked variance with the moral failings and corruption which actually characterised the institutions of Rome and their tendencies to oppress the downtrodden and sustain the powerful. Corruption, not courage, was the mark of this sinful social order.

There is an analogy between Augustine's critique of the Roman rhetoric of glory and Michel Foucault's critique of the rhetoric of scientific progress.[89] The procedures of scientific investigation involve a sense of detachment from the earth that is deeply implicated in the wider cultural and political rejection of a moral obligation to care for the earth on the part of citizens, governments and corporations. Most climate scientists work in institutions that are heavy users of power, based in Northern countries which emit many times the greenhouse gases of those in the South. Climate change research is itself climate-changing since it relies heavily on power-hungry supercomputers, on satellites sent into space with the aid of great quantities of fossil fuel, burnt in the stratosphere where it does the most damage, and on research trips to far-flung places in jet aircraft with high-level emissions far more damaging than greenhouse gas emissions at ground level. In its tone of serious attention and concern at the present trajectory of the planet, scientific power-knowledge seems to signify that this trajectory has taken place outside science, that it is an externality of which science is not the cause. As Val Plumwood suggests, the implication is that while technology may do damage or harm to the ecosystem, science itself is 'pure knowledge' and as such can do no wrong.[90] And this 'ideology of objectivity' enables the privileged and the powerful to sustain their control and their power, and to hide the acts of expropriation and the structures of injustice which sustain their superiority over the powerless. But 'knowledges that involve injustice to those who are known do not provide accurate or ethically acceptable forms of knowledge'.[91] In other words, if the scientific message is to result in new moral behaviours with respect to the biosphere, it will require not only clear talking but also truthful action, including radical measures to conserve energy and reduce wasteful consumption activities on the part of those scientists who claim that there is a problem.

Many scientists are now far more alert to ecological problems than they were even twenty years ago, and many now dedicate some of their own time to attempting to communicating the problems and perils the planet now faces in the public sphere. They sincerely believe that if they can describe these perils with sufficient clarity, and with the minimum of uncertainty, the public and politicians will respond with policies to match the seriousness of the crisis. However, this strategy misses the extent to which science and technology have already been co-opted in the quest for endless growth and consumption. Science is a core constituent of the nexus of power-knowledge which sustains modern states and corporations in their power over the earth's resources and peoples. It is unlikely that citizens or politicians will respond to the message from the planet in the language of science, for they are used to this language speaking to them of its ongoing success in bringing the planet into ever more complete subservience to the desires and devices of the growth-led technological society. Furthermore, the scientific narrative of climate change is a global narrative which, by definition, is disconnected from the practices of everyday life and their embedding in the fossil-fuelled economy.

The Social Shaping of Fossil Fuel

The pervasive reach of the fossil-fuel economy into the practices of daily life is sustained by networks and grids of energy supply systems. And these grids in turn are socially shaping, distancing citizens from the material roots of their uses of energy and their consumption behaviours, while also enabling the centralisation of economic and political life. Through these grids corporations and nation-states have abstracted material power from communities of place and from households. Much attention is given in the United Kingdom to efforts to reduce household energy consumption, and this is entirely appropriate, given that around 30 per cent of domestic production of greenhouse gases emanates from electricity consumption and heating and cooling devices in households. However, given the demise of the household economy, and the consequent alienation between consumption and making in the modern economy *as* household, it is clear that the causation of global warming is intricately connected with the modern form of political economy. Without fundamental reform of the centrifugal tendencies of economic management and corporate and technocratic rule, the actions of individual citizens can never outweigh the ecological debts being run up by multinational corporations, banks and international financial institutions. As we have seen, the offshore carbon emissions of British banks and corporations exceed all British-based emissions by a factor of six or seven.[92] A programme of greenhouse gas reduction which targets household consumption while neglecting the offshore emissions of British companies is clearly inadequate.

The paradox is that the concentrated material power that the fossil-fuel economy has placed in the hands of corporations and nation-states has accustomed them to forms of sovereign power which they show no willingness to cede back to those from whom it has been amassed. But if it is the case that climate change requires every individual and household to engage in greenhouse gas reduction behaviours, then a topdown political response to the problem is entirely inadequate. Instead efforts need to be made to empower citizens and local communities to recover control over the production and use of power, and to deliberate over ways to conserve it and to produce it sustainably. A political move towards a local and renewable economy of power will assist in creating a new moral climate. There are no governance procedures which alone can centrally deliver the radical changes in lifestyles and behavioural patterns which anthropogenic climate change necessitates if the vulnerable poor are not to be devastated now and in the future by its continuing effects. These effects represent a deeper structural injustice in the global economy, which climate change reproduces in the carbon cycle. Modern citizens consume without regard for consequences distant in space or time from the sites of consumption. Recovering agency and responsibility in the practices of everyday life will involve individuals and businesses reconnecting their production and consumption activities with the ecological and social relationships that sustain communities of place in the bioregions of the earth.

When Thomas Edison first set about providing electricity to domestic consumers and factories to power homes and machines in New York City, his plan was for locally generated Direct Current which he thought, quite rightly, would be more efficient, as well as putting control over this important new resource closer to the end users. DC was only transmissible for around one mile in the kinds of power lines that were then available, and so Edison built small micropower plants in homes and offices. These small power plants were also amenable to the energy-saving features of combined heat and power, and he built the first combined heat and power unit in America close to Wall Street.[93] However, Edison's vision of low-voltage, locally-produced electricity was soon overtaken by the vision of a national grid which would deliver high voltage Alternating Current, allowing large generators to be constructed at considerable distances from the users. The result is the networked societies that now predominate in Europe, the Americas, Asia, Australasia and Japan. The electricity grid has shaped modern society in profound ways, so much so that few in the modern world can imagine living life off the grid.

Here lies the central political conundrum of the climate change problem. The modern world has been so successful in the centralised collection, mastery and dispersal of enormous reserves of planetary power that the household, the nation-state and the corporation have been shaped by the technologies of the grid, the pipeline, and the steam generator. Sheldon Wolin suggests that technological power has played a determinative role in the epochal shift of modern America from the 'politics of citizens' to the 'economic polity', and the same is true of politics in Europe, and in other parts of the modern world. What Wolin calls the politics of citizens was a 'decentralised system of small property owners', or what Arendt calls households, who governed themselves through 'diffused powers' in 'scattered towns, villages, and settlements'.[94] The economic polity, on the other hand, represents 'an imperial system struggling to preserve its global influence while simultaneously launching its power in outer space':

> Once, power had to be operated in a context of localised politics, emphasising liberty, suspicion of concentrated power, and strict accountability of power holders to what was once called a 'body of citizens'. The politics of citizens has since been overwhelmed by an imperial system and rendered marginal and predictable, first by a politics of professional politicians and party bureaucracies, then by a politics of managers, and now by a politics of low military and high technological 'imperatives'.[95]

The amassing of state power is made possible by the industrial division of labour, and the mechanisation of work which represent a bargain between the state and the citizen in which citizens give the state a significant proportion of 'the powers and products' of their labour. Through welfare and warfare the state provides moral and material rewards which confer a kind of collective identity, and even a modicum of security, on the citizen/worker. But over time this process hollows out local politics and produces a 'political passivity' and a transformation of civil society which results in a new dependence of the citizen on giant economic corporations and on big government.

At the same time successive waves of capital formation and technological change produce anomie and dislocation, eroding even those vestigial forms of associational identity and shared moral purpose which the worker citizen had adopted. Under neoliberal economic management, this leads to social disintegration and to what Richard Sennett has insightfully called the 'corrosion of character', where the moral qualities of traditional working class communities are gradually worn away.[96] The response of the state is to

> employ the technology of modernity to amass and centralize power in magnitudes unavailable to previous ruling groups – then to use power to penetrate and change the social relationships, economy, and belief systems of civil society. The twentieth century is the century of totalizing power, of the concerted attempt to unify state and civil society.[97]

Wolin names social structures which make ordinary life technologically dependent on the state and the corporation 'the system'. He suggests that this technological shaping of modern governance has radically undermined the democratic vision of the state as a society of societies as conceived by the Founding Fathers of America, and has brought the postmodern state close to the condition of state domination otherwise associated with the fascist and socialist regimes of Nazi Germany and the Soviet Union.

The new 'megastate' is not only the product of technology but also of organisation for total war. The Second World War in particular, both in Britain and America, produced 'a mobilization of science, universities, and private industry' which made possible not only the manufacture of the greatest and most universal form of destructive power in history,[98] but also the subsequent and continual mobilisation of these institutions so as to put America (and Britain) on a footing of total and continual war, first called the Cold War, and now the 'war on terror'. We may therefore expect that the global warming crisis will be conceived and mapped in similar terms in coming decades. From denial and refusals to regulate or act the state, the corporation and scientific institutions will eventually move to declaring that a new war front has opened up with the climate and only the state and the corporation can save us.

In the United Kingdom advertisements are already being sponsored by ESSO (ExxonMobil in America), and by the partly-state owned oil company BP which claim, despite the poor environmental record of oil companies and their extensive lobbying against global warming science and carbon taxes, that they are already mapping out and investing in a new ecologically sustainable energy future. The implicit message is clear: carry on driving, filling up at our petrol pumps, growing our profits and draining our oil wells, and let the oil companies deal with the future of the planet.

What is most troubling, and yet most insightful, about Michael Crichton's novel of climate change denial, *State of Fear*, is his suggestion that the occasion for what he sees as the conspiracy of scientific and political opinion around what he quixotically insists is the bad science of global warming, is the continuing desire of the nation-state to enhance its power and disciplinary control over the life of the citizen. We can already see this process at work in the growing advocacy by politicians in Britain and elsewhere of the necessity of nuclear power as the means of addressing the fossil fuel emergency. Nuclear power is such a dangerous and expensive technology that it can only be delivered and managed by a partnership between corporations and the nation-state in which risk and security are handled and shared in ways that corrode, or are obscure to, proper democratic scrutiny. As Wolin suggests, it is this tendency towards the privatisation of state power which produces the 'end of politics', for, once the state amasses power and then consigns that power to the private corporation, politics is eliminated.[99] Arendt makes the same point in her suggestion that technocratic power leads to the rule of 'no-man'.[100] In this condition, individual economic and political actors are disabled by their sense of being embedded in technological systems of such complexity that their individual actions seem to be of no account.[101] This is the worst form of tyranny because no one can be held accountable, a familiar position in contemporary British politics where so many public services are now run by corporate technocrats that government ministers often refuse responsibility when these services fail.

As we have seen, the paradox is that this political overwhelming of the citizen by the megastate is the fruit of the promise of technology to liberate humanity from futility and toil. This promise was first anticipated in Francis Bacon's and René Descartes' insistence that scientific rationality would confer on humanity 'utility and power' and make humans 'masters and possessors' of the forces of nature.[102] This control is now so advanced that humanity has in effect *become* a force of nature: species evolve new capacities – such as drug and pesticide resistance in insects and viruses – in response to human technology, and now the climate is changing for the first time in response to the behaviour of one species.[103] But humans lack the capacity to know in advance the consequences of their interventions. Hence the expression of this force is as likely to create new pathologies as to resolve old ones. Humans are not therefore capable of directing the course of evolution, or of changes in the climate, just as they are not in control of the centrifugal forces which have seen the absorption and amassing of political power in the megastate and the multinational corporation. However, technological systems sustain the illusion of control. And this illusion comes at great cost since it is accompanied by a loss of true agency in human labour on the earth, and hence a loss of genuine moral and political agency. This is because as nature is experienced only in its anthropogenically-altered form of risks and threats created by human techniques and tools which have got out of control, the essential human need to 'trust in the reality of life' is undermined.[104] Consequently, human labour on earth loses touch with the life-giving vitality of natural rhythms and purposes and people lose the ability and the will to conserve these through toil and care. The modern desire to be free from the authority of Creation, and from suffering and pain in serving

God's creatures, produces a new form of servitude. As Arendt puts it, 'man cannot be free if he does not know that he is subject to necessity, because his freedom is always won in his never wholly successful attempts to liberate himself from necessity.'[105]

This explains the alienating forms of consumption and production in modern making. Work which takes the form of the servicing of machines, such as the flickering screens which fill so many modern offices, involves a loss of skilful and embodied engagement with the biophysical world and other people. What Albert Borgmann calls the 'device paradigm' corrodes meaningful work in the home and the workplace, and so subverts the quality and character of everyday life. This produces a real moral decline in human experience, as well as in planetary ecology, as 'the world of machines has begun to destroy the world of genuine ends'.[106] The machine world produces societies which lack the sense of endurance, place and stability that was available to humans in previous eras.[107] Consumerism advances this still further as it turns the durable artefacts of human work – clothes, cooking utensils, furniture, houses, musical instruments – into consumer objects which do not endure beyond a few seasons or years. In the resultant culture of waste things are made to be thrown away, and made in such a way that up to 90 per cent of the materials used in the process of making are wasted even before the end product appears in a store.[108] As the fashion-driven cornucopia of consumer abundance multiplies throwaway objects and waste, it produces a growing instability in the humanly-fashioned world of dwelling that begins to destabilise and disrupt the fertile cycles and energy flows, the cycles of life and death in the more than human world, on which ultimately all life depends.[109] In the case of global warming, the culture of waste and instability in the mode of human dwelling undermines the stability of the total earth system on which these cycles depend.

This culture of waste is not only destructive of ecosystems and precious natural materials. It also destroys families, small farms and the household economy: and it pervades and distorts the moral fibre of human communities. As Borgmann suggests, neither conservatives nor liberals have perceived the real threat to the moral character of human life that the modern perversion of making represents:

> In an advanced industrial country, a policy of economic growth promotes mindless labour and mindless leisure. The resulting climate is not hospitable to the traditional values of conserva-

> tives. This is the predicament of the conservatives. Nor is such a climate favourable to 'human development in its richest diversity'. It produces a wealth of different commodities. But underneath this superficial variety, there is a rigid and narrow pattern in which people take up with the world. This is the liberal predicament.[110]

The irony is that the liberal quest for a just society involves a bargain with technology in the belief that more technology will produce a better society. But this does not happen because technological rule requires that liberal democracies give up on deliberation over ends, or on what kinds of taking up with the world make for a good society.[111] Technological modernisation sustains the illusion that it is possible to create procedures and policies that ensure that such good ends as justice or prudence can be achieved without the people being good. This illusion is further sustained by the endless deferrals of agreed social goods involved in the collective pursuit of technological advancement. The rich enjoy speed and visits to a variety of geographical places far in excess of the poor. For the poor to catch up means that the rich need to slow down, but with technological efficiencies and cost reductions the poor start to own cars, or enjoy holidays away from home, and so the rich begin to travel more often on planes and own foreign homes rather than merely having foreign holidays.[112]

The point to note here is that the assumed automatic association between human wellbeing and technological progress endlessly defers real political deliberation over ends. The devotion of modern nation-states to the ends towards which energy-fuelled technological progress directs modern human life results not only in what Bill McKibben calls the 'end of nature' but in the end of politics.[113] Hence the sacrifices the empire of oil requires of those whose lands and places are subjugated to its production, and of those whose lives are blighted by the 'accidental' deaths and injuries, noise and pollution caused by the machines which it fuels, are not subject to genuine political deliberation.

If this seems too large a claim, it is worth recalling that, despite the deaths and injuries caused by road transport, just below one quarter of the energy consumed in the United States and the United Kingdom is devoted to the fuelling of this dangerous mode of transport. Vehicle emissions are killing the planet as inexorably as vehicles kill humans and other fragile creatures that get in their way. But governments do not declare 'war on automobiles' or 'war on trucks'. On the contrary, road building, along with war, is the least likely of all public services to be privatised or contracted out, since it is seen as a core function of the nation-state to keep things moving. Consequently no modern parliament has even discussed the *possibility* of outlawing the car. Only a few religious groups – most notably the Amish – consider them too morally ambiguous an instrument to be owned by right-minded people.

Perhaps we should not then be surprised that political deliberation over climate change has failed to challenge the devotion of the nation-state to economic growth, or the automatic association between growth and wellbeing, and has instead been concerned with the technological means for the continuing fuelling of the growth engine, without which it is assumed modern life cannot be lived at all. Much of the public debate about climate change in Britain and America concerns the *method* of power generation, whether fossil fuel, nuclear or renewable. But this debate ignores the *character* of life towards which the energy-driven economy directs us. As Borgmann suggests, the real power failure is the 'regardless use of power'.[114]

And this lack of care in the uses of power manifests not just in a political and moral loss but in a spiritual malaise. Technological cornucopia and the endless comforts and distractions it makes possible has produced a growing disregard for true spirituality. Limitless energy provides such apparent power and security that the idea of human dependence on God, and on the biosphere, seems increasingly alien. And behind this exchange of metaphysical sovereignty for material security lies another exchange – the god of mammon for the God of Moses. This is the figural significance of the trees of Lebanon in the Old Testament. The material exchanges of labour and land for the high-value cedars of Lebanon reshaped the covenantal and federal economy of Israel. As the Israelites abandoned the laws of the covenant so they abandoned dependence on Yahweh. Instead of devoting themselves to justice and mercy they were consumed by the dream of a grand imperial Israel conquering other nations.

Like the Hebrew prophets who denounced these imperial aspirations, Christ also protested at the bargain with empire which was the source of the centralised authority and power exercised by King Herod through the restored Jerusalem Temple over the people of first-century Palestine. In his offer of forgiveness and healing without recourse to established religious or political authority, Christ offered an alternative model of power, a model which is elaborated in the writings of St Paul about the nature and order of the local congregation. As John Howard Yoder argues, the politics of Jesus represented a paradigm of revolutionary subordination which set it apart from the politics of centralised states and empires.[115]

The Micropolitics of Energy Conservation

The empire of oil is a form of empire with an even larger global reach than those of the ancient world. But its material base and effects are analogous to these earlier empires. The sacrifices required to sustain it are not paid by those favoured citizens who enjoy the good life it proffers. And these sacrifices are so far distant from the lives of the favoured as to hardly trouble their consciences. All empires involve distance between the places they colonise and the places of governance, and between places of production and consumption. But no empire before the modern global economy could sustain these remote chains of command and control over the whole earth. Cheap and ill-gotten fossil fuel is the material base which enables this global extension of distance, and the speeding up and multiplication of its deleterious effects, on colonised peoples and places, and on local and now global, ecosystems.

The careful use of power will require not just a technological change in the methods of power production but radical change in the corporate and imperial systems which are responsible for this global distancing. These structural systems powerfully reshape the daily lives of those whose material needs depend on power grids and oil pipelines, and whose daily rituals are shaped by the consumption patterns which they sustain. If the centralising tendencies of the modern nation-state, the corporation and the global economy are to be reversed, then the surest way to achieve this is to begin to undo the dependence of all these remote command and control systems on cheap energy to sustain their production and consumption pathways. This will create space for and require a new kind of politics, which needs to be cooperative, local, face-to-face, and reorganised around the household and place if significant energy savings are to be achieved. This is because electricity grids, and other forms of centrally-directed fossil-fuelled energy supply, are vastly wasteful. A typical coal-fired power station wastes 60 per cent of the potential heat energy in the fuel, which is expended in the form of waste heat in steam cooling towers. A further percentage, between 10 and 20 per cent, is wasted in the resistance of the transmission lines and the local transformers which transfer high voltage electricity from power station to buildings and factories. While gas-fired plants are more efficient, even these emit 30 per cent of the potential energy as waste heat in cooling towers.

The most significant energy savings in terms of electricity consumption – which is the single largest source of greenhouse gas emissions – are offered by a move away from grid-based supply back to the more localised kinds of generation first envisaged by Thomas Edison. Communities and households which act together to steward a precious resource will require more micropolitical human interaction than those which are hooked up to the inefficient and wasteful centralised delivery grids of the empire of oil. And this is not just a utopian dream. There are only around 1,000 combined heat and power schemes operational in Britain, but in Europe there are hundreds of thousands of such schemes.[116] Many millions of households and firms in Europe and North America, and smaller numbers in the developing world, have also installed microgeneration devices such as windmills, solar panels, fuel cells and household-sized combined heat and power units. The University of Edinburgh, where I work, has recently installed a combined heat and power generator on its science site which is heating buildings as well as generating electricity, as a contribution to a larger quest to improve sustainability in the university's operations. These 'early adopters' are evidence of a response to global warming which also begins to address the malign effects of centralised power systems on political participation and on power relations between the energy-rich and energy-poor.[117]

Power systems which carefully use all the available kilojoules in their energy sources are good because they do not foster or rely upon a culture of waste, and because they care for the earth. Power that is generated locally also fosters the recovery of a locally-managed household economy. Power that is carefully used, locally generated *and* from renewable sources is the best of all. Power from local renewable sources enables human dwelling and economy to be properly situated in local energy flows between the sun and the earth's surface, and within local ecosystems rather than in the global extractive economy, and is thus truly anti-imperial. Such power is also more likely to be carefully used, just because its local sourcing trains households and businesses that their uses of power need to be in scale to its local availability. Equally pollution and 'waste' in local energy systems is more visible to the people who use it and so is much more easily discouraged. And hence the best kinds of local power schemes are ones which do not involve the importing of energy sources from outside the local ecosystem and region. Businesses, households, villages and cities which generate their own power from locally-available energy flows – using appropriate mixes of local biomass, ground-sourced heat, solar, wind and water power – will not only have more secure forms of energy supply than ones that depend on Siberian gas or Iraqi oil. The people who live in such places will also be learning to frame their practices of living in relation to the character of the places where they dwell. The micropolitics involved in true ecological citizenship are dramatically at variance with the remote chains of command and control of the empire of oil. They also represent a fundamental challenge to the liberal ideology of a mechanistic free market, which autonomously coordinates the actions of individual consumers into a vast nexus of production and price.

4
Climate Economics

And when you tell this people all these words, and they say to you, 'Why has the Lord *pronounced all this great evil against us? What is our iniquity? What is the sin that we have committed against the* Lord *our God?' then you shall say to them: It is because your ancestors have forsaken me, says the* Lord*, and have gone after other gods and have served and worshipped them, and have forsaken me and have not kept my law and because you behaved worse than your ancestors, for here you are, every one of you, following your stubborn, evil will, refusing to listen to me.*

Jeremiah 16:10–12

Among the first people on earth to experience cultural and ecological losses from anthropogenic climate change were the Inupiat,[1] the indigenous inhabitants of Alaska and the Arctic Northwest of Canada. They are traditionally whale hunters: whale oil, whale meat and blubber are the staples of their diet and lifestyle and sustain them through the extreme cold of the long dark Arctic winter. In the 1970s the International Whaling Commission (IWC) banned all commercial whaling in order to save whale species from what scientists at the time believed to be near-terminal decline. The ban was a tragedy for the Inupiat because their whole culture, community life, diet, economy and festivals revolve around the whale.[2] Permitted to hunt just twelve bowhead whales throughout the whole of Alaska by the IWC, they formed their own whaling investigation, recruited Western scientists, and were able to prove scientifically what they from their constant observation and deep knowledge of whales in their oceans had known all along: that there were 8,000 bowhead whales in Arctic waters and that a typical annual catch by the Inupiat of 100 or so whales therefore represented no threat to the survival of the bowhead.[3] The IWC eventually bowed to the superior local knowledge of the Inupiat and permitted an annual catch of 75 whales.

In the last twenty years a new threat to the whaling culture of the Inupiat has now arisen from climate change.[4] In a subsistence culture whose hunting and gathering practices are deeply enmeshed in extreme climatic conditions climate change makes the environment, and in particular the behaviour of other animals and of pack ice, much less predictable. The main threat to whale hunting from climate change comes from the changing behaviour of the shore pack ice. Whale spotting, hunting and killing is traditionally performed on the edge of the pack ice with the aid of small boats made of animal skins.[5] Warming Arctic seas and a warmer climate are making the pack ice less stable and hunters are at greatly increased risk of being stranded on ice which breaks off from the shore without sufficient warning. Similar problems affect seal hunting. Seals and walruses live for much of the year on old pack ice because it fosters rich krill growth in the ocean at the ice edge and hence rich fishing opportunities.[6] As the floating ice disappears, opportunities for hunting these species are also reduced.[7]

A similar problem afflicts polar bears, who traditionally stay well clear of human beings because they live on the floating ice pack where they mate, raise their young, and hunt seals. With the melting of the ice pack, bears are forced onto the shore earlier in the year. Bears normally pass the summer on shore where they mostly fast until they can get back out on the ice again, but as they are forced onto the shore for longer periods they get skinnier and hungrier, and increasingly threaten the lives of humans, whose settlements become scavenging grounds for land-locked hungry bears.[8] Some predict that polar bears may ultimately become extinct because of global warming. This possibility has forced the United States government to recognise polar bears as a species at risk of extinction because of carbon pollution, which in turn provides the first internal legal opening for the proposition that greenhouse gas pollution requires regulation by the United States Federal Government.

The warming ocean also has an effect on the teeming life of the Arctic seas. The Arctic and Antarctic oceans are the most productive on earth because the cool waters are zones of upwelling of nutrient-rich bottom waters, which stimulate production of algae and marine invertebrates such as krill, forming a rich food chain for fish and ocean mammals. But with the warming that is now taking place ice is thawing, algae production is diminishing and krill, which feed on the algae, are declining in the Arctic and Antarctic oceans. Consequently, the numbers of whales, seals, penguins and wild birds are dropping, and some Inupiat communities now go through a whole spring and summer without killing a single whale.[9] While some species are threatened by warming, others grow out of control.

Bark beetles, whose numbers are normally held in check by extreme sub-zero temperatures in winter, are ravaging boreal forests in Alaska, many of which are already reduced to eerie skeletal remains of silvery dead standing trees and melting subsoil.

Melting permafrost and increasing open water also threaten human settlements. Without the impeding effects of floating and onshore ice, waves build up over hundreds of miles and erode the shoreline. Cliffs collapse as the land edge is exposed to the action of waves for more of the year. At the same time, the frozen subsoil or permafrost is melting and destabilising the foundations of houses. Some island communities have no stable land left on which to build houses. Other coastal communities are forced to relocate further from the shore.

Inupiat elders and political leaders have tended to resist the scientific claim that changes in the Arctic and in other earth regions were novel and the consequence of an unprecedented anthropogenic influence on the earth's climate. Their folk stories recalled earlier times when the ice pack shrank, the shore ice became less stable, tundra began to melt and mosquitoes were overabundant in the summer months and they argued that these present signs were part of a long-standing cycle of warmer and colder periods. As with the whale count, the Inupiat put more faith in their folk traditions and local knowledge than in modern science, whose practitioners do not dwell in the Arctic and yet claim through their data-gathering instruments and research hypotheses to have access to superior knowledge. However, the changes now observed in Arctic climes in the last twenty years are so severe that elders and politicians alike now accept that anthropogenic climate change is a reality and is threatening the survival of Arctic settlements.[10] Recognition of the problem is made particularly ambiguous for Alaskans because they have benefited significantly from the financial compensation packages negotiated by Alaskan politicians for their communities from oil production in the state.

Climate change is manifest more powerfully at the North Pole than anywhere else on earth; scientists call this the 'Arctic amplification effect', which is in part caused by the different reflective qualities of snow, soil, shrub and open water. When solar radiation reaches the earth different quantities are reflected back out into space depending upon the albedo of different earth regions. The albedo of the earth's surface is greatest at the poles because for substantial parts of the year they are covered in snow and consequently reflect back out into space roughly 80 per cent of the radiated heat from the sun, compared to around 20 per cent for bare soil or vegetation.[11] The melting of snow because of climate heating amplifies the local effects of climate change significantly, producing a feedback loop. The North Pole is entirely constituted of sea ice and recent studies conducted for the IPCC indicate that the Arctic Ocean will be almost free of summer sea ice within fifty years at present rates of warming.[12] This abrupt change in ice conditions will have a dramatic effect on the climate of the Arctic region, and potentially on the planet as a whole, since the dark waters of the Arctic will absorb much more heat from the sun than ice-covered seas. In planetary history ice ages lasting thousands of years have been partly triggered by the increased albedo of ice sheets in the North. Glaciers spread south as they reflected more of the sun's short wave radiation out into space and drove the planet into long cool periods. But these cool periods were also related to declines in solar activity, or shifts in the amount of sunlight hitting the earth from tilts in the earth's axis. This time it is likely that the Arctic Ocean will be driving the climate, an event that will be unprecedented in planetary history.

The melting of sub-arctic permafrost also sets up another feedback loop. The mechanisms here are again complex. One of the factors is increased shrub growth during the longer periods in which the land is snow-free, which then take longer to be buried by winter snowfall, further reducing the albedo of the tundra. Raised atmospheric CO_2 also helps land plants to grow faster and tundra when bathed in increased CO_2, like other earth regions, is more productive, and absorbs more CO_2 than it did fifty or a hundred years ago. However, frozen sub-arctic tundra is also a great carbon store, after the oceans themselves one of the largest on earth. Even though through photosynthesis the tundra is greening and so fixing more CO_2, the rate of this increasing fixing, while it has tracked the increasing release of carbon from the formerly permanently frozen tundra soils, has not absorbed it completely. Arctic stations now show a gradual annual rise in the average quantity of CO_2 in the local atmosphere as the declining albedo of the tundra absorbs more of the sun's heat and releases more of the tundra's long stored CO_2. At the same time, frozen methane is also released from sub-surface soils. Russian scientist Sergei Kirpotin brought the world's attention to an abrupt shift in this process when he revealed that an area of Siberia equivalent to France and Germany had turned from frozen tundra into a wetland in the summer of 2005. Bubbling up through this new wetland was methane gas, which has, as we have seen, twenty times the warming effects of CO_2, although its lifespan in the atmosphere is much shorter. The quantity of methane in Russian peat bogs is so great that it is estimated that the West Siberian peat bog released 100,000 tonnes a day in the summers of 2005 and 2006, a warming effect greater than that of the greenhouse gas emissions of the United States.[13]

In sum, climate change is already having significant effects in Arctic and sub-arctic regions. These effects include disruption of traditional ways of life and in particular subsistence hunting and fishing, destruction of some human settlements, decline of non-human mammals, decline of floating ice and ocean fertility, increase in ocean temperatures, increase in land temperatures, melting of permafrost, melting of frozen ice, melting of glaciers, destruction of forest and excessive insect infestations. Though all of these phenomena affect the Inupiat in ways which harm their culture and welfare, they have nonetheless benefited in monetary terms from the burning of fossil fuels, which have contributed to these climate change effects. Alaska in the last thirty years has been the largest source of oil production in the United States and Alaskan political leaders negotiated very favourable payouts from oil drilling and production to native Alaskans. Like many other indigenous peoples, they have been persuaded through their contacts with Western civilisation to trade their cultural distinctiveness and isolation for cash payments from oil wealth which have bought them satellite TVs, mobile phones, ski-doos, SUVs, large freezers and centrally-heated homes. But these monetary benefits have come at a price, not only in a loss of traditional ways of life and a local and sustainable economy, but, inasmuch as oil from Alaska has contributed to global warming, in climate change which is even more threatening to their long-term survival.

The Gift Economy and Relational Abundance

The Inupiat are an extraordinarily resilient and self-reliant people and they have developed a range of strategies to deal with the changing and increasingly unstable climate. These strategies reflect both the strengths of their community life and of their local knowledge of their environment which has frequently proved to be superior to scientific knowledge. The local knowledge that characterises their symbiotic relation with the nonhuman world is dramatically at variance with the knowledge of the nonhuman world that is prevalent in Western civilisation, because it is not predicated on a radical dualism between human and other than human life. Whereas

Western climate science regards climate change as indicating the need for new efforts to control and manage the earth's climate, an indigenous approach to knowledge finds in anthropogenic climate change a need for greater humility and responsiveness to the non-human world; patient listening and observation of an array of different signals from land, ocean and sky, rather than control and management, are the innate skills of the native hunter and gatherer. This approach is evident in the ways in which Inupiat train their young people to learn to hunt and live in the extreme Arctic environment. They allow them to learn for themselves, to make mistakes and to take risks which may have life-threatening outcomes, because only when their young people are exposed to the power and precariousness of the Arctic elements will they learn the kind of humility which is required of people who live in such environments; it is humility which enables the experienced whale hunter to retreat from failing pack ice, while arrogance and over-reliance on technologies such as skidoos and satellite-assisted weather forecasts leads to situations where hunters are stranded on ice which has suddenly broken from the shore.[14]

Humility in response to natural elements and forces is matched by a form of cultural subordination, sustained by the economic and social practices of the Inupiat, that prevent extremes of inequality of the kind which would undermine the bonds of collective reliance needful for human communities in such an isolated and climatically extreme environment. For the Inupiat a whale caught through courage, skill and low-tech hunting methods is regarded as a gift from the whale species to the community, and the proper reception of the gift requires social sharing. The team who kill the whale gain prestige for their luck, skill and success but the butchering and sharing of whale meat is something that includes the whole community. The hunters get the choicest cuts of meat but every family in a 900-strong community receives a share. This practice of sharing has the effect of preventing any one individual or household from becoming so wealthy that it becomes independent, no longer reliant on the larger community. Humility before the elements and humility toward one another are intimately connected: respect for nonhuman life and a refusal to over-exploit other species or one another are therefore integral in Inupiat culture and this produces simplicity of lifestyle as well as a lack of arrogance with regard to other species or other people.[15]

The other animal to which the culture of the Alaskan Inupiat is closely tied is the caribou. The Arctic National Wildlife Refuge

(ANWR), the last part of Alaska not to have been sullied by the oil-drilling industry, is the migratory breeding ground of a 120,000 strong herd of caribou, and also home to thousands of bears, wolves, musk ox and moose. In 2005 the Bush/Cheney administration persuaded the US Congress to legalise drilling in the ANWR, and the last great wilderness in America is now laid open to industrial exploitation. Local reaction to this news is mixed, but given the already parlous state of the oil pipelines in the Prudhoe Bay area, and numerous oil leaks, the likelihood of new drilling areas being opened in Alaska spells doom for an already threatened ecosystem.[16]

Anthropologist Tim Ingold has spent many years studying human–nature interactions in the sub-arctic North and he tells the story of the Canadian Cree hunter and the caribou. When chased by wolf or man, the caribou typically runs to a halt towards the end of the chase in order to catch its breath, and as it does so it turns to look its predator in the eye. Biologists note that since it is the deer that decides when to start running again, this strategy gives the deer an advantage in the final pursuit, helping it on most occasions to outrun the wolf. However, the same behaviour when it is performed in a hunting chase by humans armed with rifles means that the deer, as it turns its head to its pursuer, provides the ideal opportunity for the hunter to bring it down. The explanation the Cree offer of the deer's behaviour is that the deer offers itself up to the hunter – its body is not taken but *received* by the hunter. Whereas the biologist sees the behaviour of the deer as an example of successfully evolved survival behaviour, the Cree see the animal standing its ground in the hunt as the caribou offering itself up as a gift of life to the hunter.

Ingold uses the contrast between scientific and indigenous explanations of animal behaviour to point up the problematic bifurcation between embodied environment and rational representation that characterises the modern attitude toward the nonhuman world. This bifurcation reflects the Cartesian-informed Western account of human perception as 'an interior intelligence enclosed by its physical container'. On this model, the world is communicated to the mind by sensory perception which responds through reason, planning and execution in the world. Indigenous accounts of perception present a valuable corrective because they recognise 'no physical barrier between mind and world' while they 'posit the self in advance of the person's entry into the world'. In an indigenous perspective 'self is constituted as a centre of agency and awareness in the process of its active engagement with an environment. Feeling, remembering, intending and speaking are all aspects of that engagement, and through it the self continually comes into being.'[17] This account of selfhood is relational rather than autonomous and mechanistic. It assumes that persons are constituted *by* their environments; personhood is then received as gift from person and place rather than, as moderns tend to conceive of it, a property of the individual which is pre-social.

The traditional perception of the situated self is radically different from modern narratives of personhood and exchange. And this contrast is richly described in Marcel Mauss's essay *Le Don*. Mauss argues that the modern understanding of 'gift' as something which is 'freely' given without expectation of return is a consequence of the dominance of capitalist concepts of contract, price and private property in industrial societies. In indigenous cultures gift is understood not in contradistinction to exchange but as a biologically and socially embedded exchange that sustains human relationships in communities of place.[18] At the heart of the economy of gift are gratitude and grace; the one who gives and the one who receives both acknowledge dependence on one another, and on the larger than human world.[19] And the gift economy is therefore an economy which creates abundance, since it depends upon and sustains both human–human relationships and human–nature relationships. In gift-giving ecological and social power is enhanced in the procedures of exchange. Primitive exchange is also situated in the biological cycles and processes through which life is sustained. It lives within these exchanges rather than seeking to control or undermine them. Hence indigenous societies thrive when they foster the fecundity and fertility of the ecological systems in which they reside, rather than relying for their welfare on stored wealth. This is why hoarding of goods in primitive societies is regarded as threatening to community and the common good, because the desire to store and secure status or the future subverts dependence on one another and on the community of all living things.[20]

By contrast, the modern economy rests upon the systemic extraction of resources from relational ecosystems and from communities of place. The paradox is that societies which rely for their security on stored property and wealth accumulation ultimately foster scarcity and waste, while societies which rely on the diversity and fecundity of local ecosystems foster abundance and sharing. Thus the energy-hungry accumulation systems of industrial capitalism have produced scarcity in atmospheric carbon sinks, while indigenous economies which rely on the abundant energies present in local energy flows rarely use such flows to the point that they are extinguished.

But this is not to say that the principles of gift-giving and relational abundance are limited to primitive societies. There are significant examples of gift-giving in contemporary industrial societies, of which parenting is the most essential, although *time* for parenting is increasingly in short supply in the most affluent industrial societies. It can also work in the realm of industrial making. Open source software is created by individuals and placed on the worldwide web for use and review by other computer users. Users make donations towards products they find work well and use regularly. Peers comment on and improve on the software and codes of others. All those involved share agency in the procedures of software making and use, while the best software writers acquire honour and fair reward for their labours, through acquiring reputations as writers of the most reliable software.[21] The procedures which govern open source software are a working industrial model of the gift relationship.

In the global market constructed under the aegis of industrial capitalism, exchange transactions are displaced from the energy flows and relational nexus of local ecosystems and human communities and subjected instead to the rule of contract and the goal of wealth accumulation. As Karl Polanyi argued in *The Great Transformation*, the market economy is a novel moral structure in human history which subjects human societies to an ethical preference for anonymous and autonomous exchanges over exchanges based on relationships between parties who are known to one another and which express justice in the parity of what is given and received.[22] The market construct has the effect of disembedding exchange relationships from human or ecological communities.[23] It thus becomes possible for the activities of exchange and making to appear healthy and flourishing while human and species communities are in fact continuously being destroyed by exchange transactions and the practices of making.

Frontier Capitalism Versus Ecological Economics

The modern capitalist concepts of contract, price and private property find a significant historical root in the first encounters between colonial settlers and indigenous peoples in Virginia and Georgia in the eighteenth century. Observing these encounters, which were sometimes violent, the English philosopher John Locke developed an account of property relations and 'natural resources' which has become foundational to modern political economy. Locke argued that the colonial settlers acquired rights to fence and farm the land because their agricultural industriousness improved the land and enhanced its ability to produce wealth, whereas the Native Americans had simply lived off the land without improving it or redeeming it by hard work.[24] Locke also imagined that work itself needed redeeming, since most of the products of daily work were perishable things such as food and fibre for heating, building and clothing. He therefore introduced the concept of money as the means by which work is preserved from decay, since it is 'a lasting thing which man may keep without despoiling'.[25] This account of nature as standing in need of improvement by human work, and of money wealth as the means to redeem work on nature from decay, is deeply formative both of modern attitudes to wealth creation and industry and to economic practices and property law. It assumes that ecosystems which are not manifestly controlled by humans are less productive than ones which bear a strong stamp of human activity. And it assumes that redemptive wealth is wealth that is stored in money systems where it is impervious to natural processes of decay and renewal.

This assumption is taken up by Smith and Marx and powerfully shapes the modern distinction between labour and work, and the whole enterprise of industrial making.[26] In this new tradition, labour becomes brutish and demeaning, an idea which in part arises from the Protestant misreading of Genesis 1 in which work is said to be imposed on humanity as punishment for human sin. However, in Hebrew tradition labour is in fact 'the blessing of life' and not a curse and hence, as Arendt puts it, the

> inevitably brief spell of relief and joy which follows accomplishment and attends achievement. The blessing of labor is that effort and gratification follow each other as closely as producing and consuming the means of subsistence, so that happiness is a concomitant of the process itself, just as pleasure is a concomitant of the functioning healthy body.[27]

The in-built logic of Lockean political economy is that labour which is determined by the conditions of life, and involves service to the natural cycles which sustain the fertility and health of the body and the soil, is demeaning and even punishing. For Locke, the cyclical nature of human labour on earth is redeemed through work which creates property or money and so can be stored and secured against the cycles of life. This attitude to labour and wealth promotes forms of work and making which refuse the natural laws by which creation reproduces and sustains life.[28] And these forms of work promote a frontier mentality in which land or natural resources exhausted to the point of destruction can always be replaced when settlers move on to appropriate new land and resource banks. Just as the American settlers went West as their animals and farming practices depleted soils that Native Americans had lived off sustainably, but not 'industriously', for thousands of years, so the in-built assumption of economists since Locke is that nature will continue to provide substitutes for the goods and services which the human economy uses to exhaustion. The logic of conventional cost-benefit calculations, as Stephen Schneider suggests, is that '"nature" is either constant or irrelevant' and that 'the economic value of ecological services is negligible or will remain unchanged with human disturbances'.[29]

The extreme form of this logic is expressed by Stephen Hawking when he suggests that global warming will force humanity to 'spread out into space' by developing new forms of space travel which challenge conventional physics.[30] As Arendt notes, space travel trains those who live in the speeded-up world of scientific discoveries and technological developments to imagine that it will be possible to escape the prison of the mortal limits of earth.[31] It is as though the modern emancipation from the patriarchal god as the heavenly father of men 'will end with an even more fateful repudiation of an Earth who was the Mother of all living creatures under the sky'.[32] The idea of the human colonisation of space is a dangerous illusion, given that the nearest solar system which shows anything like the arrangement of planets in this one, and hence at least the potential to support life, is hundreds of light years from earth.

The frontier logic is sharply opposed to the logic of the hunter-gatherer, who does not move on to alternative resources and terrains once present ones are exhausted. Instead he develops customary laws and practices which model a symbiosis between the needs of the hunter and the capabilities of the landscape, so that prey are not hunted to the point where their survival is threatened. The frontier logic is closer to the assumptions of John Locke and modern economists, while the hunter-gatherer logic is closer to the assumptions of modern ecologists.[33]

Ecological economists have tried to bring these competing logics together by means of an ecologically modified set of instruments for valuing natural resources. Constanza and Daily argue that the economic value of all the 'services' provided by ecological systems, including biomass production, nutrient recycling, waste sinks, warmth and water, actually exceeds the total monetary value of the human economy, even after a twenty-fold increase in the size of this economy in the last hundred years.[34] Attempts to value ecosystem services reflect an approach to environmental problems which highlights the intrinsic dependence of the human material economy on the natural economy of the earth, while not challenging the utilitarian account of human interests as the guiding moral frame.[35]

As Charles Pigou first pointed out, in conventional approaches to property rights common resources such as clean water and air have no market value and are therefore always at risk of abuse.[36] Thus a river whose fishing rights are traded by landowners and rented to fishermen is a marketable commodity and its pollution or otherwise acquires market value. But a river where fishing rights are not private but shared by communities through whose land it runs has no marketable value and may be polluted without direct cost to the polluter. Such pollution is called an 'externality' by economists, as it does not generate internal costs within market forces of supply and demand; it is an example of what economists call 'market failure'. The logic of this approach is that for natural resources to contribute to human welfare, and to be protected by humans, they need to be privately owned, or, if they remain in the common realm, their use needs to be regulated and taxed, hence the phrase 'Pigovian tax'.

A recent example of the adoption of a Pigovian approach by regional governments concerns watersheds. Municipalities around the world which are responsible for providing clean piped water to households are challenging the behaviours of property owners with respect to the watersheds they inhabit and which they may pollute by their activities. Part of this project involves education of householders and companies so that they understand that when they put toxic waste, such as engine oil, in a drain they pollute their own water sources. Municipalities are taking this approach in part because they are able to attribute an economic value to unpolluted watersheds which derives from the costs of constructing filtration plants. These treat polluted water from such activities as fertiliser use on agricultural land, outflows from factories into river systems or dumping of noxious substances in drains by householders. Where the watershed is protected from abuse by farmers or factory owners or householders, the need for filtration is reduced and hence regulatory mechanisms which prevent pollution, expensive as they may be for polluters, are more economically efficient than complex filtration plants; by such means unpolluted watersheds are acquiring economic value.[37]

But these reforms to the autonomous procedures of modern economic relationships, for all of their pragmatic value, fail to challenge the deeper conflict between the modern industrial economy of accumulation and waste and the economy of the earth. In the earth's economy there is no 'waste' other than the heat of the sun, which is reflected back off the earth's surface and out into space. Everything else in the earth's economy is recycled and reused. And in this way the earth fosters biodiversity and abundance. By contrast the industrial economy, because it neglects local energy flows, ultimately destroys the relational sources of the earth's abundance and diversity. And this is why the industrial economy is a frontier economy; its practices of accumulation and making rely upon waste and therefore require endless expansion of nutrient sources. It is their tendency to *waste* and wreck ecosystems which is the reason why imperial economies constantly encounter biological limits which ultimately lead to their collapse. Modern humans are the only species to regularly take more nutrients from the earth than they return to it.[38] And this is why the industrial economy is now hitting the limits of the earth system; its destructive culture of waste has reached from the ocean floor to the upper atmosphere. This economy therefore presages a more momentous – because more global – collapse than any predecessor empire or economy.

Global Warming Requires Global Mitigation

To prevent global collapse the nations of the world have begun to talk. But the nature of these talks does not begin to measure up to the depths of the mismatch between the industrial model of wealth accumulation and waste and the earth's economy of relational abundance and reuse. The primary focus of international efforts is not the redesign of industrial making and the procedures of wealth accumulation but the setting of artificially-constructed and socially-agreed limits on waste emissions. Since most of the waste carbon emanates from rich industrialised countries, the Climate Change Convention (CCC) has adopted the position that these countries should bear the costs of mitigating climate change, and be the first to reduce their consumption of fossil fuels to bring them closer to the much lower average levels of energy use among the inhabitants of less industrialised, or industrialising, countries. This approach is given the name of 'contraction and convergence', as articulated by Aubrey Meyer of the Global Commons Institute. Meyer proposes a long-term aim of equal global carbon emissions for every individual on the planet to meet the radical reduction in present fossil fuel use needed in order to prevent dangerous climate change.[39] However, as Meyer points out, the world's economies are presently on a course of 'expansion and divergence' with carbon use and per capita income increasing in industrialised countries while income and fossil fuel use have actually fallen per capita in parts of Africa and Asia under the conditions of neoliberal globalisation.

The moral case for contraction and convergence is given added weight by the observation that while gains and losses from climate change in temperate areas mostly occupied by industrialised countries may balance themselves out under some scenarios, losses will far outweigh gains for developing countries which are mostly located at or near the tropics. Thus while increased warmth in temperate zones may produce more deaths from heat exhaustion in the summer months, it will reduce deaths from cold in the winter: increased CO_2 in temperate zones also promotes faster forest and crop growth. However, at the tropics there are no such gains to counterbalance losses from climate change; as we have seen, monsoons will become less reliable, storms more frequent, malarial and other insect-borne diseases more common, and ocean levels and temperatures will rise.

Contraction and convergence has many advocates but its critics suggest that countries such as China and India, which are not required to act to reduce their energy outputs before 2012 under the Kyoto Protocol because their per capita levels of consumption are so low relative to more heavily industrialised countries, ought to be brought into the process of negotiation and technological change sooner if the problem of climate change is to be fairly and comprehensively addressed at international level. As we have seen, it is this argument which is made in defence of the decisions of the United States and Australia to withdraw from the Kyoto Protocol. Both nations have sought to establish an alternative climate change negotiation process which envisages technological transfer and efficiency gains in energy production rather than conservation or contraction as the way forward. This alternative approach to the international politics of climate change reflects the view that restraints on economic growth represented by attempts to reduce fossil fuel consumption are not warranted by the likely costs of climate change.

The case for not devoting resources to mitigating climate change also draws strength from the argument that environmental regulation and legal restraints on energy consumption represent interferences in the economic market which create 'inefficiencies'. Better, suggest mainstream economists who write on the economics of climate change, to use the market as the principal device for addressing what they describe as the 'externalities' of economic pricing mechanisms.[40] Economists therefore are finding ways to give economic values to carbon emissions and so bring carbon production into the sphere of market relations. This strategy originates from a new neoliberal approach to pollution trading developed by Canadian economist J. H. Dales as a response to pressure from environmentalists to clean up industrial capitalism. Dales suggested that pollution is an economic activity like any other and is best treated as an economic resource.[41] By turning a pollution dump into a marketable commodity it is possible to treat pollution as another wealth-creating component of the market.[42] Under the influence of this neoliberal idea the United States has established markets in pollution permits which can be bought and sold by electricity utilities and other corporations. These markets have not reduced the quantities of particulates, nitric oxide and sulphates emitted by heavy polluters such as the Tennessee Valley Authority, a coal-burning electricity company responsible for extensive pollution in the Southeastern states. But they give the impression that companies are getting serious about pollution since they are paying to pollute, and using conventional economic instruments to price pollution.

Again there is a deep irony here. Opponents of regulation are in one sense quite right. Government limits on the economy of making are inefficient because they discourage activities which generate the shared wealth on which even governments rely for their provision of public services.[43] However, the deeper inefficiency is not in the regulations. The regulations are only necessary because the industrial economy and the procedures of making are themselves more deeply inefficient because of the culture of waste on which they rely. The real solution is not *more* regulation but radical ecological reform of the culture of making. However, so powerful is the modern idealism of the market that it is proposed that the most efficient way to correct 'market failures' such as pollution is to turn these limits also into marketable commodities.

Commodifying the Climate

During the negotiations at the CCC which led to the Kyoto Protocol, representatives from the United States, with backing from Australia, Britain and Russia, proposed the creation of a new global market in carbon emissions and carbon sinks as the principal means to address climate change mitigation.[44] They also made it clear that if the Kyoto process did not introduce carbon emissions trading the United States would withdraw from the treaty process and refuse to ratify the protocol. Such is the status of the world's only superpower that this threat outweighed opposition to carbon trading from Europe and the South.[45] The irony is that, having blocked realistic carbon reduction targets for the principal polluting nations and corporations, and having then forced the inauguration of a carbon emissions trading scheme which further weakened the already diminished targets agreed at Kyoto, the United States still refused to ratify the treaty. The strategy has however been highly influential. Governments, businesses and many non-governmental organisations (NGOs) are now already involved in carbon trading, and this is widely perceived as the best procedure for addressing the moral problems of global warming even though, as Michael Grubb points out, the structure of emissions trading inaugurated by Kyoto 'makes emissions higher than they would be in the absence of trading'.[46] The principle behind carbon emissions trading is known as 'cap and trade'. The idea is that each company and organisation in countries which participate in the Kyoto process are assigned permits to emit carbon up to an assigned cap which is set on the basis of their previous emissions, a principle known as 'grandfathering'. If they want to emit more than their cap then they must go into the carbon market and purchase carbon permits.

The carbon emissions trading scheme inaugurated under the Kyoto Protocol includes two major elements. The first is called Joint Implementation (JI) and under this mechanism carbon producers can purchase permits to produce carbon in excess of agreed emissions caps by acquiring carbon credits conferred on nations or organisations which produce less carbon than their allowances under CCC agreements. Russia is the owner of the largest quantity of carbon credits in the scheme because of the shrinkage of the Russian economy since the collapse of the Soviet Union in 1990; Russia's reduced economic activity on 1990 levels gives it approximately $10 billion of 'avoided emissions' under the agreement. Consequently, organisations which plans to exceed their carbon allowances can purchase carbon permits from Russia on the carbon market. In this way, countries which are committed to real physical reductions of national output of CO_2 will actually be able to exceed these reductions provided their constituent carbon producers have purchased the proper permits. Neither Russia nor those who trade permits with Russia are incentivised by the scheme physically to reduce their carbon emissions.[47] The second element of the CCC emissions trading scheme is known as the Clean Development Mechanism. Under this scheme heavy carbon producers such as oil companies and electricity utilities can acquire carbon permits by investing in energy-reduction technologies and new carbon sinks in developing countries. Similarly, governments and organisations in developing countries can establish methane reduction plants or renewable energy generating facilities, and once these are certified they can then sell carbon credits on international carbon exchanges.

The Kyoto Protocol established only the bare bones of an emissions trading scheme and the European Union in 2005 created the first working transnational market in carbon permits, known as the European Emissions Trading Scheme. Under the scheme, any organisation which creates combustible power above 20 megawatts must have permits to emit carbon dioxide. If it emits beyond its allowances it must purchase more permits or pay a fine. Carbon permits are traded on the European Climate Exchange.[48] The scheme only targets stationary carbon dioxide emitters such as power utilities, oil companies and cement factories, estimating their emissions on the basis of metered purchases of coal, gas and oil. It excludes 50 per cent of European carbon production, including all forms of transport and domestic and small business users of electricity and gas. At the inauguration of the scheme large quantities of permits were handed out to power generators and oil companies. In the United Kingdom power generators and oil companies received around £1 billion worth of permits for free. As a result, the price of carbon has fallen to just 0.27 per tonne at time of writing because the caps were so generous that no incentives were provided for carbon reductions in the first phase of the scheme to 2012. And consequently this approach has not yet had any effect on the European output of greenhouse gases. But it does represent a significant further advance in international law of the neoliberal utopian dream of regulating all human activity through the interaction of supply and demand curves, and autonomous economic management, instead of through moral and political deliberation in parliaments, courts and local forms of governance. Carbon emissions trading also represents an attempt by governments to avoid their responsibility for legal regulation of carbon emissions which would make the polluters pay – through taxation and fines – for the damage they are doing to the climate.

Carbon trading in its various forms is a highly ambiguous development. The new global carbon market is not incentivising real reductions in emissions. But it has created tremendous new trading opportunities, and new opportunities for fraud and injustice.[49] Verification of the viability of carbon sinks established or purchased by countries or corporations would require extensive global environmental policing, and the system relies on extensive self-policing by both polluters and those who sell carbon permits to the carbon market under the Clean Development Mechanism. Thus under the Kyoto scheme some forests planted after 1989 can earn carbon credits. But this incentivises agroforestry companies to log and burn old growth forests – which are significant carbon stores – and replant them with industrial monocrops such as fast-growing eucalyptus. Theoretically Kyoto forests cannot be planted on old growth forests and must represent a real increase in forest cover. However, if an old growth forest has been cleared, used in some other activity – such as soya farming – and is then turned over to an agroforest, then it can count under the scheme.[50] This of course creates a strong financial incentive for countries and companies to clear forests and eventually draw them into the emissions trading scheme, since old growth forests are not permitted to count as sinks under the scheme even though they are far more precious – in terms of avoided carbon and methane outputs if they are not felled – than new forests.

One such 'carbon sink' project in Brazil, funded by the World Bank, has involved the clearing of thousands of acres of rainforest and the wide-scale spraying of pesticides and herbicides which adversely affected local people and domestic and wild animals. The company turned over forest land to eucalyptus production which was then turned into charcoal for use in iron ore smelting. The company then claimed carbon credits, not directly for the forest but for the use of a 'biofuel' in its iron ore smelters. The protests of indigenous groups and NGOs at the deleterious effects of this project did not prevent the company involved, Plantar S. A., from claiming carbon credits under the Clean Development Mechanism.[51] A study of drained forest lands in Kalimantan in Southeast Asia which are currently used for palm oil, and would therefore be eligible for carbon credits if turned over to monocrop forest, reveals that the drying peat in these former rainforests continues to release billions of tonnes of CO_2 and methane for years after the original forest is cleared. Consequently such areas could continue to be net carbon emitters, even though they could be eligible for carbon credits when replanted with trees.[52]

Agroforestry projects are also notorious for the ways in which they exacerbate fuel and land poverty. Where such schemes are run by corporations or state agencies with a focus on commercial species like eucalyptus, they have had very poor outcomes in terms of restoring forests or meeting the needs of local people. Another difficulty with agroforestry carbon sinks is that the new forests will only act as carbon sinks if the trees grow to maturity and are not then harvested and utilised in a similar way to their predecessors.[53] Although the emissions trading scheme cannot guarantee any of these conditions, it nonetheless treats new forest schemes as carbon sinks with the capacity to absorb real carbon emissions. Again, we encounter here the competing logics of the industrial economy of waste and wealth accumulation and the earth's economy of diversity and relational abundance.

Modern forestry is a discipline invented in the eighteenth century, and emanating from the same idealist stable as classical market economics. It is about the economic management of monocrops – that is, of single-species tree plantation. These monocrops were presumed to be more 'efficient' than diverse natural forests because it was thought they produced more timber mass than more chaotic but species-rich natural forests. However, it now emerges that the monocrop is far less 'efficient' than a naturally diverse forest. The natural forest fosters an extraordinary abundance of species: ants, beetles, worms, birds, small and large mammals, all dwell in unmade forests. And it turns out they do not just dwell there but they are intricately involved in the fostering of a rich and productive eco-system. The ants and worms break up the soil and enable it to soak up rainfall and get it down to the roots of the trees. Other insects and birds fertilise the trees, create humus which conserves surface soils, and assist fungi and moulds in breaking down dead trees, branches and leaf mulch into fertiliser which again adds nutrients to the tree roots. A diverse forest therefore produces more lumber, and utilises more carbon dioxide, than a monocrop precisely because it is home to such a great diversity of species that are all intricately involved in enhancing local energy and nutrient flows and so in sustaining the earth's economy of abundance.

By contrast the monocrop plantation involves the destruction of this prior nutrient-rich and species-abundant economy, and the release of the large quantities of carbon and methane sequestered in such forests into the atmosphere. The forester then plants his monocrop in serried ranks, and often so close together that little else can live or grow. But such a forest looks more 'efficient' in the optics of the accountant's rule. Trees on a plantation are eminently countable – the eventual lumber they will supply can be measured and their market worth thus accurately inscribed and exchanged as an economic value.[54] And when the forest is clear-cut to realise this value it leaves behind an even more degraded ecosystem than that left behind when the original forest was cleared. This countability also makes monocrop forests better candidates for emissions credits than efforts to regrow mixed-species forest, which would restore some of the lost biodiversity of the old forests.[55]

The idea, now at the heart of the Kyoto Protocol, that the spread of more corporately-owned monocrop plantations will *reduce* carbon emissions is therefore a monstrous lie. Carbon trading is leading to a new form of imperialism, 'carbon colonialism', which effectively commodifies the earth's atmosphere and forests, privatising a common resource in the monetised form of carbon credits to be traded between governments and corporations.[56]

Given the limited carbon absorption of new growth as compared to old growth forests, and the large quantities of permits given for free to carbon polluters under the grandfathering principle, the new European carbon emissions exchange has so far provided few incentives for heavy energy users physically to reduce their carbon outputs. Dieter Helm suggests that in the early days of emissions trading it should be expected that the price of carbon remains low.[57] However, if corporations reduce their emissions faster than the carbon credit scheme allows, they are actually penalised, and governments have handed out so many carbon permits to corporations that their market value is in any case too low to provoke serious energy reductions. Estimates of real reductions from Kyoto are as low as 0.1 per cent because so much offsetting is allowed under emissions trading that few countries or corporations will be forced to reduce their physical emissions to meet their Kyoto commitments.[58]

If the intention of the Kyoto Protocol was to incentivise individual firms and householders to reduce emissions through a decentralised market-driven approach, this could have been achieved far more equitably by giving carbon permits to poorer developing nations who could then have sold them on to corporations in exchange for assistance with renewable energy projects, or by giving them to energy-poor citizens of industrialised nations who, with the proceeds of the sale of these permits, could have been required to insulate their homes or buy advance quantities of energy from energy utilities to reduce their future bills. The current scheme simply rewards the heaviest polluters with a large injection of newly-created public wealth in permits to pollute atmospheric carbon sinks, even although they have already profited for decades from this same common good without having paid for it.

Anil Agarwal criticises the Northern bias of international climate negotiations which permit developed countries to 'grandfather' their emissions, while developing countries who have emissions far below those of the North are still expected to come into the Kyoto process at some point on the basis of a commitment to restrain growth in their own emissions, even though their present emissions are a small fraction of Western emissions.[59] Agarwal suggests that the poor outcomes of these kinds of arrangements reflect the inbuilt bias of the global scientific narrative of climate change which privileges industrialised and developed nations, where most of the research is undertaken and framed, over less developed peoples.[60] To counter the perceived bias to the rich, and to economic corporations, built into the practice of emissions trading, priority should be given instead to those suffering the worst effects of global warming. Whereas the aggregate approach to value confers the new monetised carbon credits on the most powerful agents of global warming, including oil and energy companies, steel and cement makers, a fairer outcome between developed and developing nations could be achieved with a conception of the human good, and of justice which reflected an 'option for the poor' and focused on the least powerful in the climate change scenario.

Personal Carbon Trading

International efforts to address the problem of global warming through cap and trade and other market procedures are replicated at nation-state level. Carbon rationing has become the favoured solution among climate change campaigners and environmental activists in the United Kingdom. Myers Hillman, George Monbiot and others propose that every citizen be given a personal carbon cap in the form of a carbon ration card, which indicates the quantity of carbon they are permitted to emit and with which they buy fuel for electricity, cooking, heating and travel.[61] Each transaction would be logged on the card in much the same way as transactions are presently logged on credit and debit cards. Individuals would also be permitted to buy more carbon credits when they had used up their allowance, just as companies are already permitted to do, while those who did not use all their credits could sell them on the carbon market. In principle this sounds like a grand idea. Poor people who use less carbon would gain financially by selling their rations to richer people. A trading system in carbon rations might even be constructed internationally alongside the inter-governmental and corporate trading system, so that consumers in rich countries could buy permits to pollute from those in the developing world who produce very little carbon. But like corporate emissions trading, carbon rationing does nothing to address the underlying conflict between industrial making and the economy of the earth. Instead it merely replicates the Kyoto solution of cap and trade at local level. Carbon rationing would also require heavy bureaucratic regulation. All transactions involving carbon would need to be monitored and registered through computerised and bureaucratic supervisory procedures. This would involve an exceptional level of government surveillance of household and personal life, even exceeding those already inaugurated by Britain and the United States in pursuit of the 'war on terror'.

In its favour, carbon rationing has a beguiling simplicity and seems to promote equity. All, rich and poor, are given the same carbon ration. As in a time of war, it is thought rationing might produce a sense of social solidarity, of shared endeavour in responding to the climate crisis. Also it brings home the problem of climate change to every citizen and household. Some are not waiting for a bureaucratic scheme to be imposed and are organising themselves in local climate action groups. Individuals in these groups voluntarily commit to a carbon target, and where they exceed the target they pay a fine to the group. Through such groups individuals and households encourage each other in living a lower-carbon lifestyle. And by embracing the practice of rationing these groups are pioneering what some believe is the solution to global warming for humanity as a whole.

There is however an alternative to either personal or corporate emissions trading, and this is the adoption by governments of a Pigovian approach, with the introduction of taxation on greenhouse gas production. Currently industrial economies tax productive activities by corporations and workers. Thus corporation taxes tax profits, while income and payroll taxes tax employment and human work. Traditional pre-modern economies by contrast taxed not work, which can be productive without necessarily burdening the planet ecologically, but physical commodities. A tax on the land and its product was the first historical form of taxation, as recorded in the Joseph saga. A shift of the tax burden from employment and profits to carbon, by means of a carbon tax, would then have significant traditional precedent. Such a shift would also be the most effective means for shrinking the carbon footprint of advanced industrial economies in the short timespan of twenty years, which is all the time many scientists now believe humanity has to mitigate climate change before the earth proceeds towards a runaway climate disaster. Carbon taxation would turn the already existing fiscal instruments of industrial societies to the moral imperative of reducing carbon emissions. And it therefore does not require the creation of new markets in carbon rations and carbon permits and the expensive and invasive forms of bureaucracy and surveillance such rations and permits involve. It would also likely produce less corruption. As Grubb notes, the carbon market is already prone to abuse, and there are few effective penalties in the system as currently regulated.[62]

A shift in taxation from profits and jobs to carbon would also have other human and ecological benefits. It would reduce the costs to employers of creating jobs since, if new jobs were carbon neutral, companies would not incur increased contributions to social security and other kinds of payroll tax when they took on new staff. At the same time it would dramatically increase the price of the most climatically harmful activities such as air travel, car travel, electricity consumption, space heating, meat eating, and long-distance transportation of food and manufactured goods.[63] Consequently it would exercise a powerful restraint on economic globalisation, which has produced dramatic growth in global movements of resources with associated increases in energy costs. A move of the public fiscal base from production taxes to carbon taxes would also mean that current cost disincentives to the installation of energy-saving insulation and energy-producing devices such as windmills and solar heating would disappear. Households and corporations would reap fiscal rewards for all such investments in energy saving and renewable energy generation. The principal downside of carbon taxation is that it would be fiscally regressive, hitting the poor even harder than conventional taxes.[64] But this is not an irresolvable problem. The most equitable way to compensate low-income households would be with fuel credits.

Comparative studies of alternative ways of managing reduced emissions by consumers and corporations indicate that carbon taxation is by far the most efficient method of reigning in consumption and production activities to the available carbon sinks of the planet, since on the one hand it promotes low carbon activities, and on the other it enables a shift in taxation away from polluting to non-polluting activities.[65] But the failure to recommend carbon taxes in the policy arena reflects the conflict between genuine efficiency and the preference of neoliberal economists for market approaches to value. Taxation is a political act which garners public resources from private economic activities for use in the public provision of services and law enforcement. But neoliberal ideology opposes taxation as an unwarranted imposition on wealth creation: hence the rush of corporations to relocate many of their activities to 'free trade zones' and low-tax regimes in developing countries. Hence also the considerable efforts of corporations and wealthy individuals to evade and avoid tax through dubious accounting devices, tax havens and foreign domicile.

Resistance to carbon taxes, and the move towards carbon trading as the dominant and so far failing method for mitigating greenhouse gas emissions, reflects the powerful corporate lobbying which has gone on around the Kyoto process. The creation of a market in carbon, for individuals, governments and corporations, represents a tragic distraction from the urgent need to re-regulate the money supply and reinvent industrial making so that they are brought back into relation to the indigenous energy and nutrient flows of the earth system.

Moral Myopia and Cost Benefit Analysis

Behind the Kyoto process lies a range of economic and political actors who are wedded to the core assumption of modern economics, which is that constant growth in the transformation of ecological resources into industrially-produced goods and services, and thence into waste to be buried in the ground or emitted into the oceans and atmospheres, promotes human welfare. In conventional economic theory, economic growth measured in terms of GDP is assumed to generate increases in human welfare regardless of the ecological destruction and waste involved.[66] It is therefore also assumed that since growth is sustained by fossil-fuel use, reductions in fossil-fuel use will produce reductions in human welfare which outweigh any potential benefits arising from the reduced costs of adapting to climate change in the future. Thus Lord Stern estimates that the costs of climate change may represent 20 per cent of present consumption levels in the future and that this will lead to a reduction in welfare equivalent to the lost level of consumption, even though empirical evidence fails to find a clear association between the levels of consumption enjoyed by industrial consumers in the last forty years and increased wellbeing.[67] Indeed, the evidence is in the opposite direction. On the analysis we have so far offered, growing alienation and a sense of futility and *ennui* in contemporary consumer societies suggests that the turn of human making to instrumental and commodified exchanges in a culture of waste produces a deep alienation between humans and their own labour, and between human work and the fertility of the earth.[68] And this alienation ultimately manifests as a spiritual loss – a sense of no longer being at home on earth.

Against the assumption that increased GDP produces increased wellbeing, ecological economists point out that many events that are extremely damaging to human beings and the planet, such as car accidents, deforestation or fishing a species to extinction, are currently counted as benefits by the crude measure of GDP. Other economists point out that beyond a certain level of income individuals or communities do not report increased wellbeing, and that societies which pursue economic growth regardless of other policy aims – such as equity – manifest declining reported states of wellbeing.[69] Despite these observations, governments and intergovernmental agencies continue to act on the assumption that increased economic growth equates to increased welfare. Even the IPCC adopts this core economic assumption in its models for estimating the costs of efforts to mitigate climate change, although it attempts to balance what it calls 'top down' estimates of the cost of mitigation with 'bottom-up approaches' which allow that fossil-fuel driven GDP growth may not equate to increased wellbeing in all households and communities.[70]

This devotion to growth regardless of costs and waste explains why economists have been particularly resistant to the policy implications of scientific predictions of climate change, and have played up uncertainty about the extent and timing of the costs of climate change. Citing the wide range of costs of climate change to developed and developing countries as estimated by the IPCC, and even larger ranges of estimates of costs of reducing fossil fuel use, McKibben and Wilcoxen argue that 'uncertainty is the single most important attribute of climate change as a policy problem'.[71] Since the estimated costs of mitigating the future effects of climate change are so great, and the nature of these effects so uncertain, many economists consequently argue that it is more economically beneficial to plan to adapt human behaviours and procedures to climate change when it occurs than to regulate economic activity so as to reduce present carbon emissions so that these potential future costs may be reduced.[72]

The problem with such economistic dismissals of the need to mitigate climate change in the present is not just that they undervalue costs to future generations of climate change, but that they involve measurements of cost and benefit which are so theoretical as to misrepresent the real world of biogeochemistry. The reason is simple; cost benefit calculations are conducted on the basis of theoretical economic rules of supply and demand, and monetary accounts of profit and loss. But these rules and accounts notoriously fail to count as costs many of the environmental and social costs, dubbed 'externalities' by economists, which economic activities impose upon individual and collective bodies in the real world. Climate change is the most dramatic and long-lasting of all such 'externalities', but as yet neither corporate nor national accounting systems include climate change effects in monetary measures of economic activities.

Economists are also resistant to efforts to physically restrain greenhouse gas emissions because of the time lag between reduced emissions and any effect on the climate.[73] On these grounds, even the minimal targets of the Kyoto Protocol for carbon emissions reduction of between 2 and 6 per cent for industrialised countries between 2005 and 2012 are criticised by mainstream economists: adaptation to climate change when it happens is the more 'rational' course to follow because it will involve fewer measurable costs than mitigation through efforts to reduce energy consumption.[74] As Cornelius van Kooten suggests, 'the general conclusion from economic research is that immediate action to avoid climate change through aggressive mitigation policies may be premature' and 'most economic models tend to come out in favour of adaptation over mitigation'.[75]

Even when economists take a stance in favour of mitigation, most tend to argue that the wise approach is to delay mitigation for as long as possible until more is known about the likely harms inflicted by climate change – in other words, until they are actually happening to lots more people.[76] This means that the only grounds on which they commend mitigation is on the basis of what is called a 'no-regrets' strategy: where there are benefits or cost savings other than unpredictable climate change costs from energy efficiency and reductions in fossil-fuel dependency, then consumers, corporations and governments should pursue these.[77] Thus less dependence on oil in the United States would mean less need to fund expensive military interventions in the Middle East, so more fuel-efficient cars – widely available in other parts of the world – might be adopted by US consumers because it would reduce the costs of US foreign policy.

Discounting the Future

There are many difficulties with conventional Cost Benefit Analysis in relation to ecological issues, and climate change perhaps more than any other problem brings these difficulties to the fore. Two of these difficulties concern social discounting and compensation. The economic practice of social discounting compares the monetary value of investment now and investment in the future. Discount rates are designed to estimate relative costs and benefits of investment between the present and the future through a computation of factors which include economic growth, monetary inflation and bank interest rates. The assumption is that because of these factors, the costs of present activities will be lower to people in ten or twenty years because economic growth will have made them more prosperous, and inflation will have reduced the value of the costs. Interest rates also impact on the calculation because money invested on a future good, as compared to money left in the bank and accruing interest, makes a good purchased now cost more than a good purchased in a year's time, because in a year's time interest rates will have increased the value of money saved by not purchasing the good now, and it will therefore be possible to buy more of it with the money saved.[78] On this basis the comparative advantages of different courses of action in the present and the future are assessed and choices made about preferable courses of action. Typical cost discount rates are around 5 per cent per annum, but given that climate change involves very long-run costs, this means that in only two decades the costs to future generations of harms from climate change are discounted to near zero. Using this logic, the benefits of economic activities which threaten harms to future generations beyond twenty years always outweigh the costs.

It is this kind of calculus which determines the opposition of bodies such as the United States government and Senate to fossil-fuel use reduction designed to mitigate climate change.[79] If the costs of climate change cannot be clearly quantified, and therefore demonstrated to exceed the costs of adaptation, then no action that would harm the US economy should be taken to reduce fossil-fuel use. However, this approach neglects the gravity of the problems that future generations will face if climate change is not mitigated by action now.

One economist who has staked his reputation on a more precautionary approach to estimating the balance between present and future costs of climate change is Lord Stern, who produced a major review of the economics of global warming in 2006 for the British government. Stern argues for a much lower than conventional discount figure of only 1 per cent, and suggests that when this figure is adopted the economics of climate change mitigation become much more favourable towards action by the present generation. This is how he arrives at his estimate that effective mitigation of climate change which reduces global emissions by 2–3 per cent a year for the next fifty years will cost only 1 per cent of global GDP. If action is delayed and emissions not reduced, Stern estimates future costs of adaptation and mitigation could be 20 per cent or higher of GDP for future generations.

Stern also makes a very telling point when he argues that the decisions economists make about discount rates and monetary values are not simply reducible to mathematical equations and statistics but reflect instead moral decisions about the scale of values in particular societies.[80] The extent to which moral decisions of a peculiarly repugnant kind have influenced costings in relation to climate change is indicated in approaches to compensation in economic approaches to climate change. The conventional position is that if group A stand to lose from the economic activities of group B and group C stand to gain from these activities, then provided group B, or groups B and C, find ways of compensating group A for their losses, then advantages for all are maximised and proceeding with the economic development will promote economic (Pareto) efficiency. But such calculations, though fundamental to all modern economics,[81] are problematic because they presume that all losses can be compensated for in monetary values. But there are many ecological and human qualities the loss of which is not reducible to monetary values; these include attachments to particular places and communities and preferences for peace and quiet, beauty or biodiversity. With climate change this is even more the case, since the possibility of life itself may be lost by future generations if drastic actions are not taken by present generations to curb fossil-fuel use.[82] A further problem with this kind of economic rationale is that the calculation assumes equal access to social power among the three groups, whereas in most social situations there are power imbalances between different groups. And such power imbalances clearly exist in the global warming scenario, since heavy carbon polluters are wealthy while those who stand to suffer loss, or even risk death, from global warming are poor farmers, fisherfolk and rural or urban squatters.

Economists attempt to resolve disparities of income in estimating the relative costs and benefits of climate change, and of attempts to mitigate it, by measuring the cost of human life itself in relative terms. Fankhauser establishes an economic ratio of 15 to 1 between the value of a human life in an 'advanced' industrial nation such as the United States and in a low-income country such as China.[83] This ratio was controversially adopted in the estimate of costs and benefits of mitigating and adapting to climate change in the IPCC's *Second Assessment Report*. The way in which these estimates of the value of life are arrived at is by the concept of 'willingness to pay' which is derived from the monetary sums paid by insurance companies in developing countries to relatives in cases of accidental death.[84] This value is then taken as the base line for comparing the likely costs arising from the deaths of individuals in developing countries from extreme weather events caused by climate change with the costs to high-energy consuming countries of reducing their energy consumption in order to mitigate climate change.

The adoption by the IPCC of these differential values of human life was heavily criticised by economists and politicians from India and other nations in the South. As the Indian environment minister, Kamal Nath said in a letter to the first meeting of the Conference of the Parties:

> We unequivocally reject the theory that the monetary values of people's lives around the world is different because the value imputed should be proportional to the disparate income levels of potential victims … it is impossible for us to accept that which is not ethically justifiable, technically accurate or politically conducive to the interests of poor people as the global common good.[85]

The core problem here is the assumption that money is always substitutable for non-monetary goods and that the money economy exists within a material frame where all goods, even life itself, are theoretically exchangeable for other goods. Thus in conventional economics if one good, such as a precious metal or a species of tree, is exploited to extinction, human creativity operating through the laws of the market will find effective substitutes as prices of such commodities rise to the point that research into alternatives becomes potentially highly rewarding. But there are certain biological materials and qualities essential to human life – carbon sinks, climate stability, oxygen, soil, sunlight, water – for which no technological substitutes can suffice. In this perspective it makes more sense to think of the human economy as a 'wholly owned subsidiary of the earth's systems' rather than as a monetary system which is physically independent from the great economy of the earth and therefore endlessly expandable.[86]

The Ecological Failings of Cost Benefit Analysis

The dominant economic model involves an implicit set of value judgments, as a result of which industrial societies undervalue ecological and social goods – what some now call ecological and social capital – and overvalue manufactured goods and trade in goods and services. The primary value that modern economists most strongly affirm in their models and descriptions of human behaviour is efficiency. The second value preference is for market aggregates of individual choices and decisions – the laws of supply and demand – as a device for pricing goods and services over collective or co-operative deliberation. As Clive Spash argues, both of these value preferences are highly questionable. The claim of efficiency in present market arrangements assumes that the present state of the economy is more efficient than one in which greater efforts are made to reduce waste and reduce greenhouse gas emissions. Estimates of the costs to world GDP of such reductions put them in the range of many trillions of dollars.[87] However the assumption that the present use of energy and other resources in the economy is *less* costly and more efficient than possible alternatives is highly dubious. It neglects the intrinsic waste and the constant expansion into new ecosystems that the present industrial economy requires. And hence it neglects the fact that a number of corporations, and householders, have already sought and found ways to dramatically reduce their fossil-fuel use without detriment to the health of the firm or the household. For example, the world's largest carpet producer, Interface has radically reduced its use of fossil fuels and its levels of waste, and hence its running costs, by radically redesigning the way it makes carpets. The long term goal is to produce carpets entirely without waste, and the planned tactic to achieve this is for the company effectively to rent carpets to their purchasers for as long as the carpet is required. When it is no longer needed the carpet is taken back by the company and used as a nutrient source for new carpets.[88]

Economists fail to appreciate that market models are covert valuing devices in which human lives and ecosystems are traded against economic growth in market aggregates of costs and benefits. And these devices are not even very good at doing what economists imagine they are so good at, which is promoting efficient behaviour in the real world.[89] For example, cost benefit approaches to value entirely neglect such issues as institutional inertia and the costs of waste. Companies whose managers have chosen to behave more virtuously with respect to energy conservation are seen by 'business as usual' advocates as mavericks, or even as acting against their fiduciary duties to maximise shareholder values. But present energy markets, and the many subsidies for conventional fossil fuel production and use which industrial states have long provided, encourage waste and destruction. So long as the fossil fuel market is regulated without regard to the true costs of use of fossil fuels for present and future generations, bad stewardship and waste of this precious resource will continue. But if the procedures of making and exchange did not rely on waste, but instead actually fostered and enhanced local energy and nutrient flows, there is no reason at all why more efficient use of capital and labour could not be accompanied by dramatic reductions in total fossil-fuel use. The issue here is that market measures define efficiency as low-cost even where this involves waste and destruction. Economists refuse to recognise that their adoption of this negligent conception of efficiency involves an implicit and foundational moral judgment about their preference for forms of making which are ultimately destructive of the economy of the earth.[90]

Repoliticising the Economy

The industrial economy of wealth accumulation is wasteful because its procedures and rules are reductionist, and so they train firms and householders to neglect the complex networks of relationships which characterise truly flourishing human and biological communities. Labouring and living in such a way as to foster and sustain these communities is the virtuous way, the way of wisdom, as the ancients well knew. Recovering this way of living will require a radical reform of the way things are made, buildings constructed, and people and resources move around, such that factories and homes, farms and offices and those who dwell and work in them respect and foster natural nutrient and energy flows rather than destroy them. However, economic policy in respect of global warming continues to follow the reductionistic logic of the economy of waste, concerning itself with *distributing* the fruits of the economy of waste while leaving its practices of making and accounting largely unchanged. Thus it is argued that if the poorest and most vulnerable are those who stand at risk of death from climate change, the developed nations who are most responsible for climate change have a moral duty to compensate the poor with resources to adapt to climate change. When the problem is put in these terms, it seems to suggest that the solution to global warming is to foster more economic growth and wasteful making to provide the funds with which to compensate the vulnerable. In other words, the issue of economic compensation acts as a distraction from the more foundational issue of the wasteful and destructive course on which industrial civilisation is set. But so long as the procedures of economic exchange are governed by techniques which are neglectful of the welfare of ecological and human communities, it will be impossible to reform industrial civilisation so as to bring its energy demands into line with the carbon sinks of the planet, and to bring justice to those nations and peoples who will suffer as a consequence.

Readers may at this point argue that calls for radical change in the procedures of industrial capitalism are utopian and unfeasible. They may also suggest that the balance of benefits over costs in the autonomous procedures of technical capitalism suggests it has done more good than harm. However, against these suggestions we must recall two facts. The first is that the harms now envisaged to the planet from autonomous technical capitalism are such that most organic life on the planet will not survive if present growth in energy demands is sustained through the present century. These potential harms far outweigh all the putative benefits – such as reduced infant mortality, extended human lifespan, and improvements in cultural and educational opportunities – which technical capitalism has realised in the last hundred years.[91] The second is that the human economy has already undergone a major revolutionary change in the last forty years under the influence of neoliberal ideology, and this is the creation of an increasingly deregulated financial regime and radically new global trading arrangements.

This is a change so revolutionary that it has had far-reaching consequences both for ecological systems and for human political governance. It may be traced in its political origins to the decision of the advanced industrial economies to deregulate the quantity and supply of money. Britain abandoned the Gold Standard in 1931. However, the most significant break with the Gold Standard was the decision of the United States to break the link between the quantity and value of the dollar and gold held at Fort Worth in 1969. This precipitated the collapse of the post-war economic regime of the Bretton Woods System, and brought about the international debt crisis which began in the 1970s. Its latest emanation is the new deregulated global trading system overseen by the newly inaugurated World Trade Organization (WTO). The dephysicalisation of money, which began with the issuing of bank notes by the Bank of England in the eighteenth century, reached its zenith with this momentous change, as a consequence of which the quantity of credit and wealth-seeking productive and consumption opportunities has grown exponentially in the last forty years, and with it the accelerated destruction of ecosystems from ocean floor to upper atmosphere.[92]

The growth in the supply of money drives banks, consumers and corporations to draw on ecological footprints far in excess of the physical size of the regions or nation-states in which they are situated. The United Kingdom, for example, claims that it has an ecological footprint three times the size of the British Isles. However, if all the overseas operations of British businesses are included, this ecological footprint increases six times. Despite no longer having an empire, it turns out that Britain continues to benefit from the use of land and labour across an area 18 times the physical size of the United Kingdom. Carbon sinks are now the most urgent example of this unequal access to ecological resources, as the lands and skies of underprivileged others are increasingly affected by the outsize footprint of the United Kingdom and other advanced industrialised countries. But the new monetised condition of 'liquid modernity' means that the human economy is now ordered in a way which entirely neglects the fact that it exists within a planetary ecology.[93]

Ferdinand Hayek in *The Road to Serfdom* suggested that bureaucratic and state planning was the enemy of human freedom.[94] Those who resist the re-regulation of the human economy in response to the threat of dangerous climate change often recall Hayek's warning of the threat to human freedom represented by the growing powers of the nation-state. But Hayek also warned of the dangers which would ensue should the supply of money be unregulated. The failure of the industrial world to prevent economically-induced climate change indicates that when human societies subject themselves to autonomous processes of either a bureaucratic *or* economic kind they deprive their citizens of liberty. Future humans are in danger of being dominated not only by these human systems but also by the extreme climatic effects which these autonomous systems will provoke if not reined in.

How though are these processes to be reined in and by whom? There is clearly need for an international global legal regime which governs activities which harm the climate. And the Kyoto Protocol, since it is the treaty of the Conference of the Parties mandated by the UNFCCC, is the only game in town at the level of formal international law and treaty. But, as we have seen, the treaty process has been radically undermined by the neoliberal invention of markets in carbon.[95]

The Perversion of Making

As Patriarch Bartholomew has suggested, modern humanity's threat to earth systems, including the climate, is the consequence of human sinfulness.[96] And the role of multinational corporations and the United States government in undermining the Kyoto Protocol, and in commodifying the atmosphere through carbon trading, is an example of what theologians call structural sin. Structural sin is a concept emanating from liberation theologians in Latin America, who reflected upon the destructive impacts of the colonial and industrial economy on the ecology and peoples of Latin America. They suggest that the traditional way of thinking about sin in terms of personal morality needs augmentation in a world which is increasingly disordered by complex intercontinental economic relationships.[97] Pope John Paul II made a similar proposal when he spoke of the 'structures of sin' that are manifest in the governing ideologies that direct the path of human development away from the authority of biological and moral laws and towards imperialism.[98] But the idea of sin as the denial of biological situatedness is not exclusively a modern one. The Fall of humanity from grace, mythically described in Genesis, is often described as originating in the sin of pride. Adam's sin is a prideful refusal of the dependence of the creature on the Creator. And this refusal has the consequence of destroying interdependence and solidarity not only between humanity and God, but between Adam and Eve, and between them and other creatures. As Russian theologian Sergei Bulgakov puts it, in the fall into sin humanity 'closes off the path to divine life' and consequently 'nature now appears to him purely under its aspect as created, no longer sophianic, as "fallen" or "darkened" Sophia, an image of non-being, sheer materiality'.[99]

In the modern economy of wealth accumulation and waste, this darkening reduction of being to sheer materiality reaches new depths. The economistic neglect of biological laws and of the regenerative ways of ecosystems arises from the exclusive devotion of modern societies to economic above moral or spiritual ends. It represents a misdirected idealism in which the material instruments of modern humanity's apparent success in tackling poverty and disease, and in enhancing the arts and intellectual knowledge, have become the exclusive ends of modern political economy. As Bulgakov suggests, 'wealth is the absolute good where political economy is concerned'.[100] And when it turns into an absolute good it becomes 'a spiritual force, influencing the human spirit from within; it changes from a source of limitation to a source of temptation'.[101]

This misplaced devotion comes at a great price in terms of the enslavement of the earth, and of billions of its creatures, to the aim of wealth accumulation. And it is not only the poor and other species who suffer: the wealthy also suffer intellectually, morally and spiritually from their misguided devotion. It is already possible to see in the West a growing demoralisation of society, a loss of interest and hope in politics, and a decline in spirituality. As Bulgakov suggests, modern societies are 'drowning in sensuality' because a 'life without ideals, spiritual *embourgeoisement,* is the inevitable logic of hedonism'.[102] The crucial marker of this logic is the growing devotion of political economy to the consumption of luxury goods. Once basic material needs have been met by political economy the logic of economic growth demands that new markets and new needs are developed in order to sustain consumption. It is in one sense hard to distinguish between luxury goods and necessity. In some contexts a car may seem an essential household possession; in others air-flown pine-apples or lettuce may seem indispensable; for the super-rich and 'celebrities' private yachts and private planes may seem essential as ways to escape the attentions of the crowd, while still displaying worldly power. Although there are no external markers which clearly distinguish a luxury from an essential good, devotion to luxury is a clear sign of the internal moral decay of contemporary Western culture. As Bulgakov suggests,

> Luxury is the triumph of sensuality over spirit, of mammon over god, whether in the individual soul, or in the whole of society. Once the cult of gratification, aesthetic or non-aesthetic, has become a guiding principle, we have luxury. Luxury is the reverse side and the constant peril of wealth.[103]

The more time that consumers have just to consume, apart from the disciplining activities of physical making, the more sophisticated their appetites become. As consumption moves on from necessities to superfluities, 'no object of the world will be safe from consumption and annihilation through consumption.'[104] As the machine age trains modern people to imagine that it is possible to design 'social systems in which people will be virtuous automatically, as it were, without any kind of struggle with themselves', both human character and biological flourishing are corrupted.[105] The traditional moral struggle of human life was to build a durable world of human dwelling in scale to the limits and natural laws of the biosphere. The paradox of consumerism is that even human dwelling, the world of human making, has become unstable, as vehicles, furniture and domestic machines become luxury consumer goods which are constantly disposed of in giant holes in the ground for newer, more fashionable or improved versions.

The perversion of making is at the heart of the ecological crisis and is rooted in Western political economy. Locke's theory of property, and Smith's theory of market values, require that more and more of nature's wealth is turned into artefacts for human use and consumption, for only in the form of tradeable artefacts, and hence of monetary wealth, can labour be reified into wealth accumulation. But paradoxically this reification of labour, far from freeing modern societies from work, turns all human activity, and human identity itself, into paid employment so that work comes to dominate all other human action, and this is particularly the case in the present day United States, and increasingly in the United Kingdom. As Arendt presciently predicted, 'making a living' becomes the end of human life in late modernity, whereas in predecessor traditions work is a means to a higher end.[106] The quest to free life and labour from biological laws, and from effort and pain, has the paradoxical outcome of subverting human life to the necessity of ceaseless activity and ultimately to more extreme forms of ecological necessity as the earth is consumed to the point of collapse:

> Painless and effortless consumption would not change but would only increase the devouring character of biological life until a mankind altogether 'liberated' from the shackles of pain and effort would be free to 'consume' the whole world to reproduce daily all things it wished to consume.[107]

The modern economy of making wears down the durability of the biophysical world *and* the stability of human dwelling.[108] The great paradox which climate change unfolds is that the historical task of freeing humanity from the physical effort inherent in manual labour, including the tilling of the soil, produces new forms of slavery; of humans and nonhuman species to other humans, and ultimately of all humanity to the burden of living in a globally warmed world.

Reordering the human economy to the biological laws and limits of the economy of the earth requires a profound re-moralisation of human making, so that honest labour is valorised again, and the blessing and joy of exhausting but satisfying work is seen as the means to care for the earth and to build artefacts which endure rather than consumer objects destined to be buried in the ground. As Arendt suggests, only when human making and the human-erected home on earth is made of things that endure can 'specifically human life be at home on the earth'. A society that builds worlds which do not endure but are constantly destroyed will not only destroy the biophysical world. Its members will also be dazzled by their own abundant making and no longer able to recognise the futility of life to which they have become devoted.[109]

In this sense, again, climate change represents an opportunity as well as a threat. Industrial civilisation is on the road to ruin, spiritually as well as ecologically, if its citizens continue to pursue endless distraction and luxury at the cost of their moral and spiritual health, and the health of the planet.

5
Ethical Emissions

Thus says the Lord: *Do not let the wise boast in their wisdom, do not let the mighty boast in their might, do not let the wealthy boast in their wealth; but let those who boast boast in this, that they understand and know me, that I am the* Lord; *I act with steadfast love, justice, and righteousness in the earth; for in these things I delight, says the* Lord.

Jeremiah 9:23–24

The remains of the earliest known ancestor of the human race, called 'Dinqenesh' (in Ethiopian 'you are amazing') were discovered in the Afar desert in Ethiopia in 1974.[1] The land which is very likely the birthplace of humanity has been plagued by increasingly frequent drought and decline of average annual rainfall since the 1970s, and the country has suffered ever since from desertification and intermittent famine.[2] Population pressures, poor farming practices, deforestation, disastrous government policies and civil war have exacerbated the problem. While in many parts Ethiopia still looks green and lush, deforestation on the hills has affected rainfall levels, reduced water retention, and led to widespread land erosion. More than a million tonnes of topsoil is washed into the rivers every year.[3] There are 55 million people in Ethiopia on a land area more than four times the size of the United Kingdom, but the proportion of fertile land for farming is gradually decreasing as the Sahel desert extends eastwards. The population has doubled in the last fifty years and will double again in the next twenty-five if poverty and poor education are not addressed.

In response to the dangers of soil erosion, the Ethiopian government has developed one of the largest tree planting schemes in the world. In the province of Tigray 40 million tree seedlings were planted in one year alone. Hillsides are also being newly terraced with small stone walls and old terracing is being repaired, to reduce erosion and to provide more land for cultivation. While planting trees can help address the local causes of climate change, Ethiopia suffers like Tanzania from the gradual warming of Africa which global warming is bringing about. Given its maritime location, Ethiopia is particularly at risk from rising surface temperatures in the oceans, which are affecting the monsoon rains and 'engendering widespread drought over land, from the Atlantic coast of West Africa to the highlands of Ethiopia'.[4] As we have seen, rainfall in the whole sub-Saharan region has dropped consistently in the last thirty years, affecting not just traditionally drought-prone regions such as the Sahel and Niger, but countries further South such as Zambia, Zimbabwe, Mozambique and Malawi which were not previously so drought prone.[5]

While poor soil management, deforestation and over-grazing are implicated in local climate change in many parts of Africa, the continent shows an overall warming in the last century because of global warming and can expect increases of between 2°C and 6°C if warming continues unabated.[6] At levels of warming of 1–2°C many Northern developed countries may actually benefit from global warming because of reduced heating costs and increased crop output. But in developing countries in the South the losses far outweigh any gains, with reduced crop outputs and risks of drought, disease and famine all increased. Bill Hare estimates that 45–55 million people will be at risk of hunger with a 2.5°C warming.[7] Sari Kovats shows a demonstrable relation between diarrhoeal diseases and rises in temperature in Africa, and estimates that 47,000 deaths from this cause alone were attributable to anthropogenic climate change in Africa in 2000, while deaths from climate-related malnutrition in all developing countries in the same year were over 110,000.[8] Robert Watson estimates a reduction of agricultural production because of climate change of around 30 per cent for the whole of Africa, with a 2°C warming between now and the end of the present century.[9] Rachel Warren estimates that Southern Africa could suffer a catastrophic decline of 80 per cent in crop yields at this level of warming by 2100.[10] Low-lying coastal areas of Africa are also at great risk from sea-level rise: 6 million people on the Atlantic and Indian coasts would be vulnerable to a 1m level rise.

Local Signs and Global Descriptions of Climate Change

Before scientists recorded the data and made their projections, local people in Africa already sensed a change in the weather. Pius Ncube, the Catholic Archbishop of Bulawayo in Zimbabwe, explains that 'there has been a big climate change within living memory, and the rainy season, which used to run from October to April, now starts around mid-November and ends in February'.[11] Wangari Maathai, founder of the Kenyan Greenbelt Movement, has demonstrated that African people are also doing a great deal to try to address the local causes of climate crisis. Women in her movement have planted upwards of 20 million trees in Kenya and beyond.[12] However, tree planting is no substitute for global emissions reductions because global climate change will, if it proceeds unabated, ultimately lead to land temperatures which destabilise forests. As the Anglican Archbishop of Tanzania, Donald Mtelemela, puts it 'the West should accept that there is a problem not just in Africa but in the world. God has given Western leaders the gift of leadership. But leadership does not belong to the West only. Their leadership is for the whole world. Today the problem is with Africa but tomorrow it could be with Europe.'[13]

The potential of climate change negotiations to link nations across the world in a collective effort to preserve the global commons of the climate is huge, and the Kyoto Protocol represents the most significant international treaty ever contracted between peoples across the globe. However, the inadequacy of the present targets and policies emanating from the process is plain to those who are suffering the effects of global warming. The peoples of Africa and South Asia are already bearing an economic and physical burden of adapting to severe climate change which threatens their very survival, while the North continues to promote schemes for trading pollution permits as a way of avoiding real reductions in carbon emissions. The gap between the air-conditioned conference rooms where the parties negotiate and the reality of a warming world symbolises the conflict between the vision of an expanding international economic order governed exclusively by market considerations and the limited carbon sinks of the earth system. While it is as invisible as the 'invisible hand' of the global market, the climate is not a social construct; instead it represents a real biophysical limit on the freedoms of corporations and consumers to maximise their preferences and profits.

This conflict between a constructed human economy and the ecology of the earth also maps onto the conflict between the values and logics with which representatives from developed and developing countries approach climate change negotiations. While the rich nations tend to emphasise efficiency and a balance of costs and benefits that is focused on the putative increased costs to Northern economies of adjusting to lower greenhouse gas emissions, developing countries prefer to talk of retributive justice for past emissions, distributive justice in relation to technologies and finances to enable people to adapt to climate change and equality between present and future generations in relation to the effects of present emissions.[14] As Jekwu Ikeme suggests, disagreement between developed and developing countries over climate change can be mapped on to a conventional modern Western account of competing ethical paradigms. The developed countries favour an approach focused on utility and end states of welfare, costed by means of supply and demand curves. Developing countries tend to favour an approach based on principled accounts of justice, rights and equity.[15]

Preference for a socially constructed approach to morality reflects the tradition of European moral philosophy since the Enlightenment, according to which concepts like justice and the good are human cultural constructs emanating from human intuition and reason; justice in this approach is neither divine in origin, nor part of the structure of the earth. The view that there is something foundational, natural, and therefore non-negotiable about justice and equity reflects the teachings of many of the world's religions which are still more widely embraced in the developing world, including Christianity, Islam, Buddhism and Hinduism. On this view every individual human being, no matter how poor, has the right to a stable climate, and to grow enough food to feed her family. Such a right is not a matter of human judgment and it may not be set in the balance against the benefits that others may derive from depriving some of this right; there are no aggregate sums which can remove this foundational right. It is the birthright of every human being. And the willingness of governments in the South to trade forests or lands or development projects for the excessive carbon emissions of wealthy Northern nations and corporations is perversely opposed to this birthright.[16]

Ecological Justice and the Limits of Liberalism

According to Jeremiah justice is part of the sacred order that God has set into the character of creation, and of human being. Human justice is first and foremost a divine gift, a reflection of the character of a good and beneficent God. The Israelites first come to know of the nature of God's justice in the story of their freedom from slavery in the land of Egypt, and this freedom is won for them in part through a series of plagues which are sent on Egypt when Pharaoh refuses to release the Hebrews from slavery. The plagues of insects, large hailstones, giant frogs and red tides indicate that God's just will to free the people he has chosen from slavery is manifest on the face of the earth, and not just in the relations between the Egyptians and their Hebrew slaves, or the negotiations between Pharaoh and Moses and Aaron. In Exodus justice is narrated as written in the stars and on the face of the earth; it is not just a human construct or a transcendent ideal. The same idea is expressed in Psalm 72; the king who rules justly rules a land where the rains come in their due season and the peoples enjoy the fruit of their crops. The king is not just because he adopts a human measure of justice, but because the manner of the king's rule is in conformity with the eternal justice of God. And this conformity between human and divine justice is manifest in the conformity between the rightly ruled society, a stable climate and a fertile land. When the king loves justice the land itself and all the creatures who dwell in it prosper. When the king and the court neglect justice, the people suffer and the land suffers with them.[17]

The biblical narrative of justice suggests that it is ecologically situated and not just a human value. Thus, as we have seen, Jeremiah suggests that the reasons that the people of Israel and Judah were exiled from their ancestral lands was because they and their ancestors had neglected the covenant, abandoned the law of God and so polluted and defiled the land. Jeremiah argues that this environmental exclusion was to punish the rich and the powerful who had taken up so much of the land that the poor were excluded even before the exile from enjoying its fruits. Greed, power and wealth had become the motive force of the people of God when they went after other gods. They neglected the laws and the character of God, who was revealed in these laws as just and fair, compassionate and merciful. Environmental exclusion in the form of exile is a core theme of the Old Testament, and it speaks to the condition of the environmentally excluded on the planet today, and in particular those millions who are already finding they are forced to migrate from their ancestral lands because of drought and flood caused by climate change.

Alongside judgment Jeremiah speaks a message of hope to the people in exile when he suggests that it remains the divine intention that they should again dwell in the land and enjoy its fruits. But this return from exile will only take place when they learn again not to oppress 'the stranger, the fatherless and the widow' and not to shed innocent blood (Jeremiah 7:6). Only then will Yahweh 'cause you to dwell in this place, in the land that I gave to your fathers for ever and ever' (Jeremiah 7:7). Love, not power, is the aspect of divine being which is manifest in the reversal of exile and the return of the people of God to the land of promise. In relation to global warming the implication is clear. The poor in the land who are already being excluded by climate change from enjoying its fruits are owed justice, but they also need mercy and compassion if they are to be enabled to adapt and survive.

The problem for many with this Israelite view of justice and mercy is that it seems to be culturally conditioned. It represents Yahweh as the god of one ancient people who has less regard for the prosperity of other nations. The Israelites are redeemed from slavery in one territory only to end up putting to the sword the peoples – Philistines, Amalekites and so on – who inhabit another one which they come to regard as the 'promised land' that God gives them to dwell in after their rescue from Egypt. But this critique neglects the fact that the Old Testament writers had already begun to universalise this early ideal of justice as the redemption of a particular people, so that in the third book of Isaiah all peoples are said to be called to know the God of Israel and to enjoy the just judgment which characterises the kind of representative rule which models the sovereign justice of God.[18] And in the New Testament the confining of the message of divine redemption to the people of Israel is finally broken altogether as the gospel of Jesus Christ is proclaimed and embraced by both Jew and Gentile.

The germ of the idea that Israel's covenantal obedience to divine justice, and in particular the divine command to respect the dignity of the poor and downtrodden, is a gift to all nations is already present in the first book of the Bible. It is indicated in the suggestion in Genesis 2 that all humans are descended from the one couple, Adam and Eve, who share in the divine image which confers on them a dignity known to no other creature. The idea that all the peoples of the earth are descended from one human couple receives archaeological and genetic confirmation in recent scientific proposals, drawing on study of the human genome, that human life actually began with a few individuals of the species *Homo sapiens* who first evolved in Northeast Africa, and quite possibly in Ethiopia.[19] This idea of a common root for all human beings has tremendous moral implications, for it indicates that all cultures, nations and peoples share a common earth story. If human beings all have a common ancestry, then the divisions between peoples erected by tribes and empires, nation-states and the paraphernalia of border controls are mere social constructs, products of the human mind with no biological basis.

Gaian science provides a significant supplement to this perception of a shared human story, for it suggests that all human beings, all creatures, are relationally interconnected by the carbon cycle of the planet. On this account there are no autonomous human actions, for in a fundamental physical sense all actions are interconnected by their effects on the carbon cycle. Each individual action is an infinitesimal element in this cycle. The geochemical interconnection of all human actions and all life is a physical analogy for the Christian doctrine of the Communion of Saints, or 'cloud of witnesses' (Hebrews 12:1). This doctrine expresses the belief in the spiritual solidarity of the people of God across space and time, a belief which is also indicated in the doctrine of a last judgment, as enunciated in Christ's parable of the Sheep and the Goats, in which every human life is weighed in terms of the relationships that each individual has nurtured or neglected (Matthew 25:31–46).

The perception of a profound interconnection of all life, both geochemical and spiritual, does not, though, involve the submergence of the value of each individual, as it does in the utilitarian and economistic worldview reviewed in the last chapter. Instead, when taken together with the recognition of the common ancestry of all human beings, it suggests that all humans share the same dignity. Another way in which the Christian tradition speaks about this shared dignity is that humans are created in the image of God, as indicated in Genesis 1:27. In imaging God they have a unique species vocation to care for the earth through labouring upon it, and to promote integrity and justice in the exchange of the fruits of their labours. The covenant with Yahweh expresses this vocation in more precise terms than those adumbrated in Genesis. As we have seen, the covenant community is one in which the sovereignty of Yahweh over the earth frames all human endeavours upon the earth. And hence in their covenant community the Hebrews were to conserve the land in their use of it, and to promote peace with one another by restraining economic practices such as usury and debt which create slavery. The biblical ban on usury expresses in the most profoundly practical way the divine law of love of neighbour in human social and political arrangements for it constrained the ancient Israelites, and Christians until the Reformation, from subjecting their neighbours to the rule of money and price.[20]

The ban on usury expressed a connection between image and stewardship in the form of Israel's covenantal society. The Israelites were commanded not to oppress one another and to respect and preserve the health and fertility of other creatures and the land. This is why Psalm 72 makes the foundational connection between the justice of the king and the stability of the rains that water the earth. And so justice in the Old Testament is cosmic and not just personal. Human actions which mirror divine justice affirm and sustain the wholeness of creation, while human injustice causes oppression and misery not only among the children of men but among the beasts of the field and trees on the mountains.

Despite their foundational place in the culture and laws of the Western world, philosophers since the Enlightenment have eschewed biblical narratives in their attempts to refound human freedom and justice on reason alone. They prefer a conception of justice that is freely chosen by reasonable people, who are conceived not as divine creatures but as originally autonomous by virtue of their belonging to a reasoning species. In place of the sovereignty of God, the modern philosopher puts the sovereignty of reason, and of the individual reasoner. The sovereignty of society then emanates not from the divine character of God but from the pooling of a fraction of each individual's sovereignty in a social contract which is designed to protect the dignity, property and rights of persons, and to punish harms and maleficence. The most influential modern account of justice along these lines is John Rawls' *Theory of Justice*.[21] The theoretical device through which Rawls arrives at his account of justice as fairness takes the form of an analogy with Plato's cave, in which the good society is constructed through a 'veil of ignorance'. Rawls proposes that an idea of justice, and a just set of social arrangements, will be best constructed by a group of individuals when they imagine themselves members of a society in the future but have no idea in what strata of that society they will live.[22] In this 'original position' they will come up with a proposal for a just society which ensures that there are some constraints on the extent to which individuals are free to amass wealth and power, because they might themselves be among those without much of either: they are therefore likely to favour a polity which places certain minimal constraints on the wealthy in order that those who are not still experience some degree of liberty. The liberal society so conceived is a society founded on a rational but essentially negative conception of liberty, where individuals are freed from interference from others by minimal political procedures which express this thin account of justice as fairness. In the condition of negative liberty, goods are said to be plural and individuals are free to choose whatever goods they prefer, save only that those worst off must not be so badly off that they lose any stake in the social contract which holds society together and so refuse to confer on it part of their own sovereign power.

The putative advantage of Rawls' account of justice over economistic utilitarianism is that it does not allow for harms to some persons to be set in a balance of aggregate benefits to others. It is an example of a 'deontological' or duty-oriented account of morality that emphasises the intrinsic good of each individual person, and the duties that are therefore owed to persons. However, as Arendt argues, the Kantian framing of value on which Rawls relies – that persons should only be treated as ends and never as means – does not escape the utilitarian infection of modern value construction. Instead it makes humanity – the only unmade being – the measure of ultimate good, thereby only affirming the utilitarian subversion of the worth of other things and species in the world.[23] As Aristotle observed long before Kant, it is 'absurd' to think that man is the highest being in the world and that everything else is subject to human instrumental purposes.[24]

The moral deficiencies of economistic and utilitarian approaches to justice arise from the valuing procedures of modern liberal political economy. As man the maker comes to dominate the social and species world, all things are degraded and lose their intrinsic worth in the creation of value in the form of exchangeable objects, and the even more durable form of money. The problem is not with instrumentalisation *per se* but rather with the 'the generalization of fabrication experience in which usefulness and utility are established as the ultimate standards for life and the world of men.'[25] Establishing humanity as the 'lord and master of all things' does nothing to repair this problem and simply extends it into the sphere of transcendental philosophy. It is the takeover of the world by the activities of marketing and trading which is the source of social alienation in modern making. As social life is devoted to the creation of stored value, so the only values which are recognised, apart from human makers themselves, are exchangeable values. But even the maker is at risk in the process for, as Marx recognised, as all the products of labour become exchangeable commodities, human beings become alienated from the objects and practices of making, and hence ultimately from themselves.

Thus the 'original sin of capitalism' is the change of everything from intrinsic worth to use value and thence to exchange value. Marx, like Rawls, attempts to repair the problem by modifications in the distribution of end values. And this is why neither Marxian socialism nor Rawlsian liberalism are able to resist the infection of the earth, and the present destabilisation of the earth system, by trade. As Arendt puts it, 'the loss of all intrinsic worth, begins with their transformation into values and commodities, for from this moment on they exist only in relation to some other thing which can be acquired in their stead.'[26] And so the growing influence of monetary exchanges in liberal and neoliberal polities, as well as in socialist ones, explains the 'loss of standards and rules' external to the instrumental relativities of the exchange market.[27]

Adequate repair of this situation necessitates the recovery of the pre-Enlightenment earth story; that the diverse species of the earth and the biological laws which direct life to its original abundance are not the outcome of instrumental and autonomous processes, but of beneficent divine Creation. As Creation the earth does not serve instrumentally as the site of human (or species) construction and making but as the theatre of God's glory. In this way the earth and its myriad species have intrinsic worth apart from use values. And this is why human making should never displace or be confused with divine Creation, and why humans are not well described as 'co-creators'.[28]

Another difficulty with modern liberalism, which emanates from Lockean accounts of human dominion and property, is that humans are increasingly described as beings of their *own* making and possession. This description is the origin of the liberal account of the diverse goals and purposes of individuals, and hence of the political impossibility of specifying positive freedoms.[29] Whereas the ancients regarded politics as the *way* in which persons are formed as moral agents who love what is good and beautiful, pursue the truth and practice the virtues, the modern liberal assumption is that persons are already persons before they are political persons. Consequently, the role of the body politic is not so much to create the conditions for the formation of individuals who can live an excellent life but rather to sustain minimal conditions of civil and political order, such that property rights are respected and crime is punished while individuals are free to pursue their differing desires and competing projects.[30] This is why exchange relations and values increasingly trounce political deliberation and intrinsic goods in modern political arrangements.[31]

The liberal answer to the problem of collective action arising from the narrative of self-creating individuals is, curiously, just to further elaborate the shift towards exchange value as the origin of moral and political order. Thus liberals conceive of desire, after Adam Smith, in terms of a market analogy. The function of the body politic is to sustain a social world characterised by the maximum degree of freedom of choice and lifestyle for individuals and businesses. The market is said to work so as to direct individual desires towards a collective state in which all achieve greater utility. In the Rawlsian restatement of liberalism, the body politic is left with the task of siphoning off a portion of market goods to enable the construction of public institutions in which property rights are respected, and where necessary defended, while minimal conditions of bodily life are granted to those who are less successful in pursuing their own projects of self construction. Consequently, provided the body politic functions minimally to sustain the conditions for respect of individuals and property, individuals do not themselves have to pursue the common good – of society or the earth – in their own projects; they can defer the political requirement of the pursuit of the common good to the invisible hand of the market, and to the public institutions and services that, via taxation, their activities fund and sustain.[32]

The liberal division of labour between individuals and the body politic is criticised by Leo Strauss, who argues that it produces the core paradox of modern liberal societies. This is the claim that it is possible to construct a good society whose institutions and procedures ensure a minimal level of justice, such that property rights are respected and the poor do not go hungry, while not requiring individuals themselves to be just or good.[33] This paradox in part explains the return of neoliberal or *laissez-faire* accounts of the market and capitalism, and how it is that public provision for education, health care and welfare has come under such sustained attack in Britain and America in the last thirty years. For Strauss, and other conservative critics of liberalism, the coercive state, and its attempts to draw competing private interests into a minimal level of public provision of care and justice, is morally hazardous; only moral individuals – acting charitably as good neighbours or, in the case of corporations, from the profit motive – are able to enact the kinds of justice which will ensure that basic social needs are met without compromising fundamental liberties.[34]

This account of the state as morally hazardous has been further advanced in the move of nation-states in recent decades towards pooling sovereignty in organisations such as the European Commission and the WTO, whose aims are to promote increased economic exchange without the constraints of national borders. However, this pooling of sovereignty has in effect given more power to economic corporations, many of which now constitute centres of economic activity larger than many nation-states. Thus five British companies produce between them 100 million tonnes of carbon, which is five times the output of a small, developed nation like Croatia, and close to the total emissions of the whole of sub-Saharan Africa.[35] But though many corporations emit more carbon than many nation-states, corporations are not formally required under the Kyoto Protocol to reduce their emissions in parallel with nation-states.

The Fall of the Common Good and the Rise of Neoliberalism

The modern Western account of negative liberty commits the body politic to an under-specified account of the common human good, and it fails to recognise the intrinsic worth of the earth and its myriad creatures. These problems originate in the infection of morality and politics in liberal societies by instrumental and exchange values. This is why it is so hard for the liberal state to resolve the increasing conflict between the exchange economy and the earth system, and why liberalism has failed to resist the rise of economic neoliberalism. Under the influence of neoliberalism even the common good of the climate system is turned into another sphere of humanly-constructed exchange values.

The global market, as we saw in the last chapter, is an aggregating device which produces an enormous amount of destruction, both of human communities and of ecosystems and biodiversity. While Hayek is right that centrally planned economies are a threat to human freedom, he and his followers underestimate the extent to which national and international market economies are also instruments of collective organisation, reliant upon a great raft of market actors and technologies, from currency dealers and corporate CEOs to the computers and controllers who drive and respond to market trends on stock exchanges and in retail stock inventories. All of these actors and the technologies they deploy act from centres of power on individuals and the masses to reduce complexity and contingency in local communities of place and in local ecosystems. Commercialisation of the social realm is, then, at least as great a threat to human freedom and ecological wellbeing as centralised state planning.[36]

As we have seen, it is this centralising and reductionist nature of the modern economy which in part helps explain the extraordinary inability of industrial societies to respond to signs of ecological breakdown. The governing practices of industrial economies – and particularly those under the spell of neoliberalism – are in some respects as unresponsive to ecological and social signals of the harms they cause both to ecosystems, and to human communities, as were the Soviet-style collectivist economies of the communist era. As with all aggregating systems of social power, the market is a device for extracting and totalising natural and social power and conferring more of it on the most powerful actors in the market – corporations and investors – than on those who merely service the market.

And here we encounter the ambiguity of the scientific narrative of the climate of the earth as a global system that collectively aggregates all individual human actions which produce greenhouse gas emissions; this narrative seems to require an equivalent form of collective action – systemically planned and ordered like a planner's idealised new town, or a global market in tradeable carbon emissions – to match it.[37] But the construction of a global market in carbon permits directly mirrors the kinds of procedures which have advanced a deregulated global market in which corporations have grown enormously in their capacity to transform natural resources and whole ecosystems into consumer products and services. The privatisation of the commons of the atmosphere in this global carbon market is a powerful example of the centralising and corporatising drift of modern political arrangements under the guiding ideology of neoliberalism. Government representatives contract together to pursue the standard national interest argument in the treaty process, and hence they put the short-term interests of their own corporations and consumers ahead of the welfare of the earth, or of Ethiopian and other subsistence farmers.

These amoral outcomes are of a piece with the growing wealth inequalities of the last fifty years, which have seen the gap in global incomes grow so dramatically that the ratio of incomes between the richest and poorest countries was around 3 to 1 in 1820, and 45 to 1 in 2005.[38] The gap in life expectancy between rich and poor countries also remains very high, so that an English child today has an average life expectancy of 78 years while a child from Ethiopia has a life expectancy of 48 years. The present extremes of global inequality mean that the costs of adapting to global warming for the poor exceed their capacity to meet them. The climate crisis also demonstrates that these extremes of inequality are ecologically unjust, since they involve physical harms, such as the ecological exclusion of the poor by drought and flood from their lands, being visited through the medium of the earth system by the extreme accumulation of wealth.

Liberal Justice Beyond the Nation-State

The moral case for righting the global wrong of extreme inequality is interconnected with the case for righting the wrongs of global warming, since in both cases benefits and surpluses accrue to the already wealthy while costs accrue to the already poor. But as soon as we put matters like this, it seems to commend a globally-planned economic redistribution of wealth of a kind that would continue to reproduce the distorting centre-periphery relationships of the colonial era, and of the present global market. And this is just how Martha Nussbaum presents things in her efforts to extend Rawlsian liberalism beyond the nation-state in her recent book *Frontiers of Justice*. Nussbuam wants to preserve what she sees as the best insights of Rawlsian liberalism while recognising that it needs modification to the extent that it does not adequately address relations between nations, between humans and other species, and the moral status of the disabled. These lacunae in Rawlsian liberalism are, she suggests, indicative of weaknesses in liberal accounts of political justice, which arise in part from their conception of an original individualism which is modified for political purposes by a social contract. They are also a consequence of the limited specifications of the human good offered by liberalism.

Nussbaum proposes to correct these through the capability approach to human development that she has articulated over a number of years with development economist Amartya Sen. This theory adopts an Aristotelian-derived position that individuals are not intrinsically autonomous, but rather are formed by and flourish in the midst of social relationships. The capability approach involves an analogous understanding of human flourishing which rests on a 'conception of human dignity and of a life that is worthy of that dignity – a life that has available in it "truly human functioning".'[39] Such a life entails access to a range of life experiences which include a healthy life of normal duration, freedom from violent assault, freedom of thought, imagination and emotions, the development and expression of practical reason, the ability to affiliate with other people and to experience social respect, relations with other species, a capacity for control over natural and political environments, and a capacity for play.

Like Aristotle, Nussbaum represents the human as an embodied and a social animal and not just a practical reasoner. The capabilities she valorises range from aspects of mental and emotional life to embodied relations with other persons and with the more than human world. Furthermore, justice is the ultimate good; as for the Hebrews and for Aristotle, who described it as the unifying virtue, justice for Nussbaum 'is the one thing humans beings love and pursue'.[40] And they do so not because of mutual advantage but because humans' fundamental sociability means that in normal life people are frequently benevolent towards strangers and to those who live in distant lands, such as famine victims. The difficulty with modern liberal political institutions is that they are not so good at translating this personal benevolence into collective action, and even worse at extending it beyond national borders and to other species.

Modern theories of international relations represent nations as independent communities of people who are engaged in a social contract which unites them more fully with each other than with peoples from other nations. In other words, modern political theory conceives of the nation-state by analogy with the fictional autonomous individual of liberalism. Thus for Rawls the nation-state is assumed to be an autonomous entity, responsible first and foremost for promoting the interests of its own citizens.[41] But, as we have seen, the climate system indicates that nations are not as autonomous as liberal political theorists imagine. If the sum of human actions in a wealthy nation such as the United States has a disproportionate impact on the climate system, then in only a very limited sense can the United States be said to be an autonomous or self-sufficient political unit. Its corporations and government agencies are in any case engaged in a neo-imperial economic strategy of harvesting oil and other natural resources on every continent, which dramatically extends the ecological footprint of the United States around the world.[42] As Nussbaum avers, the misguided liberal assumption of the autonomy of nations neglects the serious inequalities of power between different nations, and confers philosophical respectability on an 'arrogant mentality that is culpably unresponsive to grave problems', which well describes the formal position of the United States government in relation to the problem of global warming.[43]

Global Justice and the Law of Love

Western political thought about relations between nations has not always been as unresponsive to inequalities of power and flourishing as the liberalism of Rawls, or economic neoliberalism. Nussbaum's capabilities approach has much in common with a natural law account of human flourishing of the kind elaborated by Thomas Aquinas in the Middle Ages. The natural law tradition also strongly marked Christian theological and philosophical reflection on relations between the new independent nation-states which emerged after the Reformation in Europe. The natural lawyer Hugo Grotius saw the emergence of the nation-state as the gift of Christendom to the emergent secular world order of the seventeenth century. And he suggested that relations between nation-states should reflect consideration for the bonds of human fellowship which unite people across boundaries of language, territory and nationhood.[44] As we have seen, the recognition of the significance of such bonds is fundamental to the conception of humanity as created in the image of God, and it finds fuller flowering in the doctrine of the Church united across boundaries of religion, race or gender on which Grotius draws.

For Grotius, relations between peoples are subject to the 'law of nations', which is a reflection in human affairs of the rights of the Creator God in the Creation.[45] The law of nations is an aggregate of the natural laws and customs of different nations and cultures who meet together to adumbrate a shared sphere of law in those areas of human life where nations interact. For Grotius true peace between peoples cannot be achieved without some extension of the Christian account of moral sociality to relations between nations, and the gospel requirement that human relationships should be shaped by works of charity, and not just legal justice.[46] This extension will require not only avoiding the kinds of territorial infractions and attacks on innocent people which are for Grotius legitimate causes for war.[47] It will also require certain kinds of agreements over the treatment of the global commons, of which the first to be adumbrated by Grotius himself was the law of the sea. In this there is a significant precedent for the kinds of international environmental regulation that the extinction of species and now global warming require. And there is a significant relationship between the laws of war and environmental law. The laws of war that emanate from the Grotian project, including the Hague and Geneva Conventions, have imposed certain international legal restraints on behaviours and tactics in warfare which, while they are not always honoured, nonetheless act as a device for restraining the evils of war. Similarly, international environmental regulations – such as the Convention on International Trade in Endangered Species (CITES), or the Montreal Protocol on ozone-depleting chemicals – are frequently broken by corporations and nation-states. But the existence of these international agreements provides the 'international community' in the form of the United Nations with a means of judgment through which to call miscreants to account.[48]

The Kyoto Protocol therefore has important theological roots in the Western political tradition. But the neoliberal effort to turn international climate negotiations into procedures for trading carbon sinks and emission permits subverts the tremendous potential of the Kyoto process. The UNFCC has set in train a crucial international legal process to bring nation-states, and multinational corporations, to an international court of judgment where they must give account and demonstrate that they are reducing their greenhouse gas emissions. But this process can only work if the divinely mandated role of government – and by extension of international governance institutions – to promote the common good and punish transgressors supersedes the partial interests of economic corporations or individual nation-states.

Grotius' account of the quest for peace between nations, and its links to an ecclesial unity of peoples which exceeds the claims of nation-state or territory, provides a stronger foundation for the international governance procedures of the UNFCC than either Rawlsian liberalism or economistic neoliberalism, because it derives from a biblical account of the rights of the Creator over the Creation, and the restoration of those rights in the Resurrection of Christ. It also rests upon a vision of fellowship between persons who are made in the divine image and, in the redeemed politics of the Church, are drawn into a social realm where charity and the quest for the common good are fruits of a shared life. Christian communities from earliest times have been marked by a special concern for the sick, orphans, widows and the poor, hence the sponsorship by the Church in Christian history of a range of institutions of social care such as hostels for the homeless and pilgrims, hospitals, orphanages and schools. Christ's Incarnation expresses in definitive form the love of the Creator for the creature. This love involves the ultimate sacrifice in which Christ 'lays down his life for his friends' and he commends them to love one another in similar fashion. It is the law of love, not legal norms, which directs the first Christians to show respect in their assemblies to those servants, slaves and others whom Roman society placed beneath moral consideration.

Nussbaum also recognises the failure of existing international legal and economic arrangements to redress injustices between rich and poor, and between humans and other species. And so what she commends across national and species boundaries is analogous to a global law of love. But while she can explain in theory why duties are owed across boundaries, she finds it hard to explain how citizens, nation-states or corporations might in practice embrace and own such duties, and make the kinds of sacrifices in their own living standards that the recognition of such duties entails. What in effect Nussbaum does is to adopt Grotius' thick account of fellowship between peoples and nations, while detaching it from its ecclesial, metaphysical and natural law contexts, and bolting it on to Rawlsian liberalism. Consequently, she neglects to address the problem of negative versus positive liberty in liberalism, and the division between private and public morality to which it gives rise. So although in theory she commends an international conception of justice founded on a rich account of relations between persons of the kind sustained by Christian metaphysics, she proposes to enact this through liberal economic arrangements. Accordingly the *theoretical* claims of justice between nations require that developed governments and multinational corporations based in developed countries redress inequalities between nations in order that the capabilities of the peoples of less developed nations are enhanced and their flourishing advanced. But Nussbaum can give no reasons why in *practice* governments or corporations will act against their interests as these are served by neoliberal market arrangements.[49]

Economic Development and the Myth of Progress

There is a related problem in the efforts of the IPCC to address the international inequities resulting from global warming. Mobilising a similar model of international justice to that of Nussbaum, the IPCC suggest that developing nations need development aid from developed nations to assist them in adapting to global warming. Such aid will be used to help with such measures as flood defences, population resettlement, and the development of drought-resistant crops and new low-carbon renewable energy technologies such as solar and wind power. The IPCC, like Nussbaum, propose that injustices between nations continue to be addressed in the language of economic 'development' while neglecting the more fundamental problems of modern political economy we have identified above.

The idea of development takes its rise from positivist perspectives on human progress emanating from the thought of nineteenth-century social scientists such as Saint-Simon and August Comte.[50] The core assumption of development is that the technical procedures, social practices and values associated with industrialisation represent progress in the human condition. Belief in progress is the understated heart of the liberal theory of justice advanced by both Rawls and Nussbaum, and in its own way by the IPCC. Development in material and cultural accomplishments is for modern liberals the guiding *telos* of human flourishing, since it is only through material advancement in subduing nature and transforming it into human wealth and property that the competing desires of individuals, as described by liberalism, are said to be satisfied.

According to the materialist ideology of development, peoples and nations who are still primarily agriculturalists and hunter-gatherers are by definition 'underdeveloped' because they do not enjoy the cultural and material accomplishments of modern industrial citizens. Living on 'less than a dollar a day', a phrase often used in development literature, and by Nussbaum, is on this definition of the good life not only a life of poverty but also a life which is hardly worth living. Progress in the condition of the global 'poor' then requires that they be assisted in moving from their traditional ways of life, which often involve barter and gift exchange rather than cash, to the modern industrial way of living.

Guided as it is by this ideology of development and underdevelopment, the proposal of the IPCC for resolving injustices arising from global warming in developing countries ironically involves forms of development assistance which will draw more of the world's peoples out of forms of life which are well adapted to the local ecological limits of natural systems into forms of life which will represent an added burden on those systems. In other words, the remedy for anthropogenic global warming is represented by the IPCC as the same as the cause – the movement of more of the earth's peoples from an ecological economy of the household which draws on locally available biofuels and sources of food and fibre to the fossil-fuelled global economy *as* household.

As critics of the ideology of development have long observed, the ways in which this 'progress of peoples' takes place in developing countries involve a tremendous amount of cultural and ecological destruction, just as they did in the first industrial revolution. Many of the development projects which the developed nations, in concert with international financial and aid agencies such as the World Bank, have funded in the last fifty years have involved the enforced movement of large numbers of indigenous peoples and subsistence farmers, the flooding or deforestation of great swathes of tropical forest, and the replacement of sustainable peasant agriculture with fossil-fuel dependent cash crops such as groundnuts and soya, and of traditional fisheries with ecologically disastrous fish and prawn farms.[51] All of this 'development' is pursued in the name of the modern ideology of progress according to which the human condition is 'advanced' when populations move from an agrarian household economy to an industrial economy, and when human work is separated from the household and the farm and provided and ordered instead by industrial corporations and government agencies.

That the planet now seems to be at risk because of this progress does not yet seem to discomfort those who suggest it remains the only remedy for the ills of the poor. And this belief is maintained despite the fact that subsistence farmers and urban migrants in sub-Saharan Africa presently experience far greater difficulties in accessing basic foodstuffs and finding fuel and water sources than did their forbears.

The Optics of Power and the Limits of Liberalism

Ethiopia is a country that has a particularly tragic history of coercive development of this kind. In the 1980s the socialist government of Lieutenant Colonel Mengistu Haile Mariam, and his infamous revolutionary regime, named the 'Dergue', imposed a disastrous agricultural collectivisation programme on hundreds of thousands of Ethiopian farmers, who were forced to leave traditional homes and fields and move to government-designed and -built villages. These villages were constructed on a centrally-designed model which was imposed regardless of local terrain such that a government servant entering any village would already know from the plan where the village hall and the seed store stood. Farmers were forced to grow crops with unfamiliar centrally-provided seeds in areas which were often ill-suited to these crops. And they were actively prevented from growing the foods and seeds with which they were familiar. The results were disastrous and helped provoke one of the worst famines in Ethiopian history, and an eventual uprising against the Mengistu regime. As James Scott argues, these agricultural reforms, like so many imposed on subsistence farmers in the name of modernism, development and progress, were driven by an 'unspoken logic' which devalued and disregarded the local ecological knowledge of the farmers, and their local plants and seeds, and instead imposed a regime which was designed to strengthen the power of government agencies and seed purveyors. The end result was the destruction of a self-sufficient household economy and the social creation of poverty.[52]

The industrialisation of agriculture in developing countries manifests an often brutal and crushing disregard for local ecology and for forms of 'place specific' knowledge and culture. This is because it manifests what Scott calls the 'optics of power' of science-informed modernity which involve a near religious faith in the belief that 'large farms, monocropping, "proper" villages, tractor-plowed fields, and collective or communal farming' represent an aesthetic order which advances the human condition, and the surface of the earth, towards the modernist goal of quantifiable, visible and planned progress towards centrally-ordered and collectively-managed forms of economic and social existence.[53]

The optics of power are not confined to agrarian 'improvement' or monocrop forests. They are equally manifest in the procedures and practices of industrial capitalism as promoted by multinational corporations, nation-states and superstate institutions such as the World Bank and the European Commission. This collectivist approach to human progress reaches its zenith in the neoliberal project of the 'advanced' nations and the intent to draw the whole world into a corporately-dominated global economy of trade and exchange.[54] This modernist project has its roots in Western political liberalism. Since Hobbes, political liberals have assumed that individuals are originally not social and that therefore property and trade relations need to be artificially constructed and ordered in a social contract which is sustained by collective institutions such as the nation-state and economic corporations. This view of human being suppresses the traditional household economy and its associated procedures for caring for common resources and sharing social power. And it ultimately destroys the local and place specific character of the agrarian household economy which rests upon forms of ecological knowledge and moral community, which made possible sustainable uses of local food, fuel and water sources. It also destroys the common access regimes which, until the colonial era, managed shared use of fragile soils and precious water sources far better than many colonial and post-colonial regimes.

The property arrangements of the colonists, involving as they did an extractive approach to the value of land, marginalised or destroyed altogether the household economics of subsistence farmers and pastoralists and their common property arrangements. And the global market economy which is now spreading like a virus right around the earth has promoted an even more brutal and large-scale disruption as it has drawn the ecological energies and social powers that traditionally reside in place into large centrifugal societal mechanisms such as markets and bureaucratic and corporate agencies.

Behind this process of abstraction of ecological and social power is a peculiar form of disembodied rationality which, from Hobbes to Hayek, envisages biological energy, and moral and political power, in abstraction from their embodied and ecological contexts in human and species relationships and geographical terrains.[55] It is Nussbaum's failure to challenge this feature of liberal political thought which explains why she opts for development as the means for advancing justice beyond borders. And the same problem is manifest in attempts to turn the procedures of liberal capitalism and development towards a resolution of the injustices of global warming.

Those who would turn development and market mechanisms and procedures towards the problem of resolving global warming do so without realising the core connection between the problem and the modernist forms of collective organisation that have given rise it. The goal of those who fashion such procedures is to preserve industrial civilisation, and the liberal narrative of freedom associated with the present form of consumer-driven capitalism. Even as radical an environmentalist as George Monbiot manifests this tendency. In his book *Heat* he assembles an impressive range of options for reducing greenhouse gas emissions.[56] But his aim is to preserve the form of liberal capitalism which is so much in conflict with the ecological health of the planet. Thus he describes a superb set of technical proposals and reforms to patterns of mobility, dwelling and consumption which will make it possible to sustain industrial civilisation while restricting carbon emissions. But the technical proposals – which involve for example dramatic reductions in car use and the end of flying – also involve dramatic restraints on freedoms citizens presently enjoy to consume and emit carbon at will. And yet Monbiot can give no account of how it is that citizens and governments will come to embrace these kinds of collective restraints in the pursuit of international and intergenerational climate justice in the absence of more fundamental reforms to the beliefs, practices and values of industrial society.[57]

In the United States a recent survey indicates that there is large-scale indifference to the condition of the planet and to climate change in particular.[58] At best the public manifest a vague worry without informed knowledge of effective means of mitigating the problem. Most however continue to believe that the scientific evidence for climate change is still uncertain and see little connection between raised taxes on fossil fuels – such as gasoline taxes – and attempts by government to mitigate climate change. In the United Kingdom things are different. In a recent survey 91 per cent of people indicate that they believe that climate change is happening and 77 per cent that there are risks for people in Britain, though a substantial proportion of respondents have little idea of the possible causes; for example 39 per cent thought 'air pollution' was the culprit.[59] More helpfully a clear majority – 62 per cent – agreed with the statement 'everything possible should be done about climate change'. However, when asked who should actually do something about it the majority suggest the levels at which action to address climate change are global and national, with only 8 per cent believing that responsibility lies with individuals or families.[60]

This brings us back to the core conundrum of liberalism. Its advocates suggest that it is possible to construct governance procedures and property relations which enable individuals to pursue their interests and goods in the market place with only minimal regard for the moral demands of justice and the common good. Nussbaum advocates an internationalisation of liberal justice and property arrangements when she suggests that, provided entitlements are properly described and recognised across national borders, then governments and multinational corporations *will* act to right injustices arising from present economic relationships. But this approach neglects the history of the relationships between colonised and coloniser, and that modern concentrations of power and wealth rest on a history of coercion and violence in which the powerless are frequently forced to give up their local place-based biological and ecological energies. The 'optics of power' sustained by present development policies, the larger global trade project and the stances taken by wealthy nations and corporations in the discussions which led to the Kyoto Protocol involve a refusal by the principal polluters of responsibility for global warming. Nussbuam's redescription of liberalism does not suffice to correct this.

Embodied Cognition and Responsiveness to the World

Climate change represents a challenge not only to energy-led consumerism and unfettered capitalism, and its latest guise in the form of borderless global trade, but to the epistemological and ontological foundations of modern liberalism. At the heart of Rawlsian liberalism, and its neoliberal offshoots, lies the assumption that individual actors are seats of consciousness, desire and rational decision-making and are intrinsically autonomous from other bodies and the biophysical cosmos. It is this assumption which explains the liberal division of labour between individual agency and the body politic; political institutions embody morality in the relational world of public space but individuals are conceived as essentially independent of this bodily domain, their actions determined by their inner desires and rational choices rather than by their biological relations to other beings and the environment.

Criticism of this essentially disembodied conception of agency and consciousness, which is rooted in the Scottish and German Enlightenments, has come from various quarters, and nowhere more powerfully than in the ground-breaking phenomenological critique of Maurice Merleau-Ponty. A key element in Merleau-Ponty's ontological move is his attempt to re-inscribe community and time upon the acts and gestures of individual bodies:

> What is alone true is that our open and personal existence rests on a primary basis of acquired and congealed existence. But it could not be otherwise if we are temporality, since the dialectic of the acquired and of the future is constitutive of time.[61]

On this account the perception and effect of every individual action or gesture already involves an unconscious but real projection of the self into relations with other bodies in the past and the future, in such a way that they are taken up into a larger embodied pattern which Merleau-Ponty characterised as 'existential rhythm'. Anthropologists such Mauss, Scott, and Clifford Geertz recognise this in the place-specific nature of gift exchange, local knowledge and moral community.[62] As Mauss proposed in his essay 'Techniques of the Body', the body is traditionally the primary tool through which humans give shape to the biophysical world. The human social world is also constituted by the assemblage of human bodies, hence the very term 'body politic'.[63]

In the mid-twentieth century Mauss and Merleau-Ponty directed anthropologists and philosophers to a reconsideration of the place of the body in human cultural and social construction as well as in human consciousness, and their insights have been taken up in ecological thought, psychology and philosophy. Thus quantum physicist David Bohm posited what he called an 'implicate order' in the substructure of atomic life which connects individual acts and the biophysical structure of the world, while Gregory Bateson constructed an ecological account of consciousness, and John Gibson proposed an ecological account of visual perception.[64] For Bateson and Gibson human consciousness and perception are constituted by their location in the body and by relations with other bodies, human and nonhuman; this recognition involves the rejection of reductionist and atomistic accounts of rationality which have achieved widespread currency under the influence of Enlightenment rationalism and political liberalism.

The recovery by phenomenologists, anthropologists, and more recently natural scientists,[65] of the embodied and relational character of human identity and perception also finds echoes in the turn among some modern moral philosophers and theologians to narrative, story and tradition as key elements in human development and rationality.[66] The life projects, desires and decisions of individuals in a narrativist perspective cannot be understood apart from knowledge of the life-story, and cultural memes, that have shaped an individual in her development, including family relations in the household, and relations with particular bioregions, communities, places, practices and traditions.[67] Similarly the problem of global warming cannot be properly narrated or addressed without attention to the history of its causation. Theoretical accounts of justice and equity will not be adequate to a resolution of the problem, any more than cost benefit analysis. Since more than 90 per cent of greenhouse gases already present in the atmosphere have emanated from industrialised nations, it is clearly their responsibility both to radically curtail their emissions in the next decades, and to compensate the victims of climate change and help them to adapt and, if necessary, migrate to new and less climatologically threatened terrains. But for the industrialised nations to recognise this responsibility and act on it would require that they give up the thin liberal narrative of justice as the mediation of competing private interests and the consequent privileging of market relations over more moral forms of exchange between nations.

Advocates of a phenomenological approach to human being and knowing have for more than fifty years attempted to unseat the regnant constructivist and rationalist assumptions in Western political and philosophical thought which underlie the project of liberalism. But they have been largely unsuccessful in dethroning the cultural power of these assumptions, given the strength of the modern commitment to the ethereal economy of unfettered desire and the 'free' market, and the relative inattention to the biophysical economy of organic bodies in the economy of the earth.[68] It may be that climate change offers not only a decisive moment in the earth's history, but a post-industrial way of conceiving of earth–human relations where the biological energy and nutrient flows of the earth and the bodies, both individual and collective, of humans are brought into a new alignment. How, though, might such a realignment be mapped onto the kinds of moral and social relations which constitute modern liberal societies?

The Structures of Sin

At the outset of the book I suggested that witness and truth-telling were vital modes of moral response to global warming. In the

Christian tradition witness and truth-telling arise from communities of worship located in particular places. These local communities are also connected to the larger household of God in the mystical conception of the Church as a communion of saints united in the mystical body of Christ. As we have seen the doctrine of the communion of saints expresses the ontological claim that these communities are not confined by boundaries of place or culture or time. The idea of connection between past, present and future actions and generations in the body of Christ is a mystical elaboration of the Hebrew conception of the unity of all humankind not only in being made in the image of God but in sharing in the fall of Adam from the original goodness of this image. The Christian doctrine of original sin arises from Old Testament accounts of the 'sins of the fathers' which the Torah represents as being visited on 'the third and fourth generations' (Deuteronomy 5:9), and from Paul's account of Christ as the 'new Adam' who redeems the inheritance of sin from the old Adam; 'as in Adam all die, so in Christ all are made alive' (1 Corinthians 15:22). This doctrine offers a valuable metaphor for thinking about the effects of ancestral decisions to mine coal and to annex colonial lands in the pursuit of industrial wealth. These decisions have collectively had an effect on the biosphere which magnifies the injustices visited on the colonies in a new form of environmental exclusion. But the idea of sin as inherited is offensive to the modern liberal mind because it suggests individuals are not free, and are not therefore responsible for their own decisions and choices. And yet when we consider the interconnections revealed by the geochemistry of the earth system it becomes clear that in reality individuals are set in a nexus of biological and social relationships across time and space. And this means that we are not free in the liberal sense in any case. As we have seen, the denial of this situatedness is part of the pathology of global warming, resting as it does on the human refusal to *be* creature, and a refusal of responsibility for the welfare of other creatures.[69]

The Christian doctrine of original sin expresses the moral intuition that individuals are only blameworthy for actions for which they are knowingly and personally responsible. It also expresses the idea that human beings receive their moral agency in the midst of relationships and structures which are already distorted by sin, and so the fields to which the word sin refers include family, self, society and ecology.[70] The global market empire and the climate system represent the most all-embracing of these fields. The market empire sustains remote structures of domination in which human and ecological communities and a stable climate are sacrificed to individual and corporate wealth accumulation and heedless personal consumption. And individuals lose agency in these remote structures and procedures. They are often unaware that their actions in excessively heating their homes or driving their cars have consequences elsewhere on the earth or for the future by adding to the greenhouse gases which are driving global warming. The distance between action and consequences in a global economy, and the centralising power of the economy *as* household, makes knowledgeable responsibility harder. This is why liberation and feminist theologians talk of the structures of sin.

But the existence of these structures, and the passivity they create in their victims, is not to deny that their victims also collude with them. It takes two to tango. And this is why the first move in the quest for redemption from the structures of sin is for individuals to acknowledge that they are responsible in some part for sustaining these structures. Recovering agency requires individuals to retake responsibility for their modes of consuming, dwelling and moving around. Acknowledging the ecological and social situatedness of human action is at the same time a way of recovering true agency and moral responsibility. Val Plumwood talks about this move in terms of the recovery of the self-in-relation.[71] Where human beings recover a true sense of being enmeshed in a rich community of social and ecological relationships, they enrich their sense of being in the world, and their encounters with other creatures.[72]

This is why the Christian claim to find in the events of Christ's life, death and resurrection liberation from the inheritance of sin takes relational form in the 'new creation' which is the Church. In this new 'household of God' individuals are reunited in a shared quest for peace and reconciliation made possible by the forgiveness of sins. St Paul uses the analogy of the 'body of Christ' to speak of this new creation. The metaphor involves the idea that the whole of the created order is being reconstituted and reconciled in the cosmic revelation of Christ as Lord, and in the shared life that this revelation makes possible in the body of Christ. The body of Christ therefore resists the divisions between body and reason, public and private, economic and personal, local, national and international, and between humans and other creatures, that are implicated in modern liberalism. Christians are trained by their participation in the divine nature through worship, and by their recollection of the divine character in the reading of the narratives of scripture, to care for one another, and to pray and work for the coming of the Kingdom *on earth* as well as in heaven.[73] In the common life of worship in local communities of place Christians rediscover the primordial unity of all persons and creatures which is affirmed in the resurrection of Christ. From their participation in the Trinitarian divine nature through Christ and the Spirit, Christians learn to attend to the common goods and the intricate inter-relationships which constitute the diversity and the sustaining energy and nutrient flows of created order.[74]

The word which best characterises this common attention to a shared identity of all creatures in God, superseding boundaries of class or race or species being, is love. According to the Apostles love and not justice should characterise human relations in the new creation which is the body of Christ; love which puts the other before the interests of the self, love which exalts the weak above the strong. And love is the form of embodied knowledge in the Christian tradition for, as Oliver O'Donovan puts it, after Augustine, 'we know only as we love'.[75] The knowledge of love supersedes the abstract theoretical demands of justice and the negative account of liberty as freedom from harm. Instead, knowledge as love indicates that freedom is to be exercised in love so that, as St Paul suggests, the needs of the weak set the standard for the requirements of love and not the capabilities of the strong.

Love is the form of knowledge that makes it possible to reflect on past moral wrongs, which have occasioned the present atmospheric emissions of climate-changing gases, without passing judgment. For only such knowledge is able to redeem present industrial humans from the sense of being trapped by the actions of their ancestors, and from guilt at the extent of their own present contribution to the problem. The possibility of forgiveness, of making a new start, does not therefore originate from human action but rather from the compassionate character of God. Humans, in other words, cannot save themselves. Instead divine forgiveness by way of Good Friday and Easter Sunday enacts at the same time both judgment on sin and liberation from sin. Past wrongs are judged as wrongs, and yet their perpetrators and their victims are liberated by the cross and resurrection to become agents in the redemption of the earth.[76] Divine grace makes possible a new mode of being where the knowledge and love of God overcome sin and transform the lives of God's creatures.[77] And this knowledge and love, so momentously experienced at first hand by the early Christians, continue to exercise a transforming power among the saints through their contemplation of the divine nature in worship, and of the reflection of divine beauty in the 'books of God' which are Scripture and Creation.[78]

Contemplation and Action

In classical and Christian thought contemplation is superior to action because contemplation is free from the necessities of life, whether in the form of labour on the soil, or the business of making. Instead the soul in contemplation is truly free to engage with the beautiful, which is the nature of divine and eternal being. And so all actions which are truly worthy, truly beautiful, must bear some foundational connection with the contemplation of beauty. And the truth of true Being, of the divine life, can only be known in complete stillness and rest.[79] As the Psalmist puts it, 'Be still and know that I am God'. This is why, for the Hebrew, Sabbath is not just the end of the week but the day which gives all the actions of the week meaning. For on this day the community of souls rests and in rest worships, contemplates, communes with the divine. This ancient priority of rest over movement, and of contemplation over action, reflects the metaphysical belief that the things that are in the physical cosmos are always superior in beauty to the things made by human hands, since as the creatures of God they reflect more fully the divine nature. And only in perfect rest and contemplation can the creature perceive the divine nature through the things that God has made.

But the everlasting temptation of the creature is to seek to make a dwelling in the world so glorious and permanent that it seems to acquire the original beauty and permanence which are the marks of the divine. Every great imperial civilisation manifests this temptation. And some of the monuments of such empires – the pyramids of Egypt, the ziggurats of Iraq, the Great Wall of China, the amphitheatres and aqueducts of Rome – have acquired a kind of semi-permanence. But the ruin of these monuments is indicative of the judgment that awaits those civilisations which have sought to replace the Creator with the work of human hands, and to substitute centrally-stored wealth for reliance upon the regenerative capacities of local ecosystems. The Tower of Babel expresses this primordial desire in classical mythic form. As humans attempt to build up to heaven they come to imagine that they are like gods, and this hubris is ultimately the cause of destructive divisions and wars.

Industrial civilisation far exceeds any predecessor culture in its attempts to build a substitute world of stored wealth and tradeable artefacts in place of the original beauty of the created cosmos. The industrial remaking and reordering of life on earth reaches from pole to pole as the chemical wastes and machine-governed order is detectable in toxins in the fatty tissue of seals and in the deformed gonads of fish, in industrial flotsam and jetsam visible on the surface of the ocean, and now in the upper atmosphere of the whole earth. The modern age, with its invention of electric light and automatic mechanistic devices, also exceeds any predecessor civilisation in its substitution of the activities of business and making over contemplative rest. And so *having* takes precedence over *being* in modern political economy in a way that is more thoroughgoing than in any predecessor culture. And industrial humans increasingly experience their identity in terms of the things that they have acquired, instead of in their *being* creatures.

In a powerful passage in the *City of God* Augustine suggests that 'a people is a gathered multitude of rational beings united by agreeing to share the things they love' and 'the better the things, the better the people, the worse the things, the worse the people'.[80] If, as advocates of industrial development propose, welfare is advanced by progress in consuming material goods, then industrial people are defined by their love of objects such as cars, televisions and mobile phones. But if contemplation of the beauty of the divine nature is the definitive form of love, then material objects are not the proper goal for a human society. Neither is their maximisation an indicator of welfare. Instead, welfare resides in the building of richer relationships between persons and all creatures, and between creatures and the Creator. Enriching relationships, not accumulating exchange value, is therefore the true source of human and ecological flourishing.

Resituating the self-in-relation describes the core task of the contemplative community. And hence such communities in Christian history have not only been centres of prayer and worship, but also of care and compassion for the poor and the sick. And they have also been places where work on the soil, and the stewardship of creation, have been restored to their central place in the arts of human making. This is why the Benedictine rule places physical work and worship in a symbiotic relationship.[81] Against the classical demeaning of physical labour as only suitable for slaves and servants, which is reborn in the machine age, Christian contemplatives discovered that worship and work exist in a symbiotic relationship, through which both creation and Creator are given their due honour and praise.

As we have seen, this framing of work by contemplation is expressed in ancient form in the Old Testament practice of the Sabbath: 'So then, a Sabbath rest still remains for the people of God;

for those who enter God's rest also cease from their labours as God did from his' (Hebrews 4:9–10). When Jews and Christians keep Sabbath faithfully, they do no work on the seventh day. By refusing to work one day in seven they not only set aside time for contemplation and worship of God. They also indicate that they do not receive their life from their work but from the Creator and from the regenerative lifecycles of the physical cosmos. Ceasing industrial activity one day in seven would also have direct material benefits, since if all industrial devices and consumption and transportation activities were stopped one day in seven this would immediately represent a more than 10 per cent cut in human use of energy. But of course the recovery of a day of rest, and the return of that day to its original ordering towards contemplation and praise would have a more profound and far-reaching effect in re-enchanting the culture of both work and leisure.

In the following chapters I will look at three core aspects of the present disordered economy of work and leisure and its impacts on the climate in relation to human dwelling, mobility and eating. I suggest that traditional moral and spiritual practices, emanating from the prioritisation of being over having, and of love over justice, hold great potential for recovering less disordered forms of making in a globally warmed world, and for the recovery of a politics and an economy which train citizens and corporations to treat the physical cosmos with greater reverence.

6
Dwelling in the Light

Hear and give ear; do not be haughty,
for the L*ORD* *has spoken.*
Give glory to the L*ORD* *your God*
before he brings darkness,
and before your feet stumble
on the mountains at twilight;
while you look for light,
he turns it into gloom
and makes it deep darkness.
But if you will not listen,
my soul will weep in secret for your pride;
my eyes will weep bitterly and run down with tears,
because the L*ORD's flock has been taken captive.*

Jeremiah 13:15–17

The Kogi are an ancient people who have lived for millennia in the cloud-covered snow-capped mountains of the Sierra Nevada of Colombia in South America. In the last thirty years they have noticed that the clouds are thinning over their mountains and that ice, rain and snow on their lands are declining. The ice that permanently encases the high Andean peaks is beginning to break away in great chunks in the summer, while in winter the snow starts to fill the rivers with melt water far sooner than it had traditionally, and the ground is bare of snow much earlier in the year. The Kogi also observe that while these strange things are happening to their own environment, the people who live on the plains and coastal regions below the mountains are cutting the forest, burning fires and shedding light into the sky from their settlements. They begin to suspect that the declining forests, the fires and the spread of light bleeding up into the sky from the growing towns and cities below are driving the changes in their mountain world. And they decide to send priests from their community to 'younger brother' down below to give a warning about what is happening to the landscape, and to suggest that the behaviour of younger brother is changing the world. In 1991 they decided to journey to the strange and noisy environment of the cities of Colombia and speak to political and business leaders of their concerns. They also cooperated with the making of a film documenting what was happening in the mountains, and their attempts to dialogue with the people in the valleys.[1] However, the Kogi did not encounter much understanding from younger brother, and they returned to their mountain-tops, their warnings unheeded, the people they had met down below largely unmoved.

Artificial and Uncreated Light

Seen from the air, the modern planet literally bleeds artificial light into the sky through the night. The production of light through electricity is one of the great miracles of the modern age. But to the extent that the ubiquity of this light now prevents most urban peoples, who are now the majority of humanity, from seeing the stars, it may ultimately turn into a curse. The ancients believed that those who could not see the stars were cut off from a vital source of wisdom about their place in the world and would become atheists.[2] Artificial light is ambiguous in another sense, for it enables industrial humans to live their lives independently of the rhythms of the earth. It makes possible 24-hour shift working in factories and offices and so subjects time to a new industrial urgency in which 'time is money'. The commodification of time creates the drive for a pattern of work and activity which is unremitting, and which subverts traditional restraints on the exigencies of work, including the Sabbath. Cities lit with electricity increasingly become places which never sleep, where both work and entertainment proceed through the night. Revellers return bleary eyed to the artificial lights of their offices the next day, even as maintenance and shift workers make their way to bed.

Electric light, more than any other commodity, represents the energy-hungry nature of the Prometheus which is the modern economy, and both the promise and the peril of this latest form of civilisation. One-and-a-half billion people still do not have access to electric light and in many ways they are disadvantaged by this lack. And yet George Mackay Brown records with ambiguity the coming of electric light to the Orkadian villages which were among the last places in the British Isles to be connected to the national grid. He observes that the spread of the light subtly changed the nature of community life, and especially when it also fuelled the new televisions that people bought. Instead of sitting around the fire listening to stories and making music and conversation, people grew apart and lost touch with the old ways.[3]

Electricity is in many ways the foundational symbol of modern life. The first homeowners to have electric light – Cragside in Northumberland was the first house in Britain to be lit by electricity which was generated from local hydropower – were seen as the pioneers of a new way of life in which humans were at last breaking free of the limiting rhythms of the natural world and imposing civility and rationality on the darkness of the night.[4] But the waste of light in modern industrial homes and cities is also a powerful symbol of the profligate and disrespectful nature of the modern economy, and of the way in which it is warming the climate. Not only does this economy reject the divine ordering of time according to the pattern of the Sabbath; it also shows deep disrespect for the created and warming energy of the god-like orb whose brilliance shone on creation over millions of years. As we have seen, this original radiance was drawn down into the substrata of the earth by dead plants millions of years ago, locking away the carbon in the subterranean carboniferous forest. Industrial humans are releasing this stored sunlight in a way that disrupts their relation to time and space and disrespects the roots of the present in geological time, and the origin of time in the primeval burst of light energy in which Creation began.

Waste light takes a variety of forms in industrial societies. People light rooms in which they do not sit, corporations light offices and advertisements when people do not work, governments light streets when people sleep. Lights flicker on appliances and computers that use energy when not being used and are never turned off. As an instance of the waste involved, the lit digital clock on a microwave oven wastes more energy in the life of the oven than the oven will use warming food. Neon lit signs and billboards light up and uglify whole landscapes with the spread of this artificially lit industrial world. In the United States, not only cities but even settlements in the mountains and on the plains assault the traveller with tall neon signs advertising motels, gas stations, fast food joints, casinos and brothels.[5] One of the simplest ways for industrial humans to save energy is to take the trouble to turn off appliances and lights when they are not needed, and to use light bulbs – such as the low energy fluorescent kind – which use less fossil fuel to keep them lit. But so long as corporations and governments insist on pouring light into the skies as if there were no tomorrow these individual acts of human care can hardly out-balance this systemic profligacy.

Just as the waste of light reveals the wasteful and careless attitude to creation of the industrial economy, so in Jewish and Christian traditions light is the symbol of revelation of the divine wisdom that orders and infuses Creation. The creation of light marks the primeval beginning of Creation in Genesis 1:3; the text names the light Day and the darkness Night. The coming of the Day reveals a universe in which Yahweh divides land from water, and on which the creatures are set to enjoy the divine light. The middle and hence pivotal day of the seven days of Creation, the fourth day, is also the day of the creation of lights in the sky – the light of the moon and the stars – whose movements in the night sky order the seasons around which the agricultural year and the liturgical year of religious worship are ordered. The creation of light is the primeval divine action which establishes and reveals the sacred nature of space and time.[6] And as the Psalmist makes clear, the sun and the moon not only order the time of worship, but also the relations of humanity with the other species:

> You have made the moon to mark the seasons;
> the sun knows its time for setting.
> You make darkness, and it is night,
> when all the animals of the forests come creeping out.
> The young lions roar for their prey,
> seeking their food from God.
> When the sun rises, they withdraw
> and lie down in their dens.
> People go out to their work
> and to their labour until the evening.
> O LORD, how manifold are your works!
> In wisdom you have made them all;
> the earth is full of your creatures.
>
> Psalm 104:19–24

As the lights of the heavens mark the festal year and time of worship, so the daily dying and rising of the sun orders human work, keeping it in bounds so that the other species also enjoy their God-given time and space on earth. As the labour of humans ceases with sunset, nocturnal animals, birds and insects enjoy the night and live their lives while humans rest. This ordering of day and night is a constant reminder that each new day is also a divine gift whose coming invites worship of the Creator, and that the succession of days is also ordered by weeks divided by the Sabbath in which praise and worship take precedence over work and making.

Light is also a symbol of redemption, just as darkness is a symbol of sin. When Yahweh redeems the Hebrews from slavery in Egypt and leads them through the wilderness to their dwelling places in the Promised Land, the divine presence takes the form of a pillar of cloud by day and of fire by night, lighting the way of the people from bondage to liberation. God is frequently described as ineffable and even dangerous light, the brightness of whose radiance no person can look upon and live. Thus Moses is told to turn aside from the face of God on Mount Sinai. Isaiah makes extensive use of the metaphor of light for the redeeming presence of God among God's people, and this paves the way for its frequent use in the New Testament. In the Prologue of the Gospel of John Christ is described as 'the light of world' who 'lightens every human being' and which the darkness of sin could not overcome. In the Transfiguration on Mount Tabor Christ's body seems to the disciples to shine with ethereal light, which commentators read as the material reflection of Christ's primordial divine nature which he shares with the Father and put aside in becoming human. After this event the disciples come to recognise Christ as the promised Messiah.

Christ uses light as a metaphor for the moral life; 'You are the light of the world. A city built on a hill cannot be hidden. No one after lighting a lamp puts it under the bushel basket, but on the lamp-stand, and it gives light to all in the house. In the same way, let your light shine before others, so that they may see your good works and give glory to your Father in heaven' (Matthew 5:14–16). The good soul is drawn to the shining of the light and the attraction and charisma of the holy life is evidence that the saints radiate this same light beyond themselves to those around them, just as the misdirected soul seeks the darkness and obscurity where misguided motives are hidden and selfish ends obscured.

The metaphor of light takes on particular significance in the hesychastic tradition of spirituality which originates with the Desert Fathers. The hesychast seeks to align his body with the divine spirit through meditation, posture and the breath prayer, with the aim of preparing the mind and the soul to receive the gift of contemplation of the divine presence which is described by St Symeon the New Theologian as a vision of light. A monk in the thirteenth century, Barlaam of Calabria, challenged the prayer methods of the hesychasts, and in their defence Gregory of Palamas developed an important theological account of what the monks were experiencing. Palamas argued that the experience of the presence of God was a consequence of God communicating directly to the soul. In this connection he identified an important distinction between God's ineffable and unknowable essence and the knowable and hence communicable nature of God. He gave a new name to the aspect of the divine which is knowable in hesychastic prayer – he called it *energeia* or the energies of the divine. He used the language of light to further elaborate this idea, suggesting that the light which the soul perceives in contemplative or hesychastic prayer through the divine energies is 'uncreated light', analogous to, but ontologically distinct from the created light which emanates from the sun.[7]

Hesychasm as elaborated by Palamas was politically and theologically highly controversial, but in the Orthodox tradition its influence grew. It provides a valuable theological commentary on the emphasis on the body and the senses in Orthodox worship and prayer, and on the liturgical embrace of created and uncreated light.[8] At the heart of Orthodox worship is the ritual engaged in by all worshippers, as well as the priests, of lighting candles before the icons placed around the sanctuary. The icons are seen as sacramental windows on the divine and the ritual of lighting connects the worshipper and the icon in a physical way.[9] For Christians from Orthodox, Catholic and Anglican traditions, the most paradigmatic liturgical event involving light occurs in the Easter Liturgy when, on the evening of Holy Saturday, the paschal fire of scented wood is lit outside the church, the paschal candle is processed into a dark nave and the light is passed among the people from candle to candle. In the semidarkness the story of creation, sin and salvation is recounted in hymns and scripture readings. The main lights are lit at the moment of baptism of the catechumens, which coincides with the traditional hour of the Resurrection of Christ in the night of Holy Saturday. The paschal fire, like the Palamite metaphor of light as the form of divine indwelling, trains Christians to treat created light with deep reverence and respect.

Palamas' account of uncreated light as *energeia* is even more suggestive as an analogy between the created light of the cosmos, the thermodynamic energy which is the product of this light, and the ineffable light of God's own being. Another hesychastic mystic, St Symeon the New Theologian, tells of a young man who was, mantra like, repeating a version of the Jesus prayer, 'Lord God, have mercy on me, a sinner', when he suddenly felt that the room in which he prayed was flooded with divine radiance, so much so that the monk 'lost all awareness of his surroundings' and

> … saw nothing but light all around him and did not know if he was standing on the ground. He was not afraid of falling: he was not concerned with the world nor did anything pertaining to men and corporeal beings enter his mind. Instead, he seemed to himself to have turned into light. Oblivious of the entire world he was filled with tears and with ineffable joy and gladness. His mind then ascended to heaven and beheld yet another light, which was clearer than that which was close at hand.[10]

According to the hesychasts, light is the sensual form of the mystical experience of the presence of God, just as it is the form in which the scriptures describe the dwelling-place of God in heaven. From this constant theme in Christianity and Judaism we may discern that being drawn to the light is a central part of our creaturely being as embodied spirits.

If light is the paradigmatic analogy of the divine life, and of the form of the divine dwelling both in heaven and in the soul, then we would expect that the increasingly atheistic and materialistic culture of modernity would manifest a disordered relation to created light. Palamism reveals the spiritual nature of the disordered use of energy that is presently warming the planet. It also suggests that redemption from the present sensual disorder of modern culture cannot be had without a recovery of devotion to the light of God, and that when the spiritual longing for light finds its true source then reverence for created light will again find its proper place in our culture.

Energy and Conservation

Light is the primeval form of the energy that drives the universe. Contemporary physicists suggest that the universe began with a 'big bang' not unlike the explosion of a massive star. The sun itself is a giant atomic reactor in which light and heat are constantly emitted to warm the earth, the only known planet in the universe so beneficently situated in relation to a stable and radiant star. The origin of heat energy in light is described in modern physics through the first law of thermodynamics, which suggests that the quantity of energy in the universe is a constant. This law indicates that when a plant turns the sun's heat into carbon through photosynthesis, or when a coal-fired power station releases stored solar energy from long-decomposed plants laid down as coal or oil in the earth's crust, these processes do not increase or decrease the quantity of energy present in the universe. Instead energy simply changes its location and form in these processes. Paradoxically the first law describes what is said to be a universal law of energy *conservation*; the sum of energy in the form of heat/light in the universe is a constant. As Rudolf Clausius first argued, what happens in heat-derived mechanical motion is that heat exchanges energy with the machine and its surroundings. But nonetheless the quantity of energy in the world stays the same, while what he called entropy increases.[11]

But there is ambiguity in the thermodynamic description of energy, for it seems to render human activity in mobilising stored sunlight in the earth's crust in the form of coal, oil and natural gas insignificant in the great cosmic scheme of things. According to the narrative of the first law of thermodynamics the rituals of energy use on which daily life in Western societies depend, and the networked power systems which make them possible, are simply ways in which humans move energy around: they have no cosmological or moral significance. The reductionist description of energy flows in terms of a mathematical formulation of the laws of energy and entropy renders human interaction with the sum of energy in the cosmos as a mere abstraction. It is an example of what Teresa Brennan calls the 'sado-dispassionate' character of scientific rationality.[12] Such descriptions neglect the ways in which biological systems exchange energy in relational and cyclical ways, and so truly conserve it. The law of energy conservation trains engineers to design devices and systems that are intrinsically wasteful, since it indicates that energy is always conserved by nature. As Hans Fuchs suggests, people do not normally conceive of heat in the way the laws of thermodynamics describe it. Instead they think of heat as a substance which moves from one body to another – like the heat of the sun moving through space to warm the earth, or the heat of a fire moving in a room to warm the one who sits by it.[13] And this common knowledge is more appropriate to thinking about the waste of primeval light that has brought on anthropogenic global warming. The modern movement of stored energy from beneath the earth's surface to the earth's atmosphere *does* change the balance of heat energy in the earth because it alters the atmospheric heat exchange between earth and space, the one place on earth where extra-terrestrial exchange is always occurring. Everything else that occurs on the earth in physical terms is confined within the earth system and therefore may better be described in cyclical terms. Everything apart from energy moves around *within* the cosmos. And this is why the law of energy conservation is so signally unhelpful as a way of narrating the physics of energy *on earth*.

Here again we encounter the cultural power of secular scientific description to misshape the way modern humans conceive and live in the material world. Physicists conceive of energy as the root of all other values. But of course this conception of value is a highly mathematical one. It does not truly describe all that is involved in heat exchange between a body like the sun and bodies on the earth, but rather reduces such exchanges to mathematical formulae which are non-relational, just like Newton's original accounts of time and motion. In the same way neoclassical economists posited money as the root of all value in the human economy. Both descriptions fail to do justice to the relational nexus in which living beings actually interact with each other in ways that are mutually constitutive and enlivening. The cultural power of these reductionist mechanistic accounts of energy and money is such that they have both contributed to ways of dwelling, or of being-in-the-world which are also mechanistic, and hence pay little regard to the relational interactions between human dwelling and the energy flows and nutrient exchanges in which human dwelling is biologically situated.

Against the thermodynamic description of energy the theological account of energy – and its forms of heat, light and motion – suggests that bodies move in a relational field which is a divinely ordered and relational whole. The reductionist account misses this complexly holistic character of the relational cosmos, and so removes the sense that what science calls nature has authority, and expresses constraints on what humans may do to it. As Oliver O'Donovan suggests, moral authority is 'inherent in the created order', but it is not an authority that is available to our immediate experience, as for example the needs of the human body for food and drink, sleep and shelter. Instead 'we become conscious of it only as we attempt to comprehend the world as an ordered whole'.[14] The key to seeing the world as an ordered whole is not a mathematical model of energy and entropy, or a computer simulation of the climate of the planet, but the story of Creation and redemption that begins with the primeval light of the dawn of the universe, and whose destiny is revealed in the dying and rising of God in Christ, who comes from ineffable light to dwell in first-century Palestine. It is in this great story, and in the stories, rituals and spiritual practices which rehearse the origin and return of divine light in the divine Son, that Christians discover that the energies that drive the universe are enchanted and are worthy of deep respect and reverence.

Energy and Sanctuary

Reverence for the Son as the primordial source of the divine light which drives the universe is realised spatially and temporally in Christian worship in the Eastward orientation of the sanctuary to the rising sun, and in the ordering of Christian worship around *Sun*day, the Lord's Day. The early Christians described the day after the Jewish Sabbath as the *eighth* day of Creation, since this is the day on which time and bodies were reborn, reordered and redeemed from the infection of sin by the Resurrection of Jesus Christ from the dead. In this way Christians celebrate that time and space have been redeemed and liberated from their subjection to sin and evil in the empires of history.

In accord with these foundational connections between worship and reverence for divine energy, the World Council of Churches (WCC) has been prominent in the last thirty years in calling for international action to conserve energy in response to the threat of global warming. Member churches organised a global petition before the meeting in Japan which led to the production of the Kyoto Protocol in 1997. Representatives of the WCC have been a prophetic and critical presence at all the Conference of the Parties meetings, drawing attention particularly to the international justice dimensions of climate-change negotiations, and in particular the imbalance of power between developed and developing nations in the negotiations, which mirror imbalances in other global fora.[15] The WCC has been active in linking efforts to mitigate climate change in the North with projects among development and relief agencies aimed at enabling vulnerable communities in the South to adapt to global warming.[16] It has also been prominent in calling for industrialised nations in the North to begin to make adjustments to a lower carbon economy, and ultimately to embrace the 80 per cent reduction in carbon emissions which alone will be enough to stave off climate disaster by the end of the present century according to IPCC predictions. The WCC is also in a position to challenge conventional divisions between nations on the basis of 'developed and developing' or 'North and South', with its long experience of working through consensus conferences across all nations and cultures. As Lukas Vischer argues, there are many developing nations in the South which are already using considerable amounts of fossil fuel.[17] Some even exceed European levels of per capita emissions. Taiwan for example emits 11.4 tonnes of CO_2e per capita, equivalent to United Kingdom per capita domestic emissions, and Singapore 21 tonnes, which is just below the per capita domestic emissions of North America.[18] There is therefore a need in future negotiations to supersede this traditional distinction.

The important conscientising work of the WCC has promoted the symbolic action of churches and Christian NGOs in relation to climate change on many continents. Much of this action has taken up the theme of light, and the symbolic significance of the use of energy in the sanctuary or church building. There is a growing movement to 'green sacred space' in North America, where new churches are increasingly being designed and built in such a way as to conserve energy and to use natural light and ventilation wherever possible in the sanctuary.[19] Churches in Britain, the Netherlands, Germany and Switzerland have installed solar panels on their chancel roofs; some, such as a Lutheran Church in Basle, Switzerland, devote the money they save from paying for electricity to fund solar energy projects in African villages and towns. In Britain the eco-Congregation project encourages congregations to undertake an environmental audit of the church buildings and attempt to reduce their ecological footprint. The Church of England has also initiated a project to encourage all parishes to undertake an environmental audit and find ways to reduce the overall energy consumption of the Church.[20]

The focus of these ecclesiastical projects on church buildings is significant because it points to the large role that space, heating and lighting play in the production of greenhouse gases. Energy used to heat, cool and light buildings constitutes around 40 per cent of the total human production of CO_2, and this figure does not include the amount of energy used in construction, which is very substantial in buildings made from concrete and steel since these materials are made with considerable quantities of fossil fuel.

Re-enchanting Human Dwelling

Humans have been building homes for hundreds of thousands of years, and the remains of prehistoric human homes and settlements can be found on many continents. In many premodern settlements living spaces are formed not simply according to function, but reflect instead a sensitivity to place and local energy flows that is often lacking in modern forms of dwelling. This is in part because the kind of delineation between sacred and secular space which emerged in the Western Christian tradition in the late middle ages is less common in nonwestern cultures.[21] Hence in most cultures before the modern it was inconceivable that the place of dwelling should be arranged without reference to the spirits of place. The 'green' architect Anthony Lawlor gives the example of the Navajo 'hogan', which is 'a simple round structure with an opening in the top and a doorway facing to the east. This simple structure embodies the entire mythology and relationship to life – just in the layout and the use of structure'.[22] The house, like the church or temple, is in many cultures a microcosm of the universe, expressing in its shape and form the orientation of those who dwell in it toward the earth and the gods.

The medieval architect Abbot Suger, the originator of the Gothic style whose principles were first fully exploited in the great cathedral at Chartres, drew his philosophy of architecture and art from the neoplatonist treatise *The Heavenly Hierarchies* of Pseudo-Dionysius.[23] Pseudo-Dionysius argued that we could only come to understand absolute beauty, which is God, through the effect of precious and beautiful things on our senses: 'the dull mind rises to truth through that which is material'. Responsible for many of the innovations of Gothic style, including the rose window and the light-flooding effects of high pointed windows, Suger developed the use of stone and glass to create an effect of weightless height and light-filled space which from Chartres onwards was to have a huge impact on European architecture.

The reverence for light in the Gothic tradition provides a significant historical analogy for the contemporary move towards green architecture. Thus Lawlor, who has pioneered sustainable housing design, explains that one of his aims as an architect is to connect the sacred significance of light in temple buildings with the role that natural light plays in making human dwellings sensually satisfying places to inhabit.[24] Lawlor and other sustainable architects are engaged in a larger project to reform their profession, which has been infected by the same reductionist and accounting logic which has driven modern economists and engineers to design systems which neglect and disrupt natural energy flows. Modern architects are trained in a 'one size fits all' approach so that buildings are designed according to function rather than location. This approach is in marked contrast to traditional building techniques. The adobe structure favoured in Mexico, for example, is highly adapted to the extremes of heat and cold between the Mexican summer and winter. In summer the thick adobe walls keep the interior cool, while in winter the same thickness helps conserve heat from the fire. Adobe structures are also made from locally available materials and therefore have few energy costs in their construction.[25] Similarly, the Bedouin tent is a superb structure for the hot desert environment. The black exterior makes the walls of the tent very hot, which has the effect of drawing heat out of the interior air. At the same time the fabric breathes and lets in cooler air through the walls while releasing hot air through the roof.

When it rains the fabric tightens up, keeping water out from the interior. And traditional Bedouin tents are made from goats' hair, which is a locally available material.[26]

By contrast many modern houses are built, as Le Corbusier, the father of modernist architecture, suggested they might be, as 'machines for living' and without regard to locally available materials or energy flows.[27] Their cell-like structure often admits very little natural light or ventilation, requiring electric light and mechanical cooling or heating at all times to make them comfortable. Reconceiving the relationship of living space to natural light and local energy flows is central to the new kinds of sustainable and low-energy housing which are already being built, albeit in small numbers, and this approach also has significant aesthetic payoffs in terms of making the dwelling space more attractive and enjoyable to be in. Homes which are south-oriented, and use a lot of glass on the southern aspect, can absorb thermal energy from the sun and store it with heat pumps and good insulation, while also maximising the availability of natural light, and so minimising the need for electric light.

The importance of natural light is not confined to living spaces. Architect Bill McDonough describes his partnership's design of a factory which was extensively lit by natural light, and gave plentiful views from the inside to the outside. The building's owners found that those who used it were happier and more productive at work.[28] Working in naturally-lit spaces makes people happier because it helps to reconnect their working lives with the rhythms of the earth, and, as we have seen, much modern work is disordered precisely because it is disconnected from these rhythms. Connection to nature is also essential to human wellbeing. Hence buildings which cut off their inhabitants from natural light and ventilation are prone to 'sick building syndrome'.

Ecological architects design buildings so that they use local energy flows, including the sun, wind and rain to cool, heat, light and ventilate buildings, and this has ecological and psychological benefits. A building that is carefully designed in this way will not only conserve energy; it can even enhance the energy and nutrient flows in its local environment. A building with a grass roof can provide spaces for insects and birds to colonise and can absorb rainwater and so reduce storm run-off into city drains and local rivers. A building can also act as a generator of energy, passing on spare energy for heating and lighting to other dwellings in the vicinity. Thus a tall city block can be used to generate solar and wind power and pass on this power to surrounding lower buildings which are unsuitable for wind turbines or are in shadow from other buildings and so receive less sunlight.

Building As If There Were No Tomorrow

Space heating, lighting and cooling is not only a major sector of energy consumption. It is also a sector in which radical reductions in energy use are entirely achievable.[29] Existing construction, insulation, cooling and heating technologies allow the building of domestic homes which use zero energy; these homes are built on the model of the *passivhaus* or 'passive house' first designed in Germany at the Darmstadt Institute. Situated so that it collects the maximum amount of solar heat, the passive house is constructed with a high degree of insulation and carefully-controlled ventilation which draws fresh air through the soil in winter to warm it, and extracts warmth from exhaust air. The system utilises all waste heat from bodies, lighting and machines in the house. The passive house also incorporates solar devices for heating water and photovoltaics for the provision of electricity. With these measures, and others, such as low energy lighting and appliances, the passive house uses one tenth the energy requirements of the average North European house, at approximately 15 kilowatt hours per m^2 per annum.[30] Passive houses are also typically built of timber and glass instead of brick and concrete, and use sheep's wool for insulation instead of glass wool, thus lowering the energy costs involved in their construction.

Alongside the building of new low-energy homes there is great energy-saving potential in the refitting and insulation of existing homes. The Oxford Environmental Change Institute suggest in their report *The 40% House* that Britain could achieve a 60 per cent reduction in domestic energy us by 2050 if the government would fund an extensive programme of refitting.[31] The target is challenging but achievable. But such a major programme of renewal requires a collective commitment from government, householders, corporations and builders to serious reductions in energy consumption in buildings, which has so far not emerged.

The British government has considerable existing powers over the energy used in buildings, and has recently increased energy efficiency and conservation requirements for domestic buildings. However, British building regulations are notoriously weak, and even weaker in their enforcement, and so commercial builders who commission 90 per cent of Britain's housing continue to build homes which leak energy, waste water and which will cost their owners, as well as the environment, a great deal to maintain as energy becomes more expensive and its climatic costs mount.[32] Those who oppose carbon taxation do so on the grounds that it will make heating fuel unaffordable for the poor. But presently more than 3 million people already suffer from fuel poverty because the great majority of the housing stock is badly insulated. Thousands of old people on government pensions under-heat their houses and risk dying of cold in the winter.[33] Insulating homes and replacing old housing stock with passive houses is not just a priority because of climate change but also because of social justice. This requires of government a preparedness to use its legal powers to re-regulate the energy and building sectors to encourage conservation and reduce emissions instead of relying on market procedures.

In its Energy Review (2006) the British government expressed a desire to see significant reductions in energy use, of up to 20 per cent on 1990 levels by 2020.[34] In 2007 three government initiatives were announced to assist in the achievement of this albeit modest target. First, homeowners will have to commission an energy audit of their homes before putting them up for sale. This will apply to houses with four or more bedrooms from August 2007, and will be introduced for smaller properties at a later date. Second, utility companies will be encouraged to provide devices for measuring real-time energy use in homes. Such devices are a considerable aid to energy conservation and attention to energy use. In current homes energy meters are typically hidden away under the stairs or on boxes on the house exterior. When homeowners can see the amount of energy they are using through a visible device, they are more likely to turn off unused appliances and lights. And third, the government announced its intention to require the building industry to construct zero carbon homes – which utilise and conserve their own energy – from 2020.

While the British government is making policy moves in relation to energy-savings in domestic housing, the regulation of energy use in the corporate sector has been much less effective because of the move to carbon trading. Public subsidies for fossil-fuel production also continue. The total European subsidy for the industry is somewhere between €12 and €19 billion a year. The problem, as we saw in chapter 3, is not only one of finance but also of the rigging of energy production and delivery systems to centralised supply networks and grids run by large corporations which are used to receiving significant government support in the form of public subsidies.[35] At the same time, all energy technologies are taxed, regardless of whether they are renewable or not.

In the United States the situation is far worse. Housing in the United States is the most energy-hungry on the planet and tens of millions of householders cannot afford either to cool or adequately heat their homes in the extremes of temperature to which much of the continent is subject. A typical thinly-built and poorly-insulated home uses 10,000 kilowatt hours of mostly coal-generated electricity every year. Yet in many parts of the States where sunlight is plentiful homes designed with renewal energy systems, natural ventilation and energy conservation measures make it possible for such homes not to be connected to the grid.[36] In most southern states, where energy demand is especially high because of air conditioning, homes could be built or refitted to be self-sufficient in fuel or even provide renewable energy to the grid. But instead of encouraging energy-efficient housing, which innovators have been building for at least thirty years, the United States uses some of the most invasive coal mining technology in the world, married to some of the planet's most inefficient coal-generating plant. In West Virginia mining companies remove whole mountaintops to uncover coal seams; the beautiful Appalachian Mountains, much fought over in the Civil War, have been systematically violated by mountain-top removal to supply the notoriously polluting coal generators of the Tennessee Valley Authority. Advocates suggest this method 'optimises the coal reserve'; communities whose valleys and rivers are despoiled with the coal leavings, and whose towns and villages are blighted by flash floods and pollution, describe it as unjust corporate exploitation, though their efforts to force the companies that do it to be more responsible, and to compensate blighted communities, have been consistently opposed by state and federal governments.[37]

Without government leadership it is hard for citizens or corporations to make a substantial difference to total energy use. It is not as if citizens are in a position themselves to re-regulate the energy market, to redirect energy subsidies to renewables, or to tax carbon instead of human labour. In moral terms, the failure of government in addressing this greatest threat to the welfare of peoples is a clear indication that politicians and lawmakers are shirking their duty to promote the common good, and instead are continuing to put the interests and wishes of economic corporations over the welfare of future generations and present communities in parts of the world already threatened by global warming.

Housing is, however, one area where increasingly local and city authorities, as well as individual housebuilders and ecological architects, are taking a lead over national governments. Low-energy housing is already being built in Britain but it is mostly commissioned by self-builders or social housing providers.[38] And the mayors of London and Newcastle have announced programmes to reduce the carbon footprints of these cities through enhancing energy conservation in existing housing stock, and through stringent energy and water conservation requirements for new homes and offices. As one example of the effects of these programmes, under pressure from Mayor Ken Livingstone, new office buildings in London are required to include in their design novel energy conservation technologies. Similarly a number of city and state governments in North America have announced climate change action plans which target every sector of the economy, including housing.

The Quest for Sustainable Communities

The movement of cities towards taking responsibility for their carbon footprint may indicate a moral shift below the level of national government towards action on climate change. There is also a growing movement towards what the WCC calls 'sustainable communities', and this involves the recognition that for energy conservation to affect the warp and woof of everyday life, it needs to become part of the micropolitics of local communities of place.[39] As we have already seen, the centralised grid is a very inefficient way of delivering electricity and energy for heating. Moving away from dependence on the grid has tremendous implications for citizen and community engagement, for it relocates power production, and political power, from large centralised organisations to local communities and households.[40] But given the intrinsic relationship between the centralised delivery systems of the energy economy and the centralised control and service systems of the nation-state, it is perhaps unsurprising that most states, and super-states such as the European Commission and the United States Federal Government, are resistant to the new energy systems which must emerge if future generations are to have a future.

Individual citizens are, however, in a position to make significant savings in their stewardship of energy in their own homes and in the workplace, as David Reay argues in his excellent guide to low-carbon living, *Global Warming Begins at Home.*[41] If the home produces between one third and one half of an average person's carbon emissions, then savings there represent important contributions to addressing the energy budget of each individual. The first and most widely-acknowledged measure is to use low energy light bulbs, and to turn off all lights and appliances when not in use. The second is to purchase low-energy appliances when existing appliances come to the end of their life. The third is to upgrade the energy insulation of properties so that they more effectively conserve what heat is produced. Insulation, though tricky to install, is the single most rewarding technology in terms of reducing carbon use in the home. The fourth is microgeneration, where a household combined heat and power unit is used to generate both heat and electricity which dramatically increases energy efficiency. There is also a growing array of renewable energy microgeneration devices, ranging from small wind turbines to solar water heaters and wood chip burners, although their efficiency needs careful investigation. Wind turbines do not work where roof spaces are overshadowed by other buildings but they are an excellent local source of energy for tall buildings and for rural homes. In southern climes photovoltaic panels will actually replace the electricity requirements of an average home, though further north they will need supplementation by other energy sources. While these kinds of technologies may seem only to be in the purview of wealthy householders who own their own properties, this is only because of the present system of energy subsidies, taxes and regulations. The installation and manufacturing costs of these devices would fall dramatically if all new houses and offices were required to produce energy from local sources, and if electricity and gas utility companies were taxed on their production of energy and the taxes offset against their efforts in reducing demand. This approach has been adopted in some states in the United States, and has seen energy utilities getting in to the business of installing insulation and more efficient furnaces in homes and offices.

The Building As Microcosm

Energy conservation is a part of a larger project to recover the macrocosmic significance of dwelling space, energy and light and this is a project with deep roots, as we have already seen, in the spiritual traditions of the West. Some readers may, however, be sceptical as to whether this kind of detailed attention to the design, construction and regulation of buildings and energy systems is really of central moral concern for Christians. In answer, aside from the tradition of medieval cathedral building, I point the reader to strong biblical testimony to the divine significance of the work of building construction, something which was well known and embraced by medieval joiners and masons. Despite the fact that the biblical record emerges from the culture of ancient Mesopotamia in which the first cities, writing tablets, taxation systems, armies and temples were all created, the Bible does not often attend to the construction and design of material artefacts. But when it does, its descriptions are rich with moral and symbolic meanings. Among its more detailed descriptions of such artefacts are narratives of the construction of one great ship, Noah's Ark, and two buildings, the Tabernacle in Exodus and the Temple of Solomon. In all three cases, the description of the design and building is rich in numerological and cosmological significance, for these three constructions are all microcosms of the covenant relationship between the Creator and the creature, Yahweh and Israel. All three are brought into being as a result of a direct command of God, and their construction is seen as analogous to the work of the master builder in creating the universe. Thus, when the people of Israel are done with constructing the Tabernacle, Moses looks on and blesses them now that their work of sacred construction is done. As Ellen Davis argues, this blessing is clearly reminiscent of the Sabbath blessing of God on all creatures in the seventh day of creation.[42] And this correspondence between the creative powers of the Creator and human builders is also of significance in the moral vision of these Old Testament texts. However, as we have seen, the Solomonic Temple, being a regal project on a grand, even imperial, scale, has ambiguous outcomes for the covenant community of Israel, and for the forests of Lebanon. Moses and Noah, on the other hand, are described as men of particular virtues which include not only wisdom but humility. The clear implication is that good building, good work, requires persons of good moral character to carry it out. These particular virtues are also crucial to living in harmony and in proper relationship to the spirit of place and the authority of the created order.[43] It is precisely the lack of humility and wisdom manifest in the reckless burning of millennia of sunlight in the last forty years which is at the root of the climate crisis.

Good Work and Moral Character

The relation between good work and moral character goes to the heart of the larger moral problem with industrial civilisation, fuelled as it is by the profligate use of energy. This is a civilisation which valorises the vices of sloth and greed, imprudence and injustice. The house builder who skimps on the energy requirements of building regulations, and corporations which throw up millions of homes like so many ugly boxes filled with wasteful technology, display the vice of sloth. They are among the most extensive commissioners of modern human work – there are few more essential forms of work than the construction of dwelling places – but they want to do it cheaply, quickly and in order to make a quick profit. House building in Britain is the ultimate example of what Wendell Berry calls bad work.[44]

Bad work is centrally implicated in the perverse ways in which industrial capitalism redirects human life away from careful, embodied and skilful engagement with the natural world towards the alienated kinds of work which are the lot of so many industrial workers. How many workers now spend hours of their working lives staring at flickering screens, moving units of data around, breathing artificially cooled or heated air and without natural light, at the service of the great corporate economy? It then becomes necessary for them to spend time after work in an artificially-lit gym interacting with machines in order to exercise the muscles that their daily work threatens to atrophy. Equally, modern cities, particularly in North America, though often also in Europe, are so designed that it is impossible to use the body in walking or cycling to work, either because commuting distances are too great or because it is just too dangerous or unpleasant to actually walk or cycle on city streets. And not having cycled or walked for purposeful travel, it again becomes necessary to waste muscle power and energy on a cycling machine or rolling road in a gym, or drive a bicycle in a car to a 'leisure' bicycle track in order to obtain adequate bodily exercise.

Another reason for this lack of balance between work and recreation is the ideology of work without limits. It is not uncommon for individuals working for private corporations, or even public universities, to regularly spend 10 hours a day in artificially-lit offices, and another part of their day commuting to and from home. Work has become an idol, and like all idols human devotion to it ultimately involves sacrifices, both human and nonhuman. Work that is prudential in its use of bodily and earth energy is work that pays attention to the needs of other persons and creatures. Such attention is needful for the proper nurture of children in the home, or for good agricultural husbandry of animals, crops and soil. As Wendell Berry suggests, good work, work that does not waste energy, and is productive and beneficent in the place where it is practised, requires a preparedness to engage the body. And it also requires community. Such work will be at the heart of a cultural revolution towards energy conservation:

> We must learn again to think of human energy, our energy, not as something to be saved, but as something to be used and to be enjoyed in use. We must understand that our strength is, first of

> all, strength of body, and that this strength cannot thrive except in useful, decent, satisfying, comely work. There is no such thing as a reservoir of bodily energy. By saving it – as our ideals of labour-saving and luxury bid us do – we simply waste it, and waste much else along with it.[45]

The individual who did more than any other, in practice as well as in theory, to critique bad work in the design and construction of human dwellings, and the substitution of skilful and purposeful crafts for machine-based drudgery was William Morris. He thought that industrial work was bad work because, as Marx also pointed out, it alienated the worker from the product of his labours. For Morris, the reason for this alienation was not just the turn to use values and the form of industrial ownership criticised by Marx but the way in which industrial machines increasingly marginalised the character and skill of the good craftsman. Like the biblical writers, Morris understood that there was an intrinsic relationship between a society's traditions of making and the character of its people. Morris also believed that there was an intrinsic good in the participation of the worker in the creation of beautiful objects.[46] For him aesthetic beauty and delightful, meaningful, glorious living are closely connected, but this connection is not of a manufactured kind: instead of the slothful gratification of machine construction, for Morris the only way to a beautiful house, sanctuary, or sofa was the good work of a skilled craftsman.

Throughout this book I have been suggesting that the threat of global warming not only presents modern energy-dependent cultures with a serious limit problem. It also raises profound questions. What is life for? Where is ultimate meaning and purpose? Whom or what do we worship? In the biblical account of the work of God, and the analogy between divine and human construction, we discern that work in Jewish and Christian traditions is not simply the way to get to something else – creative leisure, enjoyable holidays, secure wealth, a contented retirement. Work is not instrumental but intrinsic to the vocation of the human creature; the human being, unlike say a bird or a cow, finds fulfilment in the ability of this particular creature to mirror the work of the Creator. As Timothy Gorringe argues, the design and construction of dwellings is a powerful exemplar of this vocation of the human being towards good work, for in their design, ordering and arrangement, dwellings have the capacity to construct so many other features of human and other than human life.[47] In their provision of gardens and walls, their arrangements of windows and pavements, streetscapes and neighbourhoods, collections of dwellings provide the material fabric and pathways of human living. This fabric can serve to sustain characterful and virtuous communities where citizens are encouraged by the arrangement of homes to be good neighbours and citizens. Or it can be shoddily built and foster moral shoddiness, environmental neglect and violent crime, as many an urban streetscape in modern Europe and America testifies. This fabric can be just and provide sustainable homes of reasonable size which are efficient to heat or keep cool. Or it can be unjust and provide some with 10,000 square feet air-conditioned mansions while others sweat or freeze in poky and ugly concrete boxes in apartment blocks.[48]

Domestic Comfort and Public Neglect

The still-growing material and spatial differentiation between luxury and social housing, gated community and ghetto, is indicative of a culture of luxury that drives the profligate use of energy, and which is at the same time a culture of injustice, and of social and environmental exclusion. The consequent aesthetic and moral disorder of the modern built environment, and especially its urban form, is partly responsible for a turn to the domestic interior as the site of true significance and meaning. William Morris and John Ruskin resisted the brutality and ugliness of the modern industrial city because they saw the exclusion of beauty and truth as an offence *both* against transcendent beauty, and against the humanity of working people, idealised by both Ruskin and Morris as rural artisans who had been excluded from their ancestral villages and common lands and corralled in urban slums, workshops and factories.[49] Their loss of economic independence led to the loss of craft skills and aesthetic sensibilities; their participation in industrial manufacture reduced their powers of making to alienating service to the machine. Seen in this light, Morris's attempt to reform the aesthetic of urban living is much more significant than its bourgeois outcome of interior design for wealthy churches and homeowners. Morris tried to put into effect in his workshops his belief that ugliness reduces the dignity of those upon whom it is imposed, because it also rejects the sublime in the order of nature as refracted in the skill of the craftsman or woman in transforming natural materials into useful and beautiful objects.[50]

Ruskin was to describe the theological significance of the consequent turn to the domestic interior, the home, as follows:

> This is the nature of home – it is the place of peace: the shelter,

> not only from injury, but from all terror, doubt and division. In so far as it is not this, it is not home; so far as the anxieties of the outer life penetrate into it, and the inconsistently-minded, unloved, or hostile society of the outer world is allowed by either husband or wife to cross the threshold it ceases to be a home; it is then only part of the outer world which you have roofed over and lighted fire in. But so far as it is a sacred place, a vestal temple, a temple of the hearth … it is a home.[51]

As J. K. Galbraith contends, this quest for domestic sanctuary in the midst of public squalor and social injustice is redoubled in the post-industrial cities of Britain and North America.[52] In England and Wales the urban rat population has increased tenfold since the privatisation of drains and water. Beggars and litter jostle in doorways and subway stations, while people avert their eyes from the disorder of public spaces on their journeys from their increasingly over-designed homes to constantly made-over office, restaurant and retail interiors. The designed interior cannot redeem the ecological destructiveness or social exclusion fostered by corporate greed and bad work. As Richard Sennett suggests, the 'perverse consequences of the search for refuge in secular society' are 'an increase in isolation and in inequality'.[53]

Against the aesthetic duality of spiritual or moral interior and functional and amoral exterior in the post-modern city, and the ecological destruction and material exclusion which is the economic basis of this dualism, a theological and ecological account of dwelling offers a crucial corrective, for it presents the relation of material and spiritual, outer and inner, as something which is intrinsic to being, and which is rescued from fallenness and futility in the Incarnation and Resurrection of Christ. Sennett suggests that there has emerged in the Christian West a tragic alienation between the inner spiritual life and the outer material world which he rather tenuously traces to Augustine, but which others trace to developments in medieval theology which reached full flowering after the Reformation.[54] But a truly orthodox Christianity is deeply invested in the material as the crucible of the spiritual life, as represented in the hesychastic doctrine of created and uncreated light. Orthodox theologians in both East and West have seen the human vocation to work on the divine creation by analogy to the divine work of the Creator. But this analogy does not make of humans co-creators with God.[55] Rather, it suggests that human making is rightly performed when human makers recognise the intrinsic worth of creation as divine gift and as theatre of God's glory. When human work is performed in ways that respect the divine origination of creation, then the fruits of human making are offered up to the deity as grateful and reverent response to the divine intention, revealed in the Incarnate light of God, to redeem all flesh in material space and cosmic time. In so doing, as John of Pergamon suggests, human makers are redeemed *through* their respectful interactions with the created order, and not aside from it.[56]

The thought and practices of socialists such as Morris and Ruskin provide a continuing source for the critique of the coercive maldistribution of dwelling space in contemporary capitalist societies. They point us towards an understanding of what John Milbank calls 'complex space', by which he means societies characterised by complex networks of intermediate groups and associations, or what others call 'sustainable communities'. The medieval cathedral and its urban environs represented a theological conception of such a complex space as the body of Christ:

> The church as a whole was not an enclosed, defensible realm like the antique *polis*, but in its unity with the heavenly city and Christ its head, infinitely surpassed the scope of the state, and the grasp of human reason. At the same time, what was fundamentally the same excess could be glimpsed in the single person and the Christian association (monastery or guild) whose activities are legitimised by the quest for salvation, not by human law.[57]

Milbank reconnects spatial aesthetics and making with an account of social justice. Against the claim made by some that alienated space in modernity is the intrinsic outcome of a Christian dualism of earthly and heavenly cities, or secular and sacred space, Milbank sees the alienating spaces and social exclusions of modernity as the connected consequences of the attack on associations and the household economy which the emergent nation-states of Europe sustained in the pursuit of the Enlightenment ideal of the liberty of the sovereign individual, and the fraternity and equality of which the state, and not religion, was to be the true guarantor.[58] The Enlightenment fostered the concept of 'simple space', in which both the sacred and the associational are marginalised. Simple space is open to domination and coercion by the state and the private corporation and sustains the private morality of the individual as consumer or homemaker. Simple space is also open to the architectural reduction of the complex, diverse and symbiotic forms of indigenous architecture that characterised premodern approaches to dwelling in different terrains and cultures. 'Complex space' offers a more integrative understanding of being which rests upon the relation of God to the material in the Incarnate Word, and upon the gift of the social as the body of Christ, which is the spiritual form of the church after the Ascension. It is also suggestive of the multiple ways in which dwelling space can be designed to exchange and interact sympathetically with natural energy flows and the local environment and landscape.

But there is an abiding ambiguity in a Christian account of dwelling space. In the Christian tradition, even more than the Jewish, the notion of dwelling is problematised by the 'Son of Man who has nowhere to lay his head' (Matthew 8:20), and by the New Testament account of the body of Christ which is not confined to any temple or sanctuary but incarnate in the presence of God among his people. If homelessness is the archetypal condition of the soul, then perhaps the alienation of modern forms of dwelling ultimately expresses the spiritual homelessness of Christianity in material form. And hence when the Christian tradition affirms that God does not dwell in buildings made with hands, it resonates with many moderns who, like John Muir, find that the presence of God is far more powerfully mediated by a great forest or a soaring mountain than it is by the Gothic columns and soaring spires of a medieval sanctuary.[59]

There is however no essential contradiction here when we perceive that the Incarnation works a material and a spiritual redemption. For Palamas the dwelling of God on earth is known in the energies of uncreated light. For the sanctuary stonemason, the ecological architect and the sustainable self-builder reverence for nature, humanity and spirit are expressed in the creative craft of the builder combined with fidelity to locally available materials and respectful use of local energy flows. Analogously the contemplative and liturgical encounter with uncreated energy trains the worshipper to good work, and respect for created energy, in the construction and stewarding of our places of dwelling, whether a temporary shelter under the stars or a more permanent home in the city.

7
Mobility and Pilgrimage

> *A hot wind comes from me out of the bare heights in the desert towards my poor people, not to winnow or cleanse – a wind too strong for that. Now it is I who speak in judgment against them.*
>
> *Look! He comes up like clouds,*
> *his chariots like the whirlwind;*
> *his horses are swifter than eagles –*
> *woe to us, for we are ruined!*
>
> Jeremiah 4:11–13

The steam engine which began the industrial economy was a stationary machine designed by Thomas Newcomen to extract water from England's coal mines. Coal was used principally for domestic heating and as a substitute for waterpower in the cotton mills. Nonetheless, the material power that this fuel put in the hands of industrialists set the modern world on a new path which would, as Engels saw, produce new kinds of relations between human beings in the class structure of industrial society,[1] and a new relationship between humanity and the earth. With coal-fired steam power, humanity was set free from the energy budget of biopower and was able to draw on the stored capital of carbon from beneath the earth's surface and so to develop economic activity at a pace never before seen in human history.

The other vital development which is associated with the gathering pace of the industrial revolution, and which now represents the single largest commitment of mechanical power in industrialised societies, is mechanical mobility. Initially coal was drawn out of mines by men and horses, and fairly soon it was discovered that the process could be done more efficiently with carts mounted on narrow gauge steel rails. But it was not long after the invention of the stationary steam engine that George Stephenson invented the first commercially successful mobile steam machine or 'rocket steam locomotive'. This device facilitated the development of railways from pithead or coal port to industrial city, the first such being from Stockton to Darlington and Liverpool to Manchester. Before the railway all the coal had gone by ship. But this was a slow process and the steam train dramatically increased the speed of movement of this heavy material. And this greatly increased speed was a vital factor in England's industrial revolution.[2]

Railways are extraordinarily efficient mobility devices because of the low friction of metal wheels on metal tracks (a function of the small surface area between the two), the great weight a rail axle can carry on metal wheels, and the large number of carriages or wagons that can be placed together relative to wind resistance. They rapidly spread throughout the world and transformed the industrial age into the age of mobility. Time itself was remade by the railway, as the linking of different towns and regions required the coordination of clocks; where there had once been a variety of local times the railway now required that mechanical time become universal. This collectivising process is also manifest in the nature of this technology, since railways require that people move around *together*. Because they are highly capital-intensive, they also require considerable amounts of collective investment. A rail train is consequently of no use on a farm or a rural turnpike, and so it was that Michigan farmer's son Henry Ford set himself the task of inventing a mobile engine which would make possible the application of mechanical mobility to the farm: 'to lift farm drudgery off flesh and blood and lay it on steel and motors has been my most constant ambition.'[3] But instead of steam tractors it was Ford's invention of the mass-production automobile, the Model T, and of moving assembly belts in his factories, which was to transform the form and pace of modern mobility more than any other single device, for it brought the exclusive horseless carriage invented by Karl Benz to the masses. Fifteen million Model Ts were built and car ownership and use became the defining aspiration of the new consumer society.

The Climate Change Machine

When first invented, the horseless carriage was seen as a device which reduced pollution, since cities were full of the reek of horse manure from horse-drawn carriages. But the car has become the most significant polluting device in the industrial age, and more than any other consumer object, mass ownership has made the automobile the most powerful climate-changing machine. On present trends the number of vehicles in use will increase from 500 million in 1990 to 2 billion in 2030.[4] Compared to railways, cars are ten times less efficient at moving the equivalent weight. But they have displaced railways and trams as the preferred mode of transport for those who can afford them in Europe, the Americas and Southeast Asia, and are presently displacing bicycles and rickshaws in the cities of India and China. In the United States automobile and tyre manufacturers actually bought up the railroads in order to shut them down and drive up demand for automobiles, tyres and highways. In Britain after the Second World War large parts of the tram and train networks were taken apart by city and national governments in anticipation of the growth of the great car economy.

The automobile is the single device which is immediately identifiable as the largest source of luxury emissions. Most humans will not die from hypothermia or heat exhaustion when they do not have access to a car. Access to cars, and the distribution of their negative effects – including not only global warming but air pollution and traffic accidents – is also a major source of the international inequities of global warming. Of the more than 500 million cars in the world, 75 per cent are owned in the developed world, while of the 1.2 million deaths caused annually by road accidents, 85 per cent of these occur in the developing world.[5] The WHO estimates the numbers of those injured in road accidents at 50 million a year. There are also many kinds of non-impact injury from road transport, and again these fall more on some groups than others. In urban areas asthma and coronary and lung diseases associated with vehicle emissions are disproportionately experienced by inner city residents, and those who live in lower-cost housing adjacent to urban freeways. Average private mileage per vehicle in Europe and the United States is between 10,000 and 12,000 miles per year, and many households now have more than one car. How much of this is essential to the welfare of the driver? Admittedly, in many parts of the United States public transport is unavailable, so that some of these miles represent travel to work and to shops by those who have no alternative means of transportation. However, the same cannot be said for the great majority of Europeans, who mostly have access to adequate public transport facilities but still prefer to maintain and drive their own 'private' vehicles.

The ascription 'private' is increasingly problematic when applied to automobiles. Their use requires the public maintenance of an extensive concrete, steel and tarmac infrastructure, representing one half of the built space of European and American cities. And the polluting emissions of these vehicles, together with the accidents they cause, place a considerable health burden on society as a whole, the costs of which are estimated at around $100 billion for the United States alone. With the advent of global warming the costs of vehicle pollution are dramatically enhanced. In the United States costs associated with the extended drought in the Southwestern States, which scientists believe is caused by global warming, are in excess of $200 billion, and globally the cost of adapting to global warming will rise in the present century to many trillions of dollars.

In the United Kingdom the transport sector as a whole is said to be responsible for 38 per cent of final energy consumption by businesses, government agencies and individuals.[6] In the United States domestic transportation is reckoned to be responsible for 30 per cent of total energy consumption.[7] But these figures exclude the construction and maintenance of transport infrastructure, and the costs of the extraction of raw materials and fossil-fuel use involved in the manufacture, distribution, sale, repair and maintenance of transportation devices including buses, cars, planes, ships, tractors, trucks and trains.[8] Even on conventional figures, which exclude such considerations, the Energy Information Administration estimates total world output of CO_2 in the transport sector in 2004 at 7 billion tonnes, and it is projected to grow to 14 billion tonnes by 2050.[9] Road transport contributes 80 per cent of total greenhouse gas emissions from transport. Air transport is the second largest source in the sector at 12 per cent, though it is the fastest growing; shipping contributes 7 per cent, though as we have seen some estimates put the contribution of shipping much higher. Rail amounts to 6 per cent. Air transport has a larger contribution to global warming than its CO_2 emissions indicate, since its high altitude emissions have additional climate heating effects. Water vapour from jet engines creates cirrus clouds that trap more of the sun's heat emitted from the earth's surface. And nitrous oxide emitted at high altitude has a significantly enhanced warming potential. The IPCC therefore estimates that the global warming potential of air transport is between 2 and 4 times that of physical carbon emissions.[10] However, airline industry estimates of its contribution to greenhouse gas emissions neglect this multiplier effect, and hence claim a much lower contribution to global carbon emissions at around 5 per cent.

In his famous painting of a steam train moving through the English landscape spewing flames, steam and soot, J. M. W. Turner encapsulated in impressionistic form the new relationship to nature that mechanical speed produces. The painting is of a black train crossing the River Thames at Maidenhead. On one side of the train is a man at a horse-drawn plough, and on the other are people boating on the river. These traditional and slower activities are contrasted with the darkly efficient speed of the train. Turner also depicts a hare, one of England's fastest mammals, running off the tracks to get out of the path of the train.[11] As Turner's painting indicates, mechanical speed has a powerful phenomenological effect on the human relation to the earth. The speed machine – whether a high-speed train, plane or car – confers on the driver or passenger a sense of mastery of the landscape, or in the case of a plane of the spherical globe as its curved horizons rapidly shrink through the speed of the modern jet. In moving at great speed through or over a landscape the human being loses bodily and sensual connection with the organic rhythms of life on earth. This loss is important in the construction of the modern imaginary of conquest over, and independence from, the forces of nature.

John Urry suggests that the car in particular is a crucial site of the specific pattern of domination over nature which is such a vital component of the modern technological condition.[12] The one who is master of devices which confer power and speed is likely to set reliance upon the sense of mastery such machines confer over trust in the life-sustaining properties of the more than human world. This is powerfully illustrated in a 2006 advertisement that depicts a man and a boy riding in a new Volkswagen Passat. The driver is shown using an array of remote controlled and electronic devices to manipulate the condition of the car. The car opens automatically when the driver approaches it; with a push of a button a sunshade glides up behind the back seat where the boy is sitting, and when the journey is over a remote control opens the boot. The boy is shown wondering at these seemingly magical happenings. Throughout, theme music from the Harry Potter films is playing in the background. When the boy arrives home with his father he goes through the house into his bedroom: he points to the window blind and says 'open' and of course the blind does not move. The implication is already clear before the punchline is delivered: the Passat is a car which gives the user such extraordinary control over the car's environment that its users 'could get used to it'. The sense of control and mastery over the glass and metal bubble of the car is not limited to the car itself: the driver and the passenger are trained by their use of the car to expect this kind of control in other environments.

The Inevitable Accidents of Speed

As Paul Virilio suggests, the speed of modern travel and communication is such that it has lent a peculiar quality to human experience in modern society, in effect turning human life into a quest for immediacy. The effects of this quest include the normalisation of a growing number of accidents, including accidental deaths caused by moving vehicles. And the accidental quality of a speed-enhanced culture is crucially implicated in the larger accident of geophysical collapse with which the planet is threatened.

The twin phenomena of immediacy and instantaneity together present one of the most pressing problems confronting political and military strategists alike. 'Real time' – instantaneity – now prevails above both real space and the geosphere. The primacy of real time, of immediacy, over and above space and surface is a *fait accompli* and has inaugural value, ushering in a new epoch. This is evocatively expressed in a French advertisement praising cellular phones with the words: 'Planet Earth has never been this small.' As Virilio indicates this portends a very dramatic shift in the contemporary vision of humanity's place on earth.[13]

Instantaneity not only changes cultural sensibilities in relation to time and mobility. It also breeds accidents. Accidents and speed are so closely related that it is impossible to speak of those who die on the roads as having died 'accidentally', for road deaths are a direct consequence of the extensive use of hard metal machines as transportation devices at speeds which, on impact, are detrimental to soft-bodied mammals. The seeming inevitability of road deaths indicates a larger problem with the speed of modern life which has precipitated such an array of humanly-caused accidents that death and suffering from such accidents exceeded death and suffering from natural disasters in the present decade, according to the large re-insurance company Swiss Re.[14] In this sense climate change may also be considered an inevitable accident, and part of a larger inevitability of ecological collapse necessitated by the speed of the capitalistic transformation of the earth.

The modern obsession with mechanically-derived speed is such that it eviscerates moral deliberation over its ecological or human costs. Speed produces an altered state of consciousness or what Sandy Baldwin calls 'a delirium broken only by the crash'.[15] Speed, Kundera suggests 'is the form of ecstasy the technological revolution has bestowed on man'.[16] This form of ecstasy generates detachment from the violent effects of the ownership and use of the devices of speed on the millions who have been killed and seriously injured by them, or from their systemic effects on the ecology of the earth.

Something more than distancing is involved here. Speed involves a kind of moral perversion, which is indicated in the extent to which danger to life and limb from fast cars forms part of the appeal of cars and other such devices to some of those who use them. In an ethnographic study of Norwegian young people, Pauline Garvey found that they regularly used fast or intentionally dangerous driving as a device for the expression of angry emotions or to provoke pedestrians and other road users.[17] The study records observations of young people using cars to indulge in more extreme and criminal kinds of behaviours such as driving cars on the wrong side of the road towards oncoming vehicles, driving at pedestrians, or at excessive speed. In Britain the phenomenon of joy-riding or 'twocking' – Taking Without Owner's Consent – is responsible for many accidents, deaths and serious injuries each year. A small number of young people see taking, driving and even crashing cars at high speeds as a kind of sport, and certain housing areas are regularly terrorised by individuals who use cars to indulge in such behaviours.[18]

The cultural adulation of fast cars and speed does not just infect a small minority of criminal teenagers. Such criminal uses of cars are related to a more generalised cultural obsession with speed associated with the adulation of large and extremely fast cars and other high-speed devices including high-speed trains, planes and ships. This adulation takes symbolic form in car advertisements which portray lone vehicles moving rapidly through wilderness or empty city streets, unrestrained by traffic jams or other physical or statutory constraints. Millions of people in Britain regularly drive at speeds far in excess of legal speed limits, and protest at the use by the police of speed cameras and other devices intended to reduce speeding on public roads. The car and the speed it is capable of gives the owner a sense of control over the physical environment, and a quality of freedom of which the speed camera is seen as an undue curtailment, even although death and injury on the road, and damaging greenhouse gas emissions, are associated with excessive speed.[19]

The Disorientating Desire for Speed

The desire for speed is pathological in part because it curtails moral deliberation on the effects of speed. The evident lack of deliberation over the immoral consequences of speed is part of a larger 'loss of orientation' which both physical and virtual speed visits on the individual, and this loss of an orienting sense of place has moral consequences. As Virilio suggests, 'to exist is to exist *in situ*, here and now, *hic et nunc*'. The ability to experience the moral claims of others, and the restraint such claims properly represent to the expression of individual desire, arises in part from the location of individuals in communities of place such as are constituted by families and place-based communities. The erosion of a sense of being in place is consequently intertwined with a lack of moral deliberation over the sacrifices of human, as well as planetary, wellbeing, that excessive speed involves. In this climate the victims of speed become statistical persons, or statistical beings in the case of 'road kill', in the utilitarian cost benefit calculi of the corporation and the nation-state. An analogous moral loss occurs for citizens whose communities are disturbed and deracinated by new roads and the speeding cars that use them.[20]

The moral and spatial disorientations produced by technologies of speed are the reasons many Amish communities still refuse to own cars. Amish culture is delineated against 'moral worldliness', which in the modern world indicates the complex kinds of control of natural and social power that technological devices such as cars and tractors confer on the individuals who own them.[21] Against worldliness the Amish practise a collective form of ascetic holiness which involves subordination of each to the other. Cars and like technologies threaten this subordination because the power they confer on an individual distances him from the collective deliberation of the community, and hence subverts the communal ethic. Cars represent a threat to the spatial as well as relational characteristics of Amish communities since 'ownership would intensify the pace and complexity of Amish living'.[22] The increasing speed and distance available to those who own cars would undermine the close-knit communities that are the foundation of Amish culture. The distinctive moral economy of Amish communities therefore involves the eschewal of the speed and complexity which characterise the modern technologies of cars, computers, tractors and televisions. Against these they set the communal values of family, church and community, and the virtues of faithfulness, self-denial, humility and meekness.[23] Amish bans on car and tractor ownership reflect restraints on the expression of individual desire which education into such shared virtues is said to require.

Wendell Berry observes that the refusal of the Amish to seek the controlling 'efficiency' of mechanical speed in their farming methods, as in their communities, has made them better stewards of the soil than industrial farmers, and ironically has seen Amish farms thrive while many of their machine-dependent neighbours have gone into bankruptcy.[24] By remaining rooted in place, and refusing speed and excessive mobility, the Amish have not only turned out to be among America's most successful farmers, they have also created a physical landscape which is ecologically and *spatially* oriented by their relational and spiritual practices.

The word *orientation* originates from the practice of building Christian churches that are oriented towards the East. Orientation means literally to be turned toward the orient and the orientation of the church building toward the East is an aid to the worshipper, who through this physical analogy is said to be enabled to orient his or her life toward those moral and spiritual ends commended by the Incarnate Son of God, who walked in what Jews and Christians call the Holy Land. The other form of spiritual orientation towards the Holy Land, and in particular Jerusalem, which has definitively marked Christian sacred geography and mobility, is the practice of pilgrimage. I suggest in the second half of this chapter that the phenomenology of pilgrimage provides a valuable frame for critiquing and resisting the modern pathology of speed, with its deleterious effects on the physical climate, on the human psyche and on patterns of association and settlement, but without requiring a return to the kind of extreme sedentariness associated with the Amish.

The market for cars, planes and trains in the twentieth century was driven by a number of factors other than simply the desire for speed of movement as an end in itself. One crucial driver was the potential these devices offered for householders in urban and rural settings to visit other locations, either on weekend trips or for extended holidays. The growth of tourism is such that in many cities and coastal areas it is one of the largest economic sectors, and tourism remains a prime function of the technologies of speed and hence a prime driver of the annual growth in greenhouse gas emissions. China is the latest region of the world to become a net exporter of tourists.

Pilgrimage and the Enchantment of Journeying

The desire to visit exotic places at some distance from workplace and home is not new to modernity. Pilgrimage provides a significant premodern analogy to the mobility of travel associated with modern mass tourism. While not all pilgrims were, like Chaucer's Wife of Bath, journeying just to see the sites and to get away from home, most nonetheless attended to sites along the way.[25] Unlike tourism, pilgrimage represented a form of travel that was not fossil fuel but *time* dependent, since it was traditionally conducted at walking pace, or in the case of pilgrimages involving intercontinental travel, with the aid of sail power. Christian pilgrimage was the principal reason for travel by pre-modern Europeans who did not by dint of itinerancy of profession have reason to travel beyond the environs of home and workplace. It has precursors in Jewish and Egyptian cultures, and in particular in the veneration of places associated with sacred journeys such as the Exodus journey from Egypt through Sinai, and of places of burial of venerated figures such as patriarchs and saints.[26] The first Christian pilgrimages were to Jerusalem to which, as Saint Jerome observed, 'every man of Gaul hastens hither', while the Briton 'no sooner makes progress in religion than he leaves the setting sun in quest of a spot of which he knows only through Scripture and common report'.[27] Rome also became a place of pilgrimage because of its association with the Apostles Peter and Paul. Subsequently pilgrimages to a range of other shrines developed, among the most popular of which were those of St Thomas à Becket at Canterbury and St James at Santiago de Compostella.

The conceptual root of pilgrimage is the idea of the Christian as traveller and of the spiritual life as a journey or quest for holiness.[28] The travels of Abraham and the Israelites in the wilderness, the classical account of the travels of Ulysses, and the itinerant nature of the lives of Christ and St Paul, are all important precursors of the idea of pilgrimage in early Christian thought. The Gospels record Christ valorising his own mobility as an aspect of the way of the cross when he says 'foxes have holes and birds of the air have nests; but the Son of man has nowhere to lay his head' (Luke 9:58). The writer to the Hebrews similarly speaks of the followers of Christ as those who in this world 'have no lasting city' and who 'seek the city which is to come' which is the heavenly one (Hebrews 13:14). Augustine in *De Civitate Dei* similarly speaks of the City of God as the Church in pilgrimage on earth.

It was in the Middle Ages that pilgrimage acquired widespread appeal. Many thousands in the fourteenth century regularly travelled to shrines and holy places far distant from their places of abode on a quest for transforming experiences of the holy at sacred sites and through the penitential rigors of the road. As Eamon Duffy observes, pilgrimage required the pilgrim to leave the 'concentric worlds of household, parish or gild' where he laboured and worshipped for a journey to a sacred site or shrine far distant, and such leavings 'provided a temporary release from the constrictions and norms of ordinary living, an opportunity to review one's life'.[29] The long journey with its trials and pains was also a penitential practice, a *via crucis,* which offered a redemption and a liberty of its own. The end of the journey – worship at the sacred shrine and its holy images and objects – offered the pilgrim a new perspective which might transform and redirect his life on his return home, setting work, home and worship in a larger and more meaningful symbolic and spiritual context.[30] For some this new perspective concerned the restoration of the inner life, while others sought physical restoration as the Holy Land and other shrines became associated with a range of curative properties and claimed miracles recorded by their guardians.[31] The ultimate goal of the journey was to enable the pilgrim to identify in a new way with the founder of their faith:

> For them the founder becomes a savior, one who saves them from themselves, 'themselves' both as socially defined and personally experienced. The pilgrim 'puts on Christ Jesus' as a paradigmatic mask, or persona, and thus for a while *becomes* the redemptive tradition.[32]

The Turners suggest that the form of *communitas* that pilgrims experienced on the road and at the shrine was central to the liminal quality of the experience of pilgrimage.[33] Through *communitas* individuals on pilgrimage are liberated from the mundane hierarchies they endure in secular life and connect with a deeper sense of the holy.[34] Chaucer's descriptions in *The Canterbury Tales* of pilgrims on the way to Thomas à Becket's shrine lends weight to this account. Chaucer describes a remarkably disparate group of people who nonetheless form a company and exchange stories along the way. A crucial element of pilgrimage which helped produce this kind of social solidarity on the way was the sharing in the sometimes painful ordeal of the long days of walking in pre-modern footware on unpaved paths in inclement weather.[35] This slow pace of movement is the strongest point of contrast between traditional pilgrimage and modern mass tourism, which in other respects manifests some of the same functions as pilgrimage – escape from the mundane, simplicity of clothing, enjoyment of a different locality, the *communitas* of other holiday-makers. The Turners provide a substantial excerpt from a description of pilgrimage to Guadalupe in Mexico, published by the basilica in Guadalupe, which reveals the spiritual significance of walking:

> Year after year, from 1890 onwards, the Guadalupan devotees of Querétaro have made the pilgrimage on foot from their city to the sanctuary of Tepeyac. In eight days they traverse the rough road 260km in length – the old royal road, broad and austere – which joins Querétaro to the metropolis. Eight days during which the weariness of the body is submitted to hard, voluntary

> discipline, loosening the bonds of matter to liberate the spirit. The rhythm of the march is set by collective prayers chanted in the plains under the mid-day sun, in the cool dawns when leaving towns, in the evenings at the end of a day's journey. It is a march of religious folly in which the Indian – who comes on it propelled by his strong will against his better nature – is freed from his bonds, while the rich man punishes his own softness by the austerity of prayer and walking.[36]

Walking through the landscape establishes a relationship between the worship of the pilgrims and the rhythm of day and night, and of the seasons.

Pilgrims' records of their journeys, like literary and mythic narratives of heroic quests, see the pilgrimage journey as a paradigm of life itself. The centrality of the trope of journeying in traditional as well as contemporary literatures indicates that, as Richard Niebuhr suggests, the desire to travel and to be in motion represents 'a deep characteristic of human nature', and the kinds of motion involved in this desire are psychological and spiritual as well as physical.[37] Mobility is central to human identity and even to being in place for, as Oliver O'Donovan observes, humans, unlike trees, are not stationary.[38]

The Journey as Quest

Literary accounts of heroic journeys that encapsulate the archetypal desire to travel are, like traditional pilgrimage, set apart from the rapid mechanical journeys of the modern tourist in their description of the formative role of chance and circumstance that the hero encounters on the way. At the outset of the journey the hero often has little idea of the end in view, and even less confidence in his capacity to reach it. The eventualities or ups and downs of the journey are therefore crucial to the acquisition of the virtues and skills that the hero needs in order to attain the goal.[39] J. R. R. Tolkien's *The Lord of the Rings* powerfully evokes the way in which at the beginning of many classic narratives of heroic journeys the hero is often hardly a hero at all; hobbits are at first sight too quirky, short and physically weak to perform the valiant task set before them by Gandalf. Along the way Frodo, the hobbit chosen for the quest, is plagued by self-doubt about his ability to succeed. By being open to the gifts and interventions of friends, strangers, elves, ents, and even enemies, the hobbit who becomes a hero is eventually able to complete the quest to cast the ring into the fires of Mount Doom, and so save Middle Earth from the militaristic and technological darkness of Mordor. But there is also a contrast between Tolkien's tale and other shamanic myths in which the hero gradually acquires mastery and new powers through the journey which enable him to fulfil his quest. At no point on his quest does Frodo become a master of the ring, or even master of the journey. On the contrary, he relies on Gollum to guide him for much of the way, and by the end Frodo has so fallen for the dangerous attraction of the ring that he only succeeds in destroying the ring through the intervention of Gollum, who in one last desperate attempt to wrest the ring from Frodo falls into the fire with the ring in his clutches, and so destroys it, and with it the rising and destructive power of Mordor.

With this crucial twist in the tale, Tolkien has evoked an essential feature of the Christian conception of life as journey or pilgrimage, which is that of learned dependence. Far from achieving mastery and autonomy, the Christian pilgrim goes the way of the cross and therefore seeks to remain humble and open to the gifts and trials which each turn in the road, each step along the way may bring. Mastery and control are not the ends in view, but instead encounter and submission to the Master, and a related preparedness to endure the rigors and rewards of life on the road. As Christ himself puts it, 'if any man would come after me, let him deny himself and take up his cross and follow me' (Mark 8:34). Stanley Hauerwas suggests that Christ's whole life is narrated in the Gospel of Mark as the expression of 'noncoercive power': Christ does not compel but call others to follow him 'and he does not try to control their responses'.[40] Neither does he try to control the events which lead to his eventual crucifixion, nor even the future destiny of those who followed him and who would become the founders of the Church. This element of living life 'out of control', in dependence on a higher power, is the hallmark of Hauerwas' distinctive critique of the moral pathology of contemporary culture. It is this which sets the traditional walking pilgrimage apart from the pace and control of modern tourism, and from the mastery of time and motion that the devices of speed confer.

Hauerwas's ethic of dependence is also delineated in contradistinction to the emphasis in modern consequentialist and deontological ethics on the autonomy of the self as a reasoning moral agent. The underlying assumption in both these modern styles of moral reasoning is that the individual reasoner is sufficiently in control of events, and of his or her own intentions, as to be able to rely on his or her own autonomous judgment. Criticism of the modern conception of the unsituated, and hence disoriented, autonomous reasoner involves the Aristotelian recognition of the significance of external referents in the formation of the self, in those habits and practices which Aristotle identifies as the virtues to which we have already alluded.

Walking and Ecological Attentiveness

In the practice of pilgrimage as walking environmental referents play a crucial role in the formation of the pilgrim. The companions on the way, the length of the road, the vagaries of climate, all involve openness to the influence of other creatures on the body and mind. In this intentional and embodied openness to the gifts and obstacles that the journey may bring, the pilgrim enacts a relational pedagogy of body, mind and spirit. The role of the environment through which the pilgrim passes in shaping the identity of the pilgrim is evident in the earliest records of Christian pilgrimage. Pilgrim journals reveal a progression from an almost exclusive concern with sites associated with specific biblical events to an interest in the flora and fauna, and in the landscape itself. The AD 333 journal by a Bordeaux pilgrim of a seven-month walking journey from the Bosphorus to the Holy Land includes many references to geological and biological phenomena, but these are only of interest to the pilgrim where they can be specifically linked with a sacred event, such as the almond tree where Jacob is said to have wrestled with an angel.[41] But the *Itinerarium Egeria,* which is the vivid personal journal of a woman on pilgrimage through the Sinai Desert and the Holy Land, includes extensive descriptions of desert and mountainous terrain, and of the views to be enjoyed of the Sinai Peninsula and the Holy Land from such places.[42]

The role of flora and fauna, topography and landscape vistas in the spiritual pedagogy of pilgrimage finds significant analogy in the relational accounts of interaction with the more than human world in contemporary ecological philosophy. The claim that walking up a mountain is an embodied experience that involves pain as well as pleasure is a central trope of Norwegian deep ecologist Arne Naess's account of 'ecosophy T'. Naess observes that extending and testing the body through climbing, and the 'peak experience' of reaching the top and enjoying the view, involves an engagement between the identity of the walker and the being of the mountain.[43] This engagement enhances the consciousness of the walker such that she becomes aware that in this embodied relation her identity is in part constructed by the mountain. Naess suggests that this new relation involves a new conceptualisation of human identity which is extended beyond the self so that the frame of self-identity *includes* the mountain.

Whether the incorporation of the mountain within the frame of self-consciousness is truly thinking *like* a mountain is contested by Val Plumwood.[44] In her own travel writing, and especially in her account of a near-death encounter with a crocodile, Plumwood eschews the colonising discourse of identification with and mastery of nature and instead emphasises how movement through a landscape provokes a consciousness of the alterity of other creatures and a decentering of human identity.[45] Instead of incorporating the mountain or the animal into the self, the self-in-relation is open to being reordered, even dismembered, by the more than human world.[46]

In contrast to traditional pilgrimage, and to Naess' and Plumwood's narratives of travel, mastery of time and space is central to the experience of mass tourism and mobility. Modern tourism is organised around universal and mechanical time and conducted at the pace dictated by the short break or the seven- or fourteen-day holiday permitted by the disciplining time management of the industrial economy. As we have seen, the industrial control and speeding up of time even reshaped the human experience of time because rail travel's connecting of places spatially far distant required the invention and the standardisation of universal time.[47] And just as the homogenisation of time was necessitated by the technologies of speed, so the control of space is increasingly required by those same technologies, from the airspace utilised by jet planes to the extensive land area taken up by metalled roads, car parks, airports, railways, stations and marshalling yards. The modern tourist utilises the infrastructure which has subjected time and space to ecologically destructive forms of control in order to enjoy elements of the liminal experience of the pre-modern pilgrim. But since the touristic experience is typically conducted at great speed, it cannot offer profound relief from the controlling pathology and pace of industrial time.

Recovering Contingency

Traditional pilgrimage was conducted according to local times, over many weeks and months, and at a pace set by the rhythms of the earth and the feet of the walker rather than the turning of mechanical engines. And the reliance of the pilgrim on her feet and her subjection to the vagaries of the road was itself a kind of discipline, both embodied and spiritual, which required the pilgrim to submit to a different rhythm than that of mechanical wheels or mechanical time. Walking is an activity which puts the human body in the landscape. When conducted over successive days and weeks it enables the individual to begin to live by a different rhythm than the productive and efficient time determined by 'time and motion', the rhythm of an order humans have not made. And the longer the walker walks the more this alternative rhythm – the rhythm of the earth – begins to replace the persistent and controlling consciousness of industrial time. The rhythm of walking interacts with the rhythm of the earth such that the landscape itself begins to shape both the physique and the psyche of the walker. Ascending and descending the contours of the path put muscle on the walker and burn calories as fast as they are eaten. This process of muscle-building is in reality one of a constant tearing and remaking of muscular tissue which is both painful, and at times enjoyable, as it is followed *in extremis* with a flow of endorphins which can bring ecstasy to the mind.

Such bodily and psychological transformations indicate the inner transformation that long-distance walking produces as succeeding days of arduous activity, living on the road and being in motion begin to mould the walker's body and mind to being and thinking 'at nature's pace'. The longer feet are lifted and replaced on the ground, the head held under the sky, and the face proceeds through the landscape, the greater the transformation in the sense of the phenomenological placing of the body in the environment. Long-distance walking trains the body and the mind to a new kind of submission to the forces that move and mould the landscape and its myriad other than human processes, structures and inhabitants.

This kind of walking is a physical discipline of mind and body which in pilgrimage takes on a spiritual dimension, as demonstrated in the Russian hesychastic classic *The Way of a Pilgrim*, in which the pilgrim walks for many months through nineteenth-century rural Russia while learning to use the prayer of the name of Jesus.[48] The pilgrim practises the prayer – 'Lord Jesus Christ, have mercy on me, a sinner' – as a breath prayer which he repeats every few steps on his journeying. The writer narrates how the love of the pilgrim for Christ, for his fellow countrymen and for the landscape they tend increases as he gradually learns to shape his utterance of the breath prayer to the rhythm of his breathing and walking. One writer describes this nineteenth-century Russian classic as manifesting an 'Emersonian ethereal harmony', a phrase which neatly encapsulates its remarkable description of a spiritual practice which is intimately shaped by the breath of the body and rhythm of the earth.[49]

Traditional pilgrimage offers a model of mobility and travel which is both more attuned to the fragile ecology of the earth than the fossil-fuel dependent travel of modern mass tourism, while at the same time providing a therapeutic antidote to the pathology of speed in the modern world. If walking is central to the spiritual practice of traditional pilgrimage, then we may read the growing popularity of long-distance walking as a modern touristic analogy to pilgrimage walking. Those who spend upwards of six months walking the Appalachian Trail on the Eastern seaboard of the United States certainly by the end know a good deal about the privations and joys of traditional pilgrimage. They also come to experience in their long time away from industrial civilisation something of the spiritual seclusion, the fellowship of the way, and the lightness of being that were the gifts of traditional pilgrimage.[50] Such long-distance walking offers a model for a more sustainable form of tourism that involves reduced reliance upon carbon emissions. But the reorganisation of mass tourism along these lines would require the reconfiguring of temporal as well as spatial aspects of modern industrial civilisation. In particular the ordering and quantity of available holidays would need to be revisited if the long-distance walk was to become a typical form of holidaymaking.

The Love of Slow

If a pedagogy of pilgrimage and walking has the potential to transform the desire for speed into a 'love of slow', and to enhance the human ability to live life in response to the rhythm of the earth, then this would also suggest the value of recovering forms of human settlement which involve walking, of the kind still enjoyed by the Amish, by the city centre residents of many cities in Europe, Asia and Latin America, and by the billions of people who still do not own cars. In those societies which have been transformed by car ownership, there is the suspicion that reductions in car use indicated by the need to reduce greenhouse gas emissions will undermine human freedom and wellbeing. But there is a growing body of counter-evidence to this suspicion, besides the arguments already reviewed in this chapter.

The diminishment of space given over to the rapid movement of vehicles, far from enhancing freedom, in reality reduces it for the majority of humans and other animals who in most societies outside North America and Australia are not car drivers. Even in car-owning cultures, the majority of road users are not drivers. Children, the disabled, cyclists, pedestrians, the elderly and domestic and wild animals are all at risk in car-dominated urban and rural environments from accidental death or injury from moving vehicles. The residential street dominated by fast-moving vehicles is neither a peaceable nor a secure space for children to play, or for residents and walkers to take the air or engage in conversation. This author's mother-in-law was seriously injured by a car that mounted the pavement and drove into her and she never fully recovered from the shock of this accident before she died. She never drove a car in her life and neither did her husband.

The colonisation of space by the motorcar is won not only at the cost of human health and a stable climate. It is itself the major obstacle to the enjoyment of walking in most urban and rural settings. One empirical study of the effects of increased walking activity in a controlled trial indicated that both men and women reported increased environmental awareness in the form of enhanced enjoyment of the aesthetics of the places where they lived, and an increased appreciation for the convenience of their locale when they stopped driving, and walked and used public transport instead.[51] The reason for reducing car use in the face of the threat of dangerous climate change is not about a balance of consequences for car users and others, even if a rational examination of that balance reveals that the downside to the 'freedoms of the car-owning society' is far greater than many people imagine.[52] In the end it is about whether or not the majority of species, and the majority of humans, are to continue to experience wellbeing in a planet with a stable climate beneficent to life.

Reducing car use, or even getting rid of cars altogether, does not mean that travel becomes impossible. As Christian economist Alan Storkey argues in a paper on alternatives to the car, there already exists an effective mass transit system, in the form of roads and motorways in Europe and the United States and other parts of the world. The problem with these publicly-provided paved spaces is that they are so taken up with private vehicles, mostly with one person in each, that they are an incredibly inefficient way of moving people around. If the same road space were to be dedicated to well-designed modern coaches instead of cars it would be possible to move people faster than these roads presently do because the congestion associated with private cars would disappear.[53] Radically reducing car ownership and use, reducing the collective resources and space devoted to them, and embracing new forms of collective transportation, will need to become the moral priority of every society on earth if the threat of dangerous climate change is to be avoided.

As with many other practices associated with redirecting human life to respect the energy flows and nutrient pathways of the earth, there are many human and social, as well as ecological benefits to embracing slower travel. Walking and cycling, moving from individual to collective modes of rapid transit, and from planes to trains or air ships, for international travel, will enhance community, foster a greater sense of being in place and a more respectful mode of dwelling on the earth. Most significant of all, these practices will be a collective turning away from the politics of speed. And it is precisely the politics of speed which is driving industrial humans towards a future of catastrophic climate change, which will be the largest 'accident' in human history. Speed, as Kundera suggests, is a form of forgetting driven by fear of the future, whereas 'there is a secret bond between slowness and memory'.[54] As the Psalmist avers, 'be *still* and know that I am God' (Psalm 46:10).

8
Faithful Feasting

Many shepherds have destroyed my vineyard,
they have trampled down my portion,
they have made my pleasant portion
a desolate wilderness.
They have made it a desolation;
desolate, it mourns to me.
The whole land is made desolate, but no one lays it to heart.
Upon all the bare heights in the desert
spoilers have come;
for the sword of the L*ORD* *devours*
from one end of the land to the other; no one shall be safe.
They have sown wheat and have reaped thorns,
they have tired themselves out but profit nothing.
They shall be ashamed of their harvests
because of the fierce anger of the L*ORD.*

Jeremiah 12:10–13

Climate change and agriculture are intricately related. The marked warming of the earth's climate that took place at the end of the last ice age was the beginning of the 'long summer' that has characterised the Holocene era since that time.[1] As we have seen, the stability and warmth of the climate in this period is unique in planetary history and has enabled the development of human civilisation, which rests upon the production of agricultural surpluses by farmers, which release others to carve stone, build temples, make weapons and write laws. Prehistoric humans had already learned to use grasses as cereal crops alongside fruits, berries, fish and meat and there is evidence of some minimal cultivation even before the Holocene era.

Agriculture was by no means an inevitable path for humans to go down, and indigenous peoples who have access to sufficient land still get provisions from flora and fauna in forests, rivers and savannah rather than relying on their own capacities to grow crops by tilling the ground.[2] Many such peoples have experimented with agriculture, or practised slash and burn farming as a supplement to forest-derived game and plants. But some still prefer not to become agriculturalists because it takes more time and effort to grow food than it does to gather it from the wild. As Marshall Sahlins observes, hunter-gatherers use less energy than any other people on earth, and yet when their traditional habitats have not been destroyed by loggers or farmers, hunter-gathers are easily able to meet their needs. They experience what Sahlins insightfully calls 'original affluence', the secret of not only knowing how to live from wild foods, but also being satisfied with and desiring less.[3] This is why the Book of Genesis describes the move from the hunter-gathering lifestyle of Eden to the agrarian life on the plains as a fall from grace. Hunter-gatherers enjoyed, and still enjoy, more leisure than their agrarian counterparts in which to sit under trees, eat fruits, roast meat, nurture children, play sports and tell stories.

The Origins of Agriculture

As we have seen the human move from hunter-gathering to agriculture first occurred in the Fertile Crescent between the Nile and the Euphrates River systems, and is recorded in the first books of the Old Testament. The reasons for the move were most likely twofold. The first was the dryer and warmer climate associated with what paleoclimatologists call the Younger Dryas, a warming of the climate that occurred at the beginning of the Holocene era which caused mammal numbers to fall in the region as precipitation declined and wild berries and fruits were less abundant. The consequent drop in animal-derived protein would have made it hard for hunter-gatherers to survive without identifying alternative food sources. The second reason was the flooding of the rich alluvial plains on which prehistoric hunter-gatherers in the Fertile Crescent originally lived. At the end of the ice age water equivalent to hundreds of metres of sea-level rise was still locked in ice caps and glaciers. In the first few hundred years of the warmer Holocene period this ice melted, flooding lands formerly exposed between today's continents. This ancient inundation is another of the likely origins of the flood myths found on many continents, including the Gilgamesh Epic, and the Noah saga.

Archaeologists believe that the narrative of a 'Garden of Eden' may refer to a region that was once located in the Persian Gulf in an area now under the ocean. Before it was flooded, this region would have been very rich in fruits, herbs and trees and have supported good numbers of deer and other edible creatures such as turtles and molluscs.[4] Its flooding would have caused the inhabitants of the garden to move to higher ground and to begin cultivating the mountain grasses which are the ancestral plants of the cereal crops of wheat, barley, oats and rye, while in other earth regions climate change forced a similar move to other staple crops including rice and maize.[5] So the ancient move from hunter-gathering to farming symbolised in the move from Eden to the dryer and higher Arabian Peninsula was caused by climate change.

The story of the Garden of Eden is on one level a moral story about the primeval fall from grace of early humanity. But it is also an ecological and geographical story, for the memory of the garden is of a hunter-gatherer life in which there was an abundance of fruits, nuts and plants to sustain human life with relatively little labour. The move from the garden to the plains indicates the birth of agriculture. And this paves the way for the building of the first cities. And with these moves come the primordial fratricide and violence of Cain, and the hubristic building projects of the Nephilim, who with their Tower of Babel sought to recreate the high places in human form, reaching from earth to heaven. Like the inhabitants of the modern towers of Los Angeles and London, São Paulo and Singapore, the builders of this primeval tower forgot their dependence upon the surrounding hills and wilderness for the flows of energy, nutrients and water to grow the agricultural surpluses that made their building projects possible.

Wilderness and Dependence

The dependence of agriculture on wilderness is something that agrarians often memorialise in their rituals and traditions. The ancient tribes from which the Hebrews were descended knew and revered the 'high places' because they were the source of the rivers and the rains, and the original source of the grasses on which their agrarian lifestyle relied. This explains the prominence of high places and mountains in the sacred geography of the first agricultural civilisations in Mesopotamia. Whereas Moses meets God on the holy mountain and so owns this primordial dependence of humanity on the wild, the builders of Babel and of Babylon's ziggurats were constructing artificial high places which neglected the dependence of life on the creation, and substituted things made by human hands – idols – for the sustaining power of the Creator.[6] And so the ancient narratives of Genesis and Exodus indicate that the Mesopotamians in their agrarian prowess were the first people on earth to forget their dependence on the wild and to invent a cosmology centred on the powers of human growing and making. And they were the first to build cities in which the gods of the high places were set in temples on the plain, domesticated and serviced as the legitimators of the imperial civilisation of Ugarit.[7]

Just as the story of Genesis is that of a Fall from the Garden to an imperious and idolatrous urban culture, so the story of redemption in Exodus is of an urban prince who leads his people in a revolt against the slavery imposed by the city, back out to the levelling nomadic lifestyle of the wilderness. And on their journeying from wilderness to the Promised Land the people of Israel are constantly reminded that they owe to Yahweh their redemption from slavery to the imperial building projects of Egypt. In the Promised Land the people are ordered to avoid the cities which dominated the land they were to colonise, and commanded to establish a more egalitarian agrarian lifestyle on the hills and the plains of Canaan which are said to be 'flowing with milk and honey', and hence in no need of the deep plough of the Egyptian.

The contrast between a nomadic hunter-gatherer idyll in which the ancestors of Israel encountered and worshipped Yahweh in the wilderness, and knew their dependence upon the wild, with the idolatry, plagues and slavery of the agricultural and urban civilisations of Egypt and Ur is not only central to the great story of Fall and Redemption of the first books of the Bible. It also accords very well with the growing understanding we now have of the origins of agriculture, and of the continuing contest between agricultural and urban civilisation and indigenous hunter-gatherer groups in Asia, Africa and the Americas to this day. It may be no consolation to them, but the Yanomami in Amazonia and the Penans in Borneo are just the latest victims in a 10,000 year history of struggle between planters and hunters.[8]

The journeys that Abraham and Moses took, away from the cities on the plains and back to the hills, are in some ways an analogy for the path humans will need to take in the course of extreme climate change. The residents of the Pacific Island of Tuvalu, whose water sources are being poisoned by sea water and whose homes are regularly flooded by annual spring tides, are already beginning to move to the hillier terrain of New Zealand. The IPCC estimates that in the worst-case scenario by 2050 26 million people will be displaced by coastal flooding.[9] Millions more refugees from the interior of Africa and South Asia will follow as lands become so parched that they cannot sustain agriculture and aquifers and rivers begin to run dry.

As we have seen, the geographical distribution of the worst effects of climate change and the regions where greenhouse gases are mainly emitted are inversely related in hemispheric terms, and this is particularly the case in relation to agriculture and food security. Climate change will produce a longer growing season in parts of North America and Europe. And with a warmer climate European farmers will initially see increases in the production of crops such as wheat, onions and cauliflower. But raised temperatures will also be accompanied by changes in rainfall patterns and more extreme weather events including floods, storms and droughts, and there will also be decreases in yields for crops grown in Southern Europe such as sunflower and maize.[10] Some estimates of the effects of climate change on agriculture in the United States are even more positive. Reilly and Tubiello find that while there will be reductions in crop outputs in the Southern states, the principal cereal growing areas, the mid-Western and Canadian prairies, will see significant rises in yields of wheat and maize, even at the high end of predicted temperature increases.[11] However, not all agree with these optimistic predictions. Lobell and Field show that since 1980 warmer years have produced statistically significant drops in total United States' output of major cereals, including maize and wheat, which outweigh any regional advantages.[12]

While farmers in Northern Europe and North America may see some benefit from climate change in terms of agricultural yields, the prospects are much more troubling for farmers in much of Africa, South Asia and Central and South America. As we have seen, African farmers are already coping with long-term drought and significant reductions in crop yields, and consequently suffering from malnutrition and famine.[13] Asian agriculture is also threatened significantly by climate change because of its reliance upon the monsoon. If the melting of Greenland reduces the circulation of warm water from the tropics, and hence cools the North Atlantic, this will lead to a reduction in monsoon rains because it will lengthen the snow season on the Tibetan plateau, so reducing precipitation across Asia.[14] With the burgeoning populations of China, India and Indonesia, the diminishing of monsoon rains would precipitate a food crisis of unprecedented scale in which present global agricultural surpluses of rice, maize and wheat would rapidly turn to deficits. The prospect that humanity will find it harder to grow enough food for the 9 billion humans who are likely to be alive in fifty years' time ought, perhaps more than any other prediction, to cause developed and developing countries alike to seek strenuously to reduce their production of greenhouse gases in order to mitigate the likelihood of dangerous climate change.

Greenhouse Gas Emissions from Crops and Animal Husbandry

Crop growing is one of the biggest sources of agricultural greenhouse gas emissions. Rice paddies are a major source of humanly-generated methane, although different rice growing systems vary dramatically in the damaging emissions they produce. Methane emissions from rice can be mitigated through crop rotation, careful soil and water management and tillage practices, but the green revolution, in which agronomists encouraged the widespread adoption of new high yield rice varieties which are highly dependent on energy-intensive fertilisers and pesticides, also discouraged these kinds of traditional paddy management practices.[15] Other cereal crops are implicated in climate change both because of the emissions involved in the energy-intensive nature of modern industrial farming, and because of the annual release of carbon from the soil caused by deep ploughing. There are, however, sustainable alternatives. Low till agriculture and organic crop husbandry both result in substantially-reduced greenhouse gas emissions from cereal growing. Similarly, organic low-intensity rice cultivation in Bangladesh, China and Latin America is around twenty times more energy-efficient than intensive chemically-farmed rice paddy in the United States.[16]

Animal husbandry is another major source of agricultural greenhouse gases, directly responsible for between 5 and 10 per cent of global greenhouse gas production. If the full climate costs of cereal growing and energy inputs from petroleum for growing animal feeds are added, this figure rises substantially. It takes 28 kilocalories of cereal and petroleum to produce 1 kilocalorie of animal-derived protein.[17] The principal direct source of greenhouse gases from domestic animals is enteric, or from animal stomachs.[18] There are approximately 2 billion domesticated cows on the planet and the combined methane output of their stomachs is a fart of monumental proportions which contributes 70 per cent to the total of 94 million tonnes of animal-derived methane emitted annually into the atmosphere. By contrast, human stomachs contribute only 1 per cent of enteric methane. Methane not only has twenty-two times the global warming potential of carbon dioxide. Its atmospheric levels are also growing at double the rate for CO_2 and some scientists now believe that methane represents a more likely trigger of rapid and dangerous climate change. Even more significant than the direct emissions from animals is the effect of the clearing of forests and savannah to grow animal feed crops such as soya and maize. As we have seen, large areas of the Amazon rainforest are devoted to soya farming, and the majority of this soya will be sold to Western consumers in the form of meat, since soya will have been fed to the animals they eat. And the Western predilection for meat is fast becoming a global phenomenon with the move in many cultures, backed by the food industry and advertising, from meat as a supplementary element in the diet to it becoming the major focus of many meals.

The moral case for vegetarianism is that it reduces the need for killing, and therefore reduces cruelty in human relationships with their fellow creatures. In the Garden of Eden Adam and Eve are only permitted to eat plants, because the shedding of blood in Hebrew culture was perceived as polluting and dangerous. In Hebrew the word for blood is *nephesh*, which carries the meanings of blood, breath and spirit. For the ancient Hebrews the divine spirit was understood to be present in a special way in the life-blood of humans and other animals. Shedding blood was therefore seen as a sin which polluted the community. The sacrificial system was inaugurated in Israelite culture as a way of dealing with this pollution. Those who killed meat were set apart as priests in the temple and all animals were initially offered as sacrifice to Yahweh, and so the dangers of shed blood for the community were assuaged.

But the sacrificial system was not only a means for sanctifying animal husbandry, slaughter and meat eating. It was also the central device in Old Testament religion for atonement for sin, and for healing and restoring the disturbance that human sin and wickedness caused both in divine–human relations and in relations with other creatures, human and nonhuman. Sacrifice was the divinely-ordained mechanism for maintaining the 'eternal covenant' between God and Israel and for recovering the promise of cosmic harmony between God, people and land. As Margaret Barker suggests, the eternal covenant was 'the system of bonds which established and maintained the creation, ordering and binding the forces of chaos'. Disregard of the divine statutes therefore risked returning the world to its pre-creation state as 'waste and void', as Jeremiah suggests. It even risked extinguishing light in the heavens. By contrast, the restored creation is described as being returned to its Edenic state where peace, justice and righteousness reign.[19]

In the Old Covenant the means by which this restoration takes place is atonement, which literally means to cover or repair a hole or cure a sickness.[20] The *place* of restoration was the Temple, 'the meeting place of earth and heaven', where each action and ornament was said to have a heavenly counterpart. Not only the high priest but the Lord was understood to perform the acts of atonement which repaired the ruptures to creation caused by sin, and hence Jewish tradition has it that the priest in effect 'transferred the atoning power to God'.[21] And since it is the Lord who performs atonement, then for Barker the Hebrew word for atonement, *kpr,* 'has to mean restore, recreate or heal'.[22]

This account sheds important light on the significance of the 'New Covenant' for creation care because it brings the tradition of the eternal cosmic covenant, and the interconnection between physical and moral laws in Hebrew cosmology, into the heart of the New Testament. Christ was Incarnate on earth in an era when, according to Isaiah, God had abandoned the Noahic covenant: the sacrificial system of the Israelites was no longer effective because they had given up on the righteous demands of the law, without which neither blood spilt nor burnt flesh could atone for sin. The Lord therefore no longer honoured the atoning blood shed in the Temple. Hence Isaiah envisages a new atonement ritual, enacted by the 'suffering servant' who shall 'sprinkle many nations' with his atoning blood, carry the sicknesses of the people, and remake 'the covenant bond of peace'.[23] The Suffering Servant is not only the one who makes atonement by the pouring out of his blood, which heals the rift caused by sin, but he also becomes the scapegoat who carries away the sins of the people and who is, like the scapegoat 'pierced for our transgressions'.[24]

If Christ is the suffering servant, as the Evangelists and St Paul believed, then the New Covenant which Christ sets forth in the fellowship of the bread and the cup at the Last Supper with the disciples is truly a cosmic covenant for, as with the Day of Atonement, the blood of Christ atones for the polluting effects of sin on the creation. This is the metaphorical significance of the piercing of the body of Christ by the Roman centurion: his blood is poured out on the earth, as was the blood of all sacrificed and slaughtered animals in Israel, and so the earth is healed. This is why the writer to the Ephesians can say that it is the divine plan hidden from before the foundation of the world 'to unite all things in him, things in heaven and things on earth' (Ephesians 1:10) and why St Paul explains in Colossians that 'in him all things hold together' and that through him God reconciles 'all things whether on earth or in heaven'

(Colossians 1:17, 20). The atonement which Christ first sets forth at the Last Supper is truly a cosmic event, and hence, as the Gospel writers have it, not only is the meaning of his death affirmed in the Temple itself when the veil of the Temple is torn in half – Christ having become the great High Priest who has passed into the heavens and hence there being no more need for a Holy of Holies – but the earth also witnesses to its anticipated restoration with an earthquake and a partial eclipse of the sun. As Barker suggests, this understanding of the New Covenant is the essential background to interpreting the new life and new creation imagery of the New Testament – the New Covenant heals the earth.[25]

For the more radical of the early Christians, this understanding of the cosmological implications of the New Covenant meant that abstaining from eating meat was a sign of the new order of being inaugurated in the coming of Christ and his founding of the Kingdom of God. Those who abstained from meat in the early church did so because they believed that Christians are called to live a new and more perfect life in which the sinfulness of the heirs of Adam is purged. Since the Book of Genesis describes meat eating as a concession made by God to the descendants of Noah, a number of the early Christians therefore believed meat eating should not be part of the Christian life.[26] Therefore meat eating forms no part of the sacred meals of the early Christians. Images of fish, grapes, bread, wine and water recur throughout early Christian art but there are no images of the early Christians eating meat; the only meat-related image is that of Christ as the good shepherd, who is not killing but rescuing a lost lamb, carrying it on his shoulders. Some early Christian theologians, including Basil the Great, consequently understood that human relations to animals had signally changed in the light of the new covenant. They believed that God intended to save animals, and not only human beings, in the redeeming events of the life, death and resurrection of Christ.[27] The emergence of vegetarianism as a modern moral ideal in Western societies thus has significant theological roots. As Andrew Linzey argues, a vegetarian diet is a diet that is closer to the reign of God as described in scripture and tradition.[28]

The growth of vegetarianism as a moral ideal among a growing minority of Westerners in the twentieth century reflects not only traditional proscriptions of meat in religious traditions but the growing awareness among Western consumers of the cruelty, ecological harms and social injustices engendered by the industrialisation of animal husbandry.[29] Many animal-related human health threats, including the deadly new strain of avian flu H5N1, and novel and potentially harmful bacteria such as *E. Coli* O157, are associated with factory farming and industrial food production, though government scientists and food corporations have sought to blame such problems on wild birds, rogue traders or free-range animal husbandry.[30] The increasing reliance of the industrial diet on meat, and particularly the high fat and protein content of the fast food industry and supermarket ready meals, is implicated in a range of other human health problems including dramatic growth in rates of obesity, diabetes, high blood pressure, cardiovascular disease and cancer.[31] Factory farming is also responsible for levels of cruelty to animals unimagined by predecessor cultures. While the reservations of some at the cruelties and ecological harms of the meat industry are growing, nonetheless industrially-produced meat and feed crops for animals are more than ever dominating the human diet and agriculture, with deleterious consequences both for human health and for the health of the biosphere. Eating disorders have become national obsessions in the United States and Britain, while at the same time rates of obesity have reached 40 per cent in the US and are approaching this in Britain.[32] The spiralling health costs of the disordered culture of food in these two countries are yet another reason why addressing the deeper roots of the ecological crisis in the disordered culture of consumption and making will actually *reduce* the costs of the present consumption economy and enhance human welfare.

The turn to meat is a central feature of the global collectivisation and industrialisation of food production in the last fifty years. This turn is advanced by agronomists, civil servants and farmers, and by international development agencies and multinational food, seed and chemical corporations. In the pursuit of fossil-fuel driven industrial development, airports, motorways and industrial monocrops such as palm oil, soya, cotton and coffee displace the forests, subsistence farms, bicycles and carts of pre-industrial and sustainable village economies. And as motorways and coal-fired power stations spread through Asia and Latin America, the dream of a systemic cornucopia would seem to be being spread from North to South. And yet, as this transformation gathers pace, planetary limits to the ecological demands of a meat-oriented industrial food economy have begun to bite back. The Gobi desert already blankets Beijing in clouds of choking dust, even as Beijing's rulers plan to multiply tenfold China's greenhouse gas production in the next three decades. In India coastal floods and extreme heat threaten the capitalist and state-driven collectivisation of farming and industrial production which the elites of the Indian state believe will make them even more prosperous, while forcing the Dalits or untouchables into continued mute servitude to their high-caste rulers.

Industrial Oils and Tropical Soils

Another central feature of the industrialisation of agriculture is the production of novel foods which are grown for export in many earth regions, and introduced into human diets as manufactured and processed foods. Principal among these novel foods are industrially-produced food oils.[33] Fifty years ago soya hardly featured in the human diet in the raw form in which it is now widely eaten. Similarly, until the middle of the last century most fat in the human diet was derived from animal byproducts such as milk and butter, from fish and meat, or from sunflowers and olives. But industrial food companies have adopted the fatty and protein-rich palm and soya oils as the most ubiquitous ingredients in their manufacture of fat and protein-laden industrial foods, and in their redesign and repackaging of the human diet. Soya has been part of the Chinese diet for at least a millennium but only in fermented form. The extensive use of raw soy protein in Western manufactured foods has introduced an entirely novel dietary staple into industrial food. The health implications of this have yet to be realised as the Western consumer takes part in a gigantic dietary experiment. Dozens of medical research papers now indicate that soy interferes with thyroid function and the human reproductive system, and disrupts hormonal function in men and women.[34] Various food allergies and intolerances and stomach disorders are growing significantly among the populace of those countries, including Britain and America, which have the most industrialised food.[35] There is also growing evidence of a link between consumption of industrial cooking oils and growth in crime, violence and mental disorders, as these new oils replace traditional fish and animal-derived fats which were better matched to the human brain.[36]

Both soy and palm oil are deeply implicated in the systemic destruction of tropical rainforests in Southeast Asia and Amazonia. Soya farmers are clearing land in the Amazon rainforest at a considerable rate, ripping apart an ecosystem millions of years old and which carries 40 per cent of the fresh water of the planet into the atmosphere and the oceans. They replace this beautiful and species-rich environment with monotonous plantations of soybean plants.[37] Eighty per cent of global soy production is fed to intensively-reared animals, and this is yet another reason why farmed meat is such a climate-endangering dietary predilection. The farms keep spreading, not just because of rising demand for animal feed, but because the plants are very demanding of nutrient-poor tropical soils. Most of the nutrient cycle in tropical forests takes place in the moist atmosphere of the forest itself, and the soil beneath the shallow tree roots is relatively poor and thin compared to that of boreal forests. Once the land is denuded of trees, the rain leaches out nutrients from the bare earth and these areas can only sustain intensively-farmed and nutrient-hungry crops like soy for a decade or so. And as the soy farmers move on to new forest areas, they leave a near desert behind. Soya farming in Brazil is in part a consequence of poverty, as peasant cultivators who are shifted off good agricultural lands by wealthy landowners travel along the logging roads to slash and burn their own patch of land to grow crops for their families. But it is also driven by affluence as meat eating is the principal driver of growing global demand for soya. Soya is also being developed as a biofuel to replace fossil fuels which will place further pressure on the Amazon and other bioregions. Sixty per cent of the soy farming in Brazil is funded by three American agricultural corporations. This is another reason why the claim that the United States is responsible for just one quarter of the world's greenhouse gas emissions is a significant underestimate of the true global warming effects of the United States' economy.[38]

Palm oil is the crop of choice for replacing the ancient rainforests of Malaysia, Borneo and Indonesia. Demand for palm oil as shortening in industrial foods and as the fat base for the majority of cosmetics, soaps and cleaning products is high. One in ten products in British supermarkets contain palm oil.[39] And a new use for palm oil, as biofuel for cars and trucks, is now being piloted to meet the rising demand for biofuel in countries seeking to reduce their dependence on fossil fuel oil for transport and heating. Indonesia has indicated an intention to devote 40 per cent of its growing palm oil crop to the production of biodiesel. Indonesia built 11 biodiesel plants in 2006 alone, and the Malaysian government has similar plans.[40] The European Union has set a target that biofuel should make up 10 per cent of ground transport fuels by 2020. But all this does is to displace the need to reduce mobility with an unsustainable contest between land for food and forest and fuel, in which the losers will be ecological refugees and extinguished species in other parts of the world.[41] Biofuel may ease the consciences of Western car drivers, but these new fuels spell disaster for ancient forests, subsistence farmers and indigenous peoples whose lands are already subjected to enforced industrial 'development'. Though the intention behind the dash for biofuel is to reduce carbon emissions from cars, biofuel derived from forest lands, which are among the largest carbon sinks on the planet, is fuel which is *more* polluting in terms of greenhouse gas emissions than the fossil-derived petroleum it is replacing as a supposedly 'greener' alternative.

Since the mid-1990s the pace of palm oil planting in Southeast Asia has grown dramatically and large areas of Borneo and Sumatra have been set alight illegally at the behest of palm oil companies. As a result of this annual burning of the forest, the whole region is now subject to polluting haze and smog in the warmer months of June, July and August. In 1997 conditions were so bad that an emergency was declared in cities such as Kuala Lumpur and Singapore, where the air was so thick with choking toxic smog that residents were forced to stay at home, and there were many smog-related deaths and health incidents.[42] As smog blanketed the region from Jakarta to Bangkok, the whole of Southeast Asia went into a serious economic recession as stock values fell and currencies dropped in value. Suddenly the booming economies of the region looked very vulnerable to the foolishness and corruption of business and political leaders who had colluded in creating this ecological disaster. And the effects of all this burning of ancient forests and peat bogs are not confined to Southeast Asia. In 2001 so much forest and peat bog was set alight in Indonesia that not only was the region blanketed once again in toxic smog, but scientists on Mauna Loa recorded an unexplained spike of 20 per cent in global carbon dioxide emissions for that year which, they now believe, was caused by burning forests in Southeast Asia.[43]

Fossil-fuelled Farming

Industrial food production has not only had deleterious effects on the ecology of tropical forests and soils. It has also seen a tremendous shift in energy use on farms throughout the world from naturally available energy – such as that provided by the sun in photosynthesis, by human and animal muscle and animal waste – towards dependence on fossil-fuel based fertilisers and fossil-fuel driven machines. Again, the oil industry is at the heart of this transformation as industrial chemists in Germany in 1909 discovered that it was possible to fix atmospheric nitrogen as a white powder with the use of crude oil in what is known as the Haber-Bosch process, after the names of its inventors.[44] This process uses heat in fossil-fuelled chemical 'crackers' to synthesise nitrogen from the air, turning it into ammonia which is then used to produce fertilisers. Significant amounts of fossil fuel are involved in this process and millions of tonnes of fertilisers are now spread on fields every year, promoting the gradual abandonment of traditional crop husbandry, such as crop rotation.[45] While the availability of artificial fertilisers has significantly increased food surpluses[46] and made possible the growth in meat-based diets in the twentieth century, its extensive use also has deleterious environmental effects, poisoning water tables and reducing soil quality. Farmers are trained by the availability of artificial fertilisers to neglect biological processes for maintaining the condition and fertility of the soil because they know they can use fossil-fuel derived chemicals as substitutes. It has also enhanced the problems of pollution from intensively reared animals. On a traditional mixed farm, animal waste was part of the farm's nitrogen cycle and utilised on the land. In a factory farm animal waste has no such beneficent function and is often stored in ponds where it emits methane and pollutes ground water.

The extent of industrial food's fossil-fuel dependence is powerfully illustrated by a modern loaf of factory-made bread.[47] The farmer who grows the wheat first prepares the field with tractors and other factory-made and fossil-fuelled devices. The seed is delivered by truck to the farm and then dispersed mechanically on the land. The farmer applies fossil-fuel derived fertilisers to the crop. Pesticides and herbicides, also emanating from fossil-fuelled chemical plants, are subsequently sprayed on the crop, either by tractor or aerial spraying; these new synthetic chemicals are made in refineries which are also major energy users, and delivered to the farm by truck in thick fossil-fuel derived plastic bags. After application fertilisers and pesticides leach into ground water and must be removed by electrically-driven filters and pumps to make the water safe for human consumption. The farmer harvests the wheat with a fossil-fuel driven combine harvester, trucks the grain to a silo where it is sprayed with fungicides and stored, with the aid of mechanical blowers to prevent it spoiling, for months and sometimes years. Eventually the wheat is driven to a large flourmill where electric motors turn plates that grind it into flour. It is then refined, treated, and transported to flour storage facilities, and thence to a bread factory. Here the flour is made into bread by what is known as the Chorleywood process, which turns flour, brewer's yeast and water into bread on a conveyor belt which takes the ingredients from mixing to proving and cooking in one continuous movement. This process is itself highly dependent on fossil-fuel derived mechanical power, and has also transformed the chemical constituency of bread since it allows less time for fermentation.[48] Once out of the ovens and cooled by electrically-driven fans, the bread is then mechanically sliced, placed in fossil-fuel derived plastic bags and transported to a supermarket to which customers mostly drive themselves in their vehicles to purchase their bread and other supplies. At the same time as all this is going on, manufacturers are using the cathode ray tubes on the entertainment devices in peoples' homes to display advertisements, made not in bread factories but in large film studios or pretty rural villages, to entice the consumer to purchase one brand of bread over another. And at the end of each day considerable quantities of industrially-produced bread are discarded because sell-by dates are exceeded before all the brands and types of bread can be sold.

As with the introduction of raw soy and palm oil into the Western diet, not only does the industrialisation of bread have significant implications for the ecosphere, it also has health implications. Some studies indicate that the growing rate of wheat intolerance in Western societies is related to the raw state in which gluten is now found in industrially-produced bread.[49] In traditional fermentation gluten is broken down over a number of hours so that it is easily digestible in the human gut, but because of the speed at which dough is warmed and proved in the Chorleywood process, not only does the gluten in the wheat not have enough time to ferment properly, but raw gluten is added into the mix after proving to aid in its final rising. To make the bread more digestible, industrial food chemists also add a range of enzymes; the effects of these on the human gut are again unpredictable.

The quantities of carbon and methane involved in industrial food production mean that industrial food and consumption is as significant a source of greenhouse gases as electricity generation, space heating and mechanical mobility. Agricultural land use also represents one of the crucial tipping points in climate change scenarios because with raised land temperatures it is likely that the soil, which is presently a significant carbon sink, will turn into a net emitter of carbon. Potentially soils have the capacity to offset between 5 and 15 per cent of global fossil-fuel emissions, but as land temperatures warm soils are turning from carbon sinks to carbon emitters, as we have already seen.[50] Careful management of topsoil is therefore a crucial contribution that farmers can make to the mitigation of global warming. Farming practices which enhance carbon storage in the soil include no-till and low till agriculture, crop rotation and covering of soil with biomass between plantings, restoration of woodlands and the use of animal waste and biomass to restore lost topsoil. Ironically, the trend in farming for the past sixty years has been systemically against soil conservation, and farms from the American mid-West to central China are now suffering seriously from topsoil erosion. Currently cereal crop production in the American mid-West is causing the loss of millions of tonnes of topsoil every year while it also relies on unsustainable use of underground aquifers. In the giant Mississippi river system which drains the mid-West the extent of soil erosion has raised river levels, just one of the many factors implicated in the destructive effects of Hurricane Katrina. Topsoil carried into the ocean has also produced a dead zone hundreds of miles around the Mississippi Delta in the Gulf of Mexico.[51]

Wes Jackson's work at the Land Institute in Kentucky on perennial cereal crops is perhaps the most visionary response to the destruction of soils and water sources involved in industrial tillage. Jackson has demonstrated that wild varieties of wheat and other grass-based cereals can be grown productively without ripping out the roots and deep ploughing the land between harvests. These perennial varieties have deep root systems and are therefore much better at retaining topsoil, while needing less irrigation.[52] Perennial cereal crops would also significantly reduce carbon emissions from soils and fossil-fuel dependency in farming, by reducing the need for mechanical tillage, mechanical irrigation and artificial fertilisers.

The fossil-fuel dependence and excess methane generated by modern food production systems is a direct consequence of the industrialisation of agriculture in the last one hundred years, which has seen a global move from traditional agriculture to industrial monocrops and factory farming. Industrial agriculture is in many ways the last area of human activity to be subjected to the revolution inaugurated by the human release of the earth's store of sunlight in the form of fossil fuels. Henry Ford developed his tractor and light truck to free the farm-worker from hard labour but he could not have foreseen the extent to which these freedoms would be won at the cost of sustainable rural communities and sustainable tillage of the soil. The connection between human culture and agriculture, and the dependence of both upon climate and soil, is as ancient as agriculture itself. What is novel in the modern situation is the use of fossil fuels to short-circuit these connections. The consequence of this short-circuiting is not only a considerable increase in the greenhouse gases emitted by the growing of food but a disconnect between the cultivation of food and the culture of the city, to the extent that many people in cities have little knowledge of where their food actually comes from, and experience a correlative loss of connection between their welfare and the welfare of the earth.

The Moral Economy of Food

Anthropologists Claude Lévi Strauss and Mary Douglas argue that dietary practices and meal rituals are culture-shaping and enact core elements of the beliefs, rules and symbol systems which constitute the structure of different societies and people groups. Rituals of eating and food preparation act in many cultures to connect people with the life of the gods. They also provide powerful pathways for sharing and sustaining beliefs, and for passing on and retelling memories, narratives and traditions. Eating is also crucial in cementing social bonds, and meals can be sources of discrimination and exclusion, as well as solidarity.[53] Douglas argues that power relations and the structure of a society are both manifested and sustained in patterns of eating and food growing. On this account, an industrial economy which fosters the consumption of manufactured food is also likely to be an economy in which small family farmers find it hard to get a fair return from supermarkets and food suppliers, and where damage from industrial farming techniques to biodiversity, the climate and the soil is discounted as economic 'externalities'. Equally the ritual structure of a meal may manifest power relations in the family or beyond the home. If, as Mary Douglas argues, meal rituals and dietary practices are important symbolic representations of underlying social structures, then in societies where a growing number of meals are manufactured from factory-farmed food and eaten by individuals sitting alone in front of electronic entertainment devices, we would expect to find an analogous individualism in society and growing disorder in familial and community structures.[54] The human moral economy, in other words, breaks down as the relation of food to farming becomes ecologically dysfunctional.

This conception of a moral economy is crucial to understanding the impact of the ministry and teaching of Jesus in the social context of Palestine under Roman occupation in the first century of the Christian era. The phrase 'moral economy' was first coined by E. P. Thompson in a paper on eighteenth-century food riots in England. Thompson suggests that these riots, far from being disorderly events, were evidence of a moral economy of the poor, inasmuch as those who took part in them were responding to the attacks on common customs and social traditions associated with the enclosures of common land, which prevented the poor from growing their own food and forced them either into vagrancy or wage labouring in industrial factories.[55] First-century Palestine under Roman rule manifests a not unrelated set of circumstances. There were food riots at many places in the Roman Empire and in their efforts to sustain and extend its rule Rome's Emperors made strenuous efforts to extract agricultural surplus from the far regions of the Empire to keep the price of bread low in the imperial cities, and to provision their legionnaires. Palestine was one such marginal province; its place in the imperial agricultural economy was the provision of surplus. The remains of large wine and olive presses, and of bunk houses suitable for large groups of farm workers to sleep in, have been found on sites south of Galilee. Archaeologists have also uncovered evidence of fish drying and bottling facilities in the region of Tyre, the deep Mediterranean sea port, where products such as fish paste, olive oil and wine would have been exported to Rome.[56]

The parables of Christ concerning food-growing and harvest take on a new significance when set in the context of the threats to Israelite smallholder agrarianism from the combined effects of the imperial taxation system and an imperial food economy which relied on the extraction of surplus food from its colonial holdings to maintain its growing armies and urban centres. Christ's parable of the rich landowner who had acquired so much land that he needed to build bigger barns to store all his surplus is indicative of the way in which wealthy Jewish landowners were cleaning up as small farmers went to the wall in first-century Palestine (Luke 12:13–21). Christ characterises the landowner as a rather self-satisfied and self-concerned individual who plans, having stored his surplus, to rest from his labours, secure in the knowledge that while others go to the wall he has enough laid up so he can relax, eat and drink and be merry. But, as the parable indicates, his life will be required of him before he gets to build his barns. He is, quite literally, storing up judgment for himself even as he rejoices in the surplus he has hoarded from the land. In the Gospel of Luke this parable is immediately followed by one of the paradigmatic meal incidents of the Gospel, which is the feeding of the five thousand. This meal is portrayed by Luke as analogous to the feeding of the people of Israel with manna in the wilderness.

Landlessness and food poverty are closely related in agrarian cultures and, as Dominic Crossan points out, many who were drawn to Christ's preaching and ministry were ill from sickness and diseases associated with malnutrition.[57] In the context of parables and teachings which critiqued privilege and excess, the many meals and banquets in which Christ participated with his disciples had a messianic significance, and enacted a subversive recovery of the traditional moral economy of people and land. These meals recalled the egalitarian moral economy of the ancient Israelites because they refused the distinctions between righteous and sinner, and even between Gentile and Jew, which were sustained by Second Temple Judaism. Just as Christ announced in a pivotal saying that 'many will come from East and West and will eat with Abraham, Isaac and Jacob in the Kingdom of Heaven' (Luke 13:28–9), so he ate and drank with those who were considered far from grace including not only tax collectors and sinners, but prostitutes, Samaritans and the sick. The many meals at which Christ welcomed the 'unclean' to eat with him were essentially missionary meals in which the proclamation of the Kingdom of God became a lived reality; those outside the orbit of grace were invited in and through their participation in table fellowship with Christ they were offered forgiveness and redemption, and a new society was in the making.[58]

The story of Zacchaeus is the classic account of this constitutive and revolutionary function of the meals of Christ, but other less obvious exemplars include Christ's drinking water with the Samaritan woman at the well, and his feeding of the five thousand. In all cases Christ's choosing to eat with those whom debt, poverty, prejudice and illness, or their association with the Romans, had placed outside the sphere of the righteous causes embarrassment to his disciples, and outrage among the rabbis. Such meals, as well as Christ's teachings, were interpreted as an implicit attack on the Temple, blasphemy against the Temple being the principal accusation brought against him at his trial. Offering open access to grace without the mediation of the Temple religion, Christ was viewed by the Jewish authorities as a dangerous subversive. This is because the Temple was at the heart of the extraction of agricultural surplus of the Roman imperial economy in Judea. Judea and Galilee were self-governing provinces and the Temple treasury was the place where the poll tax and imperial tribute were gathered before being handed over to the Romans. Thus Christ's frequent denunciations of the religious mediation of the Temple system were also attacks on the imperial economy itself and his meals, and his teaching about the messianic banquet open to the poor and even to Gentiles, were equally implicated in this attack.[59]

The Eucharist and Christian Eating[60]

Given the importance of meals in the life and ministry of Christ, it is no surprise that eating and feasting also played a central role in the worship traditions of the early Christians. When the first Christians began to meet for worship on the day they called the Lord's Day, which was the day of the Resurrection of Christ, they met not in synagogues or Temples but in the dining rooms of Mediterranean houses, where they reclined or sat at table and shared a meal together. In so doing they were following the common practice of their Mediterranean contemporaries in the first century who, whenever they gathered for religious, social or political purposes, organised their meetings around a two course meal.[61] The meal was invariably followed by entertainment or debate known as the symposium. In the case of the Christians this would have taken the form of hymn singing, reading of scripture, oral testimony, intercession and exposition of scripture. In centring their worship on table fellowship they adopted the common custom both of their Jewish and Gentile contemporaries and of Christ and the disciples.

This recognition suggests that the Eucharist in the early Church was also associated with redressing the wrongs against the poor and the land in the food economy of imperial Rome.[62] Just as the table fellowship of Christ was the means to redemption for sinners, so Christian eating becomes an acted parable of a moral economy which recalls the idealised moral economy of the Torah. In the Kingdom, and hence at the Eucharist, the poor no longer have their land expropriated from them for the benefit of the tables of the wealthy, but instead are welcomed to the messianic banquet alongside the rich, where they find not only a place but a voice in worship after the breaking of the bread.

On this account, the now dominant practice of Eucharistic worship as a token meal which symbolises the death of Christ fails to represent either the original meaning *or* practice of this meal for the early Christians. As Dennis Smith argues, the New Testament itself unambiguously describes the first Eucharists not as token meals but real meals. And early Christian art reveals that worship in the early church was around tables in the dining rooms of the houses where the first Christians met for worship.[63] We know this also from 1 Corinthians, which indicates that the association between Christian worship and eating was not without its problems. Paul's foundational description of the Eucharist in this epistle is preceded by his extensive discussion of the problem of meat offered to idols. As N. T. Wright suggests, this account is not epiphenomenal to his account of the Eucharist. On the contrary, it provides the crucial context. The economy of eating in Corinth was problematic because it was a pagan economy in which almost all meat eaten in Corinth was butchered at pagan temples.[64] How to eat meat without the infection of idolatry was the pressing moral question that St Paul addressed in this letter immediately before his description of the Eucharist itself. His concern was that the Eucharist was marred at Corinth by its infection over the arguments about meat offered to idols, and the larger context in which rich had meat and the poor merely bread-crumbs. Bread for the poor, wine for the rich – the very Eucharistic elements had become the occasion for division in the church at Corinth.

Paul was writing to the Corinthians to put them right on their practice of the Eucharist because they had turned it from the meal of common sharing as practised by the disciples in Jerusalem into a meal more like that of a Roman symposium, in which the rich ate first and had their fill of luxury foods, while the servants cleaned up on the crumbs when the wealthy were done. This was not the practice of the common meal as Jesus had shared it with his disciples before and after his death. Paul begins by recounting the story of the feeding on manna in the wilderness, a feeding which initially went wrong because the people of Israel neglected the two rules that God established for receiving the manna which were, first, that the people should gather only sufficient for what they needed and second, that they should not hoard or try to store up the manna. The Corinthians were doing anything but this in their practice of the common meal – instead of sharing together equally some had far too much while others went without.

This problem with the form of the meal at Corinth provides a vital clue to the meaning of Christ's words. 'do this in remembrance of me', which St Paul sets centre stage in his own account of the Eucharist. Christians today tend to hear these words in the light of a later individualistic and rationalistic sacramental theology. But when St Paul writes these words he is referring to the common practice of Christians breaking bread together in the name of Christ. This common meal tradition established a foundational connection at the heart of the Christian religion between the moral economy of food and of divine grace, a connection first enacted in the messianic meals of Christ and the daily community practices and worship of the first Christians. Another key phrase in St Paul's account of the

Eucharist is the language of 'discerning the body': these are interesting words which again Christian history has turned inward. Aquinas' sacramental theology has trained Christians to think of these words as indicating some inner sin which if unconfessed will lead to the communicant being condemned when she receives the sacrament. But the meaning to the first readers of this epistle could not have been plainer. The body of Christ was an alternative political order to the imperial polity of Rome – this is why Paul uses the traditional classical metaphor of the body. And so to discern the body is to discern the significance of the way of Christ: the way of common sharing in which the weak are respected alongside the strong, the rich eat and drink alongside the poor. Discerning the body refers to the alternative moral economy of revolutionary subordination inaugurated by Christ. The body of Christ, as Paul goes on to explain in 1 Corinthians 12–14, realises this moral economy on earth in the common meal, and in the acts of worship and ministry which follow it when the strong give honour to the weak, and those with less respect in society are given voice.

Modern Christians have two economies in mind when they hear or read this passage – the economy of salvation, which is taken to be the sacrifice of Christ through which sins are forgiven, and the economy of food, which is subject to secular political and economic arrangements. But for Paul there is no such distinction between politics and religion, or nature and culture. To confess that Christ is Lord is to confess that Caesar is not – it is a far more profoundly political confession than it is for Christians after the conversion of Constantine. Similarly, to break bread blessed in the name of Christ was a profoundly political action which modelled in the first Christian communities a different moral economy to the pagan economy of imperial Rome.

All of the butchers in ancient Corinth were servants of the Roman pagan economy – all of their meat was offered to the Roman imperial cult as a part of the process of its slaughter. Eating meat offered to idols was therefore unavoidable for Corinthians. The motive context of this first narrative of the Last Supper in 1 Corinthians is Paul's concern that the Christians in Corinth were in danger of turning Christian eating into meals which were like pagan symposiums, manifesting the dubious commitments and social relations of the Roman moral economy. Thus when Paul suggests that Christians ought mostly to desist from eating meat in Corinth, because for the weaker and poorer brethren the sight of the rich eating pagan meat would be offensive, he is clarifying the distinctiveness of the Christian ritual meal from pagan ones. Eating meat offered to idols risked the Christians becoming embedded in an idolatrous and pagan worldview.[65]

Paul's narrative of the Eucharist in 1 Corinthians, like other accounts of meals in the New Testament and in the early fathers, indicates that the early Christians did not sustain a ritualistic demarcation between the Eucharist and other forms of eating. The tradition attests to a variety of ritual meals, with elements ranging from bread and wine, bread and water, bread and fish, and bread, water and fruit or vegetables. These meals occur in a variety of liturgical contexts. Some took place in a distinct and formal act of worship in which the Eucharistic meal was preceded by the kinds of prayers preserved by Hippolytus and the *Didache.* Some were more along the lines of what it has become customary to call *agape* meals – that is to say a meal which was in every sense of the word a meal, where participants literally sat down and ate together, having first blessed all the food that was set before the faithful. Prayers, scripture readings, homilies, and hymns would also have been part of these meals, much as they were in the Roman symposium.[66] What set these meals apart from pagan meals was that the food offered and consumed was seen as the gift of God, Creator of the universe. The blessing and breaking and sharing of bread and wine recovered the creation ethic of the ancestors of Israel who ate manna in the wilderness – it reconnected the gifts at the table with the gifts of God in creation, and celebrates the dependence of the people of God on God's generous gifts of spiritual and physical sustenance. In this way the Eucharist meal becomes a microcosm for the divine plan to redeem the whole creation from the effects of sin: physical food becomes spiritual food as bread and wine become the body and blood of Christ.

This recognition is doubly important in relation to the morally and theologically objectionable supercessionist readings of Christian salvation history in subsequent eras. Christian freedom in eating foods that Jewish law proscribed did not mean that the sacred meanings of the old covenant were simply set aside or supplanted. One of the most extensive treatments of Jewish dietary laws in the early fathers is Novatian's essay *On Jewish Foods*. Novatian treats of the dietary laws as allegories of the moral and spiritual life, so that for Christians the true meaning of the proscription on certain foods as unclean is that they represent human vices, while restraint from eating such foods represents the virtue of temperance.[67] For Novatian the 'consummation of the law' which Christ realises means that, as St Paul has it, 'to all who are pure themselves, everything is pure' and that 'every creature of God is good, and nothing is to be rejected that is accepted with thanksgiving'.[68] But while this means that all foods can be enjoyed by Christians in 'evangelical liberty', as blessings received from creation, it does not mean that Christians should eat with an excess of sensuality or that they must not guard against greed or gluttony. On the contrary, the Jewish laws are a continuing reminder that food is a source of temptation and that desires are still prone to corruption even for those who are in Christ. Christian eating is still subject to the rule of virtue, and the one who eats and drinks in moderation and with a clean conscience 'eats with Christ'. Christians may not be drunkards or gluttons because Christian eating is subject to 'the law of frugality and moderation' which St Paul commends to Timothy.[69] And there is one more rule that Christians share with Jews, and this is the prohibition on eating food offered to idols, for such food, since it has been offered to demons, 'nourishes the one who partakes of it for the devil, and not for God, and makes him a table-companion of an idol, not of Christ, as the Jews also rightly hold.'[70]

Novatian's account of the importance of moral virtue and spiritual worship in Christian eating indicates that for the early Christians *all* eating is potentially eating *with Christ,* and that there is a moral and ritual continuum between the Lord's Table and the household meal. This recognition is an important corrective to much Eucharistic practice and theology where the Eucharist has become a token meal, and as a memorial rite of the death of Christ conceived as a sacrifice. It indicates that not all ritual meals were associated in the Christian mind with sacrifice. Some Christian meal practices were open to sacrificial interpretation, but sacrifice is not the only way in which ritual eating is understood in the early Church. Equally strong are metaphors concerned with mission to the world beyond the Church,[71] with the binding of the fellowship in the shared act of eating and drinking bread that has been broken and from a cup that is shared, and with blessing and gift giving. Furthermore, in certain crucial respects Novatian's account indicates that there was no sharp distinction between the sacrality of a ritual meal and the profane eating of the household. All meals are indicative of the relationship of the guests with God and with one another; all meals constitute community and sociality, and offer opportunities for conversation and table fellowship.

The elements and the rituals of Christian meals, then, are *all* theurgic (and hence the tradition of Christ as the unseen guest at every meal) and all are constitutive of Christian identity, and of the boundary between *ecclesia* and *seculum* in a pagan world. Analogously, food growing and the social economy of food are caught up in the divine economy of giving and receiving which Christ's many acts of blessing and breaking of bread set forth. We see this in so many places in the Gospel – for example when he is walking in the fields on the Sabbath and he and his disciples pluck kernels of wheat to eat as they walk. The scribes condemn him for lack of respect to the Sabbath but Christ's answer is, 'the Sabbath was made for man not man for the Sabbath' (Mark 2:27). Religious boundaries may represent divine mandates, but ultimately they are to serve humanity's imaging of God. Where they obstruct that image they can be abrogated.

Modern European culture has been profoundly shaped by the sacred–profane distinction which emanates into Christian theology from the medieval practice of the Mass. The medieval focus on priestly or cultic objects and sacred space paved the way for the emergence of the non-cultic as secular in succeeding centuries. In post-Christendom Europe and North America, profanity in relation to creation and food does not take the form of disrespecting the Host or refusing to honour the priest. In a sense both Host and priest are sidelined in modern secularism. It is now the creation itself which is profaned, while the sacred Host is set aside, replaced with the sacred dollar or euro. The contemporary form of idolatry is the devotion of the culture of food to money, from crop to table. This produces a new kind of profanity in which industrial food is debased while the soil is eroded, the land poisoned and the climate changed. By participating without critique in the fruits of modern agronomy, do Christians analogously 'eat meat offered to idols' and imbibe a worldview which is contrary to the Gospel?

This point is sharpened by the recognition that most celebrations of the Eucharist in Europe and North America occur not in rural areas but in cities where the resonance of the agrarian references of the elements of bread and wine may be lost, either consciously or materially.[72] Thus congregations that decide to use a real loaf of bread rather than communion wafers may use a loaf which has been made from petrochemically farmed wheat and then manufactured by the Chorleywood process. When a loaf which is the product of this novel structure of food growing, making and marketing is presented at the altar with words that declare that it is through the goodness of God that the people 'have this bread to offer', and moreover that it is 'fruit of the field, work of human hands', credibility is being stretched.[73] Did the goodness of God intend that bread should be wrung from the land at the expense of the rape of the soil, the poisoning of ground water and the wasteful expenditure of fossil fuels, or that it should be manufactured in such a way as to turn it into the ambiguously convenient consumer product which is modern sliced bread? While it is the case that there is a foundational agrarian reference in the use of bread and wine as the elements of the Eucharist, it is questionable whether this agrarian reference is truly honoured and proclaimed in these circumstances.[74]

The related claim I am making here is that food *is* politics, and that the Eucharist is therefore political food. Just as Eucharistic practice in Europe, both before and after the Reformation, represented a very different kind of politics than the revolutionary subordination of Christ and the apostles, so a Eucharistic practice which fails to challenge the profanation of the creation represented by modern agronomy is equally flawed. Profane eating, eating food offered to idols, has become the norm in the twenty-first century and like earlier cults, this idolatry involves sacrifices – of the soil, healthy rural communities, and now of the climate of the earth.

In the shadow of the Second World War C. S. Lewis asked whither chemical agriculture would direct the modern world?[75] Lewis did not know of the technical problems which now beset modern agronomy including pesticide residues, groundwater pollution, decline in bird and other wildlife populations, diminishment of soil quality, an increasing treadmill of chemical inputs, the bankrupting of small farmers, the decimation of rural communities and global warming. But Lewis was trained by his reading of medieval literature to believe that there is wisdom in the prior order of the biophysical world, even though this is an alien conception to the modern mind.[76] According to Lewis chemical agriculture was wrong, not because of its consequences but because it was intrinsically in conflict with the wisdom of nature's order, which he describes in the medieval language of natural law and which in an ecumenical spirit, he also calls the 'tao'. For Lewis, the industrial farmer who neglects the tao of the earth invites the judgment of the earth.[77]

Wes Jackson and Wendell Berry analogously suggest that in their neglect of the laws of ecology modern farmers, and the scientists, politicians and corporate managers who guide them, are not just myopic but morally deficient. By neglecting the wisdom of conserving soil quality for the next generation the modern farmer, and the politicians and consumers who collude with him, display 'moral ignorance and weakness of character'.[78] Science-informed agriculture, far from liberating modern humans from natural necessities, has subjected them to a new form of slavery – the slavery of the machine and of untutored instinct. This mechanistic slavery takes a number of forms in the global food economy. And this enslavement is not of course confined to humans. Billions of animals are now caught up in a horrendous factory farming industry which is not only one of the largest single sources of greenhouse gas emissions but also a form of collective cruelty which is unique in human history. Never before did a civilisation rest upon such systemic cruelty to so many living creatures as does industrial civilisation. That this immoral treatment also has serious health implications for human beings in the form of novel viruses and toxins in our diet is still not seen as a cause for governments or food scientists to press the industry into more humane and sustainable approaches to animal husbandry. In Britain the present Labour government spent two years and 2,000 hours of parliamentary time discussing the ending of hunting foxes with hounds. But the same government oversaw a mass culling of hundreds of thousands of cattle in response to an outbreak of foot and mouth disease, against the advice of its own vets that vaccination would be kinder and less destructive to the environment. And it continues to permit movements of meat in and out of the country which are the prime cause of the ingress of this and other infections, such as bird flu. The reality is that the industrial food economy is the most unholy form of agriculture invented by any civilisation.

Citizens are however beginning to resist the destructive effects of the global food economy. Increasingly people want to know where their food has come from, and, in the case of meat, what animals have eaten and how they lived. But so long as governments remain committed to a 'free market' in food – and to global movements of animals, and animal feeds – people will remain at risk of eating meat which has been grown in immoral conditions, and at the cost both of extreme cruelty and of immoral threats to non-Western peoples and to the climate. The food industry is now the largest manufacturing industry domiciled in the British Isles and on the mainland of North America, since so much of industrial manufacture has been exported to factories in China, Mexico and elsewhere. In terms of domestic greenhouse gas emissions, the food sector is then the largest single sector, apart from electricity generation, transportation and housing, in which governments can regulate for a more sustainable economy. However, the power of the food companies is such that in Britain and the United States their governments resist regulating the production and sale of industrial foods so as to promote human health and environmental sustainability.

Once again, we are faced with the systematic conflict between the neoliberal, corporately-driven economy of endless growth without moral limits, and the ecological and moral structures of the biosphere and of ecological and human community. The deregulated industrial food economy promoted by neoliberals advances a fundamental disconnect between the growing and eating of food. So remote have global food chains become that the *average* item of food in an American supermarket has travelled 1,500 miles. And while some European countries, such as France, have strenuously sought to preserve regional food cultures and food supply chains, others have foods which travel at least as far as those in American supermarkets. The author's own unscientific survey of a local Scottish supermarket indicated that 80 per cent of its fresh food had been air-flown, shipped or trucked from places as distant as Chile, Kenya, Israel, New Zealand and Spain, and this included most of the organic produce, which flies in the face of the claim that organic food is more ecologically sustainable. One major consequence of these remote food chains is that it is not only Western animals that are fed on crops grown at the cost of ecological destruction and environmental exclusion in the Southern hemisphere, but Western consumers too. A growing proportion of the food consumed in the rich North is grown in sub-Saharan Africa, Asia and Latin America on lands once tilled by smallholders for local consumption and regional trade. With the turning over of this land to cash crops for export, millions of small farmers have been forced to migrate to other areas; either to cities where they eke out a miserable living as squatters, or to forests and hillsides where their desperation drives them to displace ancient forests with their crops, or farm marginal slopes, and so precipitate flooding and soil erosion.

For many Western citizens their remoteness from food production and agricultural practices also maps onto an increasingly attenuated engagement with traditional approaches to food preparation and to eating. For a growing proportion of people in Britain, including many children as well as adults living alone, eating has become an individualised meal experience which often involves pre-prepared and pre-packaged food from a factory or fast food outlet, whether of the baguette or burger variety. As a result, the traditional social context of the meal at which families gather around the table to talk over the events of the day, and to *be* family together, is increasingly a counter-cultural experience. Not only are the kinds of foods eaten in individual meals less nutritious and healthy than traditionally cooked meals, but there is also evidence that meals taken together are healthier precisely because they are moral and social experiences and not just individuated acts of energy consumption or body fuelling.[79]

This turning of eating into an individuated and denatured process is indicative of the perverted spirituality of the powers that direct the global food economy. The climatological and cultural destructiveness of the industrial food economy are both exemplars of what Marx classically identified as the guiding spirit of capitalism, which is commodity fetishism, and the associated alienation between the producer and the consumer.[80] As industrial agriculture turns food growing and eating into processes of value extraction and profit maximisation, the cultures of food and farming, and the ecosystem of the earth itself, are sacrificed on the altar of the global economy. The profanity of this economy turns food into a fetish, and hence the diseases and dietary disorders associated with food in modern consumer societies, from anorexia to obesity.

These disorders and the disordered rituals of modern food consumption and modern animal and crop husbandry reveal the cruel and unjust nature of the global food economy. The modern economy of eating is as imperially disordered and profane as the extractive economy of imperially-governed Palestine at the time of Christ. Imperial value extraction and wealth accumulation sustain, and are sustained by, forms of fetishisation and ritual behaviours which are deeply alienating both to those who practise them and to the biosphere. And just as industrial foods and eating rituals are fetishised in perverse ways as they are detached from the farm or even food preparation, so food growing has been subjected to increasing commodity fetishisation and manipulation by scientific researchers and technologists. This manipulation has had extremely ambiguous results, as we have seen in the case of industrial oils. The nutritional value of chemically-grown food has declined in modern industrial food systems, and essential minerals like magnesium, iron and calcium are present at lower levels than they were in traditionally-grown foods seventy years ago.[81] At the same time the preparation of food has also been increasingly professionalised, either by food scientists working for food manufacturers, or by professional chefs who view the preparation of food as high art, an art which is presided over by an 'academy' of elite Michelin-starred restaurants and famous chefs. Food preparation and eating become ends in themselves, detached from the *habitus* of everyday living and the culture of farming. One of these elite chefs was even driven to suicide under the stress of maintaining this temple to his trade.

Christian Worship and Spiritual Eating

If Christian worship is effectively to resist the devotion of modern civilisation to an increasingly heedless abuse of the precious resources of this good earth, and if it is to resist the particular threat posed to all life on earth by climate change, then there is an urgent need to recover the full agrarian and social significance of Eucharist in Christian worship. Breaking bread was the paradigmatic meal of first-century Palestinians and Romans. The early Christians worshipped sitting at tables where they blessed the cup and broke bread because it was a common meal, not just a 'sacral' act. The social meanings attached to the meal traditions in which the Eucharist began have considerable resonances with the witness of the Church today in a situation where once again a global imperial economy is the source of oppression of both humanity and the land. The creatures which are the elements of bread and wine are transformed in the Eucharist into the new creation which is the body of Christ, the Church, and in this transformation lies the possibility of the end of the groaning of creation and the remaking of all things towards their peaceable destiny. The Eucharist does not only remake human politics and spaces but the politics and spaces of all creatures, human and nonhuman.

The central claim of this chapter is that the Church-constituting ritual meals of Christians are paradigmatic places where the modern nature–culture distinction is refused and the modern industrial rupture of human–nature relations is healed. In the messianic banquets of Christ described in the Gospels, in the meal traditions of the early Christians, and in much of the subsequent Eucharistic tradition in Christian history, the ritual meal involves acts of blessing and sharing. In this way the whole creation is drawn representatively into the atonement, reconciliation and divine–human fellowship which are the fruits of the work of Christ and which constitute the priestly calling of the people of God to continue Christ's pioneering priestly work, in drawing the creation into praise of the Creator. For the first Christians, their ritual meals were crucial events in establishing the boundary between pagan and Christian in cities where meat and wine were intricately connected with pagan sacrifice. For Christians in the twenty-first century, paganism takes the form of consumerism and devotion to the money economy. This modern paganism has subverted the human and other than human food chain and hence the health of soil and air, forest and mountain, river and ocean. In the context of ecological breakdown Christian eating, in Church and in the home, is central to the Christian witness to the human mandate to care for creation and to resist its heedless destruction, including that advanced by modern agronomy.

For 12,000 years farmers have been guided by two fundamental beliefs: that humans are part of the web of life, and that the web of life is cyclical, communal and mutually reinforcing.[82] For most of that time these beliefs shaped not only the farming year and farming practices but human civilisation in general. The Bible is full of advice to farmers that they should attend to the ways of the natural world in their own engagement with it. It is impossible to underestimate the influence of the agricultural practices of the ancient Israelites on their religious and moral traditions.[83] And we have seen that the same influence is there in the parables of Jesus. He uses the metaphor of feasting to describe the new order which the Incarnation inaugurates on earth, and the metaphor of propagation to describe his own and his followers' struggle in birthing that new order against those who opposed it in first-century Palestine. Agrarianism in the Old Testament is focused on the need to preserve the fragile land of Palestine from the threats of soil erosion and salinity. The Sabbath provided a significant limit on Hebrew use of land – moderated demands on land and animals. The neglect of the Sabbath is also the occasion of divine judgment; the Prophets interpret the long exile in Babylon not only as judgment on Israel for neglecting the Sabbath but to help the land recover from hundreds of neglected Sabbaths.

The sacrificial system was crucially implicated in the connection between Israelite agrarianism and the divine will to preserve the web of life. Sacrifice finds its first explicit mention in the sacrifice that Noah offers to God of a burnt offering of exemplars of the animals and birds which he had saved in the Ark.[84] This primordial sacrifice attempts to restore the fallen creation and to bring it back from its path toward destruction, on which the Flood is seen as the judgment of God. Noah recovers what others in his generation had lost, which is humanity's priestly vocation to offer up from the abundant gifts of creation sacrifices which atone for sin, which are pleasing to the Creator, and in which creation is restored to the order which is the work of its maker. And so in the Noahic covenant God promises to desist again from so nearly destroying all life on earth (Genesis 8:20–21). But it is also to this paradigmatic first sacrifice, and the great banquet which no doubt followed it, that is traced the tradition of the proscription on drinking the blood of the animal which represents the spirit, the life force, and is therefore to be respected. The blood must be returned to the earth as a sign of respect for the origin of all life in the divine creation.

The offering of food in the Eucharistic feast is equally significant for Christians in connecting the Church-constituting rite of Eucharist with the whole of the divine creation. The Eucharist is in effect a microcosm of the history of creation–redemption as it finds its completion in the Incarnation of Christ. The claim that the Eucharist makes the Church, which in the twentieth century has been advanced by the Catholic theologian Henri de Lubac, needs also to be read in the light of the moral economy of food.[85] Here both Catholic and Protestant may learn something from those Christian farmers who departed the shores of Europe in the face of bloody and extreme persecution and established the Anabaptist culture of North America, which still maintains a close relation to the land today. Just one of their major disagreements with Catholic and Protestant theologians was over the theology of the Lord's Supper. They claimed that the Lord's Supper was the constituting action of the body of Christ. For the Anabaptists it was not the priest or the altar, nor their mystical association with Christ, which made the elements holy but rather the gathering of the whole people of God and the holiness of lives.[86] The crucial element in a proper invocation of the body of Christ *is* the body of Christ, where the body is conceived not as the material or mystical body of Jesus Christ but, as St Paul has it in 1 Corinthians 12, the community of the local church. As Yoder points out, Anabaptist ecclesiology broke with Reformation and Catholic theology over the question of the visibility of the Church and its capacity to act.[87]

In Anabaptist perspective the Lord's Supper has no objective power, no mystical significance, apart from the participation of the people of God in the action of breaking bread.[88] There is no invisible Church which resides behind the real, presently-existing Church as represented by the local congregation. Instead, the Church and the Lord's Supper are mutually constitutive because the local church in its gathering to break bread *is* the agent that breaks bread. The celebrant acts on behalf of the community, and the act of breaking bread creates and sustains the community. This approach places ethics and holy living at the centre of the Eucharist–Church relationship, because it also requires that those who gather to break bread are living in communities whose morality and order are modelled on the teachings of the Gospels. There is no objective sacramental mystery here. Only when the people of God act morally and live holy lives do they constitute a body of people who can break bread and so be the body of Christ.

Here we have a recovery of the early significance of the important words, 'eat and drink worthily', in 1 Corinthians 11. Fear of individual unworthiness was a major motive for the infrequency of lay participation in the Eucharist for more than a thousand years, and for some remains a motive to this day. But in the Anabaptist reading unworthiness is not an individual matter. It is the revolutionary subordination of the people of God to one another – and in particular of the strong to the weak – which is the mark of the true body of Christ on earth and provides the ecclesial link between holy living and holy eating at the Lord's Supper.

The relation of holy living to holy Eucharist in Anabaptism recalls in a powerful way the same connection in the Old Testament sacrificial system – when Israel's agrarian economy was ordered after the commands and ordinances of God the Temple sacrifices were pleasing to Yahweh. They were just sacrifices because they represented an agrarian society in which the people were just in their relations to one another, to animals and to the land. When they acted without restraint, amassing wealth and agricultural surplus at the expense of the land and of the poor, they ceased to live justly and their sacrifices became offensive. The Anabaptist recognition of the foundational significance of Christian eating is surely not unconnected with the fact that they are unique among Protestant Christians in North America in resisting modern agricultural practices. Instead of pursuing the model of power over nature, and of agriculture as an extractive economy, they have sustained a model of holy agriculture in which communal dependence and the conservation of the soil go hand in hand. Instead of debt to banks and government to sustain the high inputs and machinery of chemical farming, they have been content with growing enough to sustain their communities, and this restraint, and freedom from debt, has made them not only the best conservers of the soil in North America, but, in recent decades, the most successful farmers, if success is measured in quality of life as well as quantity of production. As Wendell Berry says, only the Amish

> as a community, have carefully restricted their use of machine-developed energy, and so have become the only true masters of technology. They are mostly farmers, and they do most of their farm work by hand and by the use of horses and mules. They

> are pacifists, they operate their own local schools, and in other ways hold themselves aloof from the ambition of a machine-based society. And by doing so they have maintained the integrity of their families, their community, their religion, and their way of life. They have escaped the mainstream American life of distraction, haste, aimlessness, violence, and disintegration.[89]

There will be those who at this point will say 'ah, but we cannot turn the clock back: global food is the future, local food the past.' While this opinion still dominates the food industry, the European Commission, and the United States Federal Government, there is thankfully a growing recognition among many American and European consumers and farmers of the need to restore the relationship between the culture of eating and the culture of growing, and to recover a more sustainable relation to the land. The rise of farmers' markets and organic box schemes, the growing demand for organic food, increased concern with food miles, and the rising market in fair trade food products all indicate a counter movement to the dominant trend of the global food economy. If the connections I have sought to establish in this chapter between Eucharist, eating and ecology are to be recovered in the Christian practice of worship, Christians will be among those who will seek in future to know where their food has come from, how it was grown, by whom, and at what cost to the climate, farmers, species and soil. In this way Christians will begin to recover the revolutionary nature of the first Christian communities and their acting out of that spirit which 'turned the world upside down' in their own moral economy of food.

There are a number of practices which would foster the reconnection of Eucharistic worship with the food economy. The first would be to recover the sense in which the Lord's Supper is a holy meal that fosters and sustains the holy lives and holy eating of Christians in their day-to-day conversation and walk. One means of remaking this connection would be to revive the medieval practice described by Eamon Duffy of the presentation of a 'holy loaf' as a Eucharistic practice, and as still performed in some parts of Europe.[90] According to this tradition a different household from the congregation makes the bread by hand each week, from flour that has been sustainably grown, and brings it to church to become the one loaf which is broken and shared in the Eucharist.

The second would be to return to the early Christian norm of worship *around* the Lord's Table and to find ways of reversing the historical trend from a real meal to a token meal. This would mean recovering the use of real loaves of bread, and not wafers or even a token small loaf, and of drinkable quantities of wine and water as central elements of a regular Eucharistic feast in Church. It would also mean the recovery of the Eucharist as the central act of worship in all Christian churches, after the problematic break with this practice among many Reformed churches.

A third means may be indicated in the extent to which in many churches in Europe and North America there is a practice of the 'potluck' lunch or supper after worship on Sundays, which for many is as constitutive of the community life of the Church as the service of the Word which precedes it.[91] But the food eaten at such gatherings is neither considered holy nor is it any different from the food sold in industrial food stores and grown unsustainably on industrialised farms. Resituating the meal which follows worship to an *agape* meal at the heart of worship and organising the service of the Word *around* the meal, rather than tacking the meal on at the end once the liturgy is over, would offer another powerful way to restore the connection between Christian worship and the fruits of the earth.

There are three ruptures which need healing and restoring. First, there is the rupture between the service of the Word, sometimes with the Sacrament attached, and the potluck meal which follows. Second, there is the rupture between the holiness of the Sacrament and the unholiness of the way in which food presented at a potluck supper has mostly been grown. And third, there is the rupture between holy eating in Church and holy eating in the home. If, as I am suggesting, faithful feasting is paradigmatic of the Christian relation to creation, and of Christian worship, then all meals are holy meals. In the home as well as in Church, food should be sourced with this in mind. All meals should be accompanied by prayerful recognition of the gifts of creation which are most manifest to us on a daily basis in the sustaining power and the communal enjoyment of the gift of good food. In such ways it turns out once again that acts of resistance to the global economy and its destruction of a stable climate through global warming are not a costly burden, but instead involve the joyful recovery of forms of life that industrialism has sacrificed on the altar of surplus value.

9
Remembering in Time

They shall come and sing aloud on the height of Zion,
and they shall be radiant over the goodness of the LORD,
over the grain, the wine, and the oil,
and over the young of the flock and the herd;
their life shall become like a watered garden, and they shall never languish again.
Then shall the young women rejoice in the dance,
and the young men and the old shall be merry.
I will turn their mourning into joy,
I will comfort them, and give them gladness for sorrow.
I will give the priests their fill of fatness,
and my people shall be satisfied with my bounty, says the LORD.

Jeremiah 31:12–14

The Church at the Eucharist is gathered by the memory of an apocalyptic moment in human history when the redemption of all things is anticipated in the Crucifixion and Resurrection of Christ. The Eucharist is a shared meal through which Christians are 're-membered' or joined as members of the body of the risen Christ. In the resurrection of Christ the world is renewed, and the physical cosmos saved from its infection by sin. As St Paul indicates, Christ is revealed in these events as the Lord of the cosmos 'in whom all things hold together' (Colossians 1:17).

Just as the climate represents a physical record and memory which connects all our carbon emissions, so Christ as cosmic Lord is the original source in the divine mind of those fecund interconnections in which life is sustained, and hence only Christ can redeem them. Remembering in the Christian tradition is therefore the wellspring of hope for the redemption of human life and creation, for God alone can redeem the complex systemic relationships of the structures of sin which will see the present 'sins of the fathers' (and the mothers) in the form of excess greenhouse gas emissions 'visited on the third and fourth generations'.

At a pivotal point in the liturgy of the Eucharist Christians confess their sins to one another. In this regular ritual act they engage in another kind of remembering – a remembering of the ways in which their lives still fall short of the standard of Christ and the hope of the Kingdom. And after confession Christians share words of forgiveness with one another and words of peace. For Christians, then, remembering is ultimately peaceable – it has peace as its *telos* – for through remembering Christians find that they are forgiven, their alienations healed, and peace restored.[1]

The story of anthropogenic climate change is also about remembering. The quantity of carbon now in the oceans and atmosphere is a physical footprint, a living memorial, to the industrial revolution and its many victims. These victims include the peoples and other creatures who lived and live in or on the terrestrial and subterranean forests which are being burned to sustain the fossil-fuelled era. They include car accident victims, the victims of fossil-fuelled aerial bombers, and the victims of the fossil-fuelled trains and ovens used in the Holocaust. They include those enslaved in fossil-fuelled industrial factories, first in Victorian England and now across many parts of the Southern hemisphere, where lives were and are foreshortened by industrial pollution and human dignity is degraded in servitude of machines. They include destroyed agrarian communities, lost topsoil, extinct species, wrecked ecosystems. They include flood and drought victims, and those who die, and will die, trying to escape from climate-stressed continents and inundated islands.

The carbon cycle is earth's way of remembering all of these people. But it is a peculiarly indifferent kind of remembering. Global warming induced floods and droughts and storms fall on the just and the unjust, the poor and the rich, though they are currently wreaking more havoc among the former than the latter. The planetary memory of all those burnt forests is not a redemptive memory, and is more like a judgment. Perhaps Lovelock is right and Gaia will judge industrial civilisation in the end, sloughing billions of humans off the planet so that she can recover.[2]

In the tradition of stewardship that occurs at a number of points in Western environmental traditions, including the Old and New Testaments, and more recently in accounts of corporate and scientific 'earth management', humans are stewards or keepers of the earth's resources, empowered to use and improve them for human benefit.[3] But industrial humanity has a very bad record of earthkeeping. If contemporary humanity was to be struck down by a deadly global pathogen and industrial civilisation ended, the power lines would rapidly fall silent, planes and cars would no longer move, bulldozers and chainsaws would no longer subdue the cacophony of life in the forests, and the cities would eventually give up their paved and concreted spaces to plants pushing through the cracks and rivers bursting out of their humanly-enforced channels. In a decade trees would begin to repopulate the denuded land, rivers would be replenished as more rainwater stayed in the soil, and so mother earth would begin to draw down the excess carbon from the atmosphere that industrial systems have put there. And with their spread myriad species presently threatened with extinction from global warming would not only survive but again multiply. And the oceans too would again begin to teem with abundant life as the trawlers and dragnets sank to the ocean floor, harbour walls collapsed and the overheated waters began to cool.[4] Depressing as such a thought experiment might seem, it is a potent reminder that what stands in the way of the healing and recovery of the biosphere is not mammalian life, nor subsistence farmers or peasant fisherfolk, but industrial civilisation and its associated ideologies, practices and values. Accounts of the global warming crisis that refuse the fundamental conflict between the imperial global economy and the health of the biosphere and its inhabitants cannot do justice to the real roots of the problem, and hence to its solution.

Without a mass die-off of human beings, can the earth be healed from anthropogenic climate change, or is it already too late? If it is, will future generations forgive present industrial consumers? We can imagine what they might say to us should they meet us in heaven: 'But you *knew* the science was coming right, you knew the predictions were real, and yet you carried on burning this stuff and wrecking the earth. Why? Why did you not change before it was too late?' One of the reasons we would need to give in response is that we were 'in denial' about our addictions.

Contesting the Evidence

There are two kinds of denial in relation to global warming. The first is that global warming science is not trustworthy. The second is that science can fix the problem without radically changing the industrial way of life. The first kind of denial has been promoted by the public relations departments of oil companies, who for more than thirty years have funded a host of denial groups which has included lobbyists, politicians, journalists and writers and even a few climate scientists. Even as the IPCC's *Fourth Assessment Report* was released in Paris in February 2007, the ExxonMobil-funded Fraser Institute was organising a breakfast for British parliamentarians to explain the 'other side' of the argument on global warming, as proposed by industry-funded scientists. According to the summary of the IPCC's report issued by the Fraser Institute at a day conference in London, 'there is no compelling evidence that dangerous or unprecedented changes are underway', and the changes recently observed in sea level and atmospheric temperatures are likely due to 'natural climatic variability'. Claims that fossil fuels are responsible are unreliable because the computer models on which such claims rely 'do not take into account the basic uncertainty about climate models or all potentially important influences'.[5] One of the scientists at the event was former university botanist David Bellamy, who still believes that recent changes in the climate are caused by natural variability, and that in any case many of the world's glaciers are 'not shrinking but growing'.[6] In America, Senator Inhofe, former chairman of the Senate Environment and Natural Resources Committee, declared on the floor of the Senate that the science of global warming still remained hypothetical:

> After more than a century of alternating between global cooling and warming, one would think that this media history would serve a cautionary tale for today's voices in the media and scientific community who are promoting yet another round of eco-doom.[7]

Inhofe insists that the media are not 'balanced' in their portrayal of a scientific consensus on global warming.

The claim that the science of global warming is unreliable has for many years been sustained by the public relations departments of the world's largest oil companies, who have clear interests in denying global warming. Their profit margins might be seriously affected if governments drastically reduce fossil-fuel dependence. And as the largest producers of fossil fuels, they might be held legally liable in an international court at some point in the future for compensating victims of climate change. The world's largest company, ExxonMobil, is the principal source of funding for the scientists and lobby groups who continue to feed global media with the claim that there are two sides to the global warming debate, and that a 'balanced' presentation of the evidence ought to represent the 'believers' and the nay-sayers. The company funds groups which suggest that climate change science is not 'sound science', and insist that considerable uncertainties remain in the outcomes predicted by climate scientists. Its own internal Global Climate Science team suggest that if the science can be shown not to support the case for reducing emissions then the moral case for the United States supporting the Kyoto Protocol will be undermined in the public's mind.[8]

A small number of working university scientists also continue to challenge the scientific consensus that global warming is being driven by industrial gases. Patrick Michaels at the University of Virginia argues that the measurements of land temperatures used by the IPCC are biased by growing urbanisation, and hence local heating, which, he claims, affects many weather stations. He also suggests that most of the detectable warming has in any case only occurred in cool latitudes in the Northern hemisphere where it can do no real damage.[9] In his claim about land temperature inaccuracies Michaels ignores the strong reliance of climatologists on satellite and other data which are not amenable to interference from local urban heating. Michaels also does not consider the phenomenon of global dimming, which indicates that particulates in the atmosphere from fossil-fuel burning, jet engines and other industrial activities played an important role in the mid-twentieth century in suppressing temperature rises by reducing the amount of sunlight which reaches the earth's surface.

Danish scientist Henrik Svensmark argues that global warming in the last century is driven primarily by increased solar activity.[10] Sunspots produce bursts of energy which drive away from the earth cosmic rays that bring atomic particles from exploded stars into the earth's atmosphere and interact with water vapour to form clouds. More clouds cool the earth's surface by reflecting back sunlight. Thus, when the sun is more active there are fewer clouds and the earth's climate warms. Svensmark suggests that sunspot activity declined in the 'little ice age', but has grown since as the world has warmed and that sunspots provide a better correlate to temperature changes than atmospheric carbon dioxide, an argument that seems to be given credence by paleoclimatological studies which correlate glacial advance and retreat with sunspot activity across millennia.[11] However, peer-reviewed studies indicate that Svensmark's claimed correlation between sunspot activity and warming in the twentieth century are not supported by the data, and that there has been no measurable increase in solar activity in the last forty years to account for the recent marked and rapid warming trend.[12]

As we have seen, the Holocene era was characterised by a unique relationship between the earth and the sun in which the earth was both warm enough and its climate stable enough for human civilisations to spread across the planet. The clear scientific consensus is that in recent decades human forcing of the climate is beginning to overtake solar forcing, particularly since 1970, when the global temperature has risen sharply, in line with sharp rises in CO_2 emissions. Andranova and Schlesinger helpfully describe the balance of the varying forces of sun, volcanoes, human activity and other factors including ocean circulation in a recent paper.[13] Their data clearly show that there is a correlation between anthropogenic forcing and recent warming trends, and much less significant correlations between sunspot activity and warming. Their data do, however, reveal a background natural variability, also recognised by the IPCC, which is not correlated to either sunspots or human emissions. It is likely that this indicates the influence of other climate drivers, and especially the oceans, which as we have seen represent a kind of long-term planetary memory of climate events. However, Andranova and Schlesinger conclude that although the sun and background variability have traditionally been larger drivers than human greenhouse gas emissions, as most recently as the cooler period of the 1940s to the 1970s, nonetheless the data clearly show that in the last forty years human emissions are now driving the climate by a factor of three times larger than the sun, volcanoes and natural variability put together.

Svensmark and Calder argue that the reason the IPCC and world governments have adopted the consensus position that greenhouse gas emissions are driving global warming is political rather than scientific.[14] Like Michael Crichton and Richard Lindzen,[15] Svensmark and Calder believe the hypothesis is adopted by people who want to sponsor a climate of fear and so enhance the regulatory powers of the state over corporations and individuals, while at the same time preventing Third World countries from enjoying the fossil-fuelled development which has lengthened the lives and improved the health of individuals in the West.

The challenges made by Svensmark, Calder and others to the widespread consensus of global warming are salutary if for no other reason than that they remind us that science has imposed consensus on societies in the past, with often troubling consequences. In the 1920s and 1930s eugenics was a major influence in the United States and Europe. Its advocates suggested that there was something unique and superior about Caucasian, and in Nazi Germany Arian, genes.[16] The scientific ideology of eugenics was linked with the Holocaust, with denials of civil rights to black people, and with the continuing colonial exploitation of Africa. Consensus, even when backed by science, can be a dangerous thing, especially when it seems to indicate that argument and dissent are no longer allowed. How then do we respond to the possibility, however remote, that climate change scientists might be wrong in their foundational claim that human greenhouse gas emissions are now driving the planet more than the sun or the ocean?

Many take comfort from the kinds of claims advanced by Svensmark, Michaels, Lindzen, Bellamy and others, who seek to debunk mainstream climate science. The dissenters argue that they represent a minority view which is being unjustly suppressed, though many journalists, and a recent Channel 4 documentary, are happy to give credence to their views. But the reason this small minority of scientists attract so much media attention is because their dissenting voices represent *mainstream* opinion about the relationship between the imperial global economy, fuelled by burnt terrestrial and subterranean forests, and human welfare. The deviant scientists are in this sense not so much dissenters as conformists, while it is the scientific consensus that is at odds with mainstream opinion. The global warming deniers suggest that it is possible for industrial humans to carry on burning fossil fuels and forests and wrecking ecosystems while not seriously endangering the future of the planet. They also suggest that industrial humans can continue to reap where they do not sow, and continue to threaten the extinction of large numbers of species.

Even if the deniers were right – which is impossible to credit on rational grounds – the core argument of this book is that the fossil-fuelled global economy is dangerous to planet earth and to human life, not just because it is warming the climate of the earth but because it is deeply destructive of the diversity and welfare of the ecosystems and human communities from which surplus value is extracted and traded across highways, oceans and jetstreams. The rituals encouraged by the recognition of global warming – turning off lights, turning down the heating, cycling or walking instead of driving, holidaying nearer to home, buying local food, shopping less and conversing more, addressing the causes of fuel poverty locally and internationally – are good because they are *intrinsically* right, not just because they have the consequence of reducing carbon emissions. Such actions correct modern thoughtlessness. They sustain the moral claim that it is wrong to live in a civilisation that depends upon the systematic enslavement of peoples and ecosystems to the high resource requirements of a corporately-governed consumer economy.

And there is a further powerful argument to put in the balance against climate change denial. The French philosopher Blaise Pascal developed a famous response to arguments for and against the existence of God. He suggested that if, against the claims of atheists, God does exist and atheists live morally heedless lives, then they risk judgment and even eternal damnation.[17] However, if they take a wager and live *as though* God exists, they have little to lose since the kind of life that Christians traditionally commend – a life which attends to the needs of others, which fosters companionship and civic responsibility and a concern for the common good – is in any case the kind of life that leads to flourishing according to the ancient philosophers. And so even atheists will do well to live as though God does exist. Stephen Heller proposes the same 'Pascalian wager' in relation to global warming: action to stem climate change would be prudent even if certain knowledge that it is happening, or about the severity of its effects, is not available or not believed.[18] If global warming is humanly caused, then these actions will turn out to have been essential for human survival and the health of the biosphere. In the unlikely event that it is not, then these good actions promote other goods – ecological responsibility, global justice, care for species – which are also morally right.

The modern quest for certainty, as we have already seen, involves an unreasonable standard of proof; it also represents, as Heller suggests, a confusion between faith and action. In real life people make decisions all the time based on partial evidence as to outcomes; if this were not so, how many would ever choose to get married or to have children. The roots of moral action do not lie in the certain ability to calculate consequences, but in a coherent relationship between the inner world of thought and emotion and the outer world of bodies, relationships and species. As we have seen, the strangely disembodied thought world of fossil-fuelled modernity is characterised by the disruption of the coherence of inner and outer worlds in the lives of modern humans, many of whom become addicted as a consequence to bad work or drugs or shopping or some combination of the three.[19] Actions which will have the effect of mitigating climate change are also actions which reaffirm the embodied relationship between inner desire and the outer world of what Christians call Creation. For this reason, such actions are intrinsically good, and will promote flourishing even if, as a minority of dissenters suggest, greenhouse gases are not the primary driver of global warming.

Technological Fixes

There is a second kind of denial associated with global warming, and this is not that the climate is not warming, nor even that it is caused by industrial gases, but that it is possible to fix the problem without redirecting the course of industrial society. Some adopt this approach because they do not think it wise to re-engineer the industrial economy, because so doing would disrupt the principal source of human welfare, which in their view is economic growth. Others put forward technological fixes because they do not believe their fellow citizens, or the world's corporations, are ready for the kind of radical changes that getting off their addiction to fossil fuels will require. Climate change scientist Paul Crutzen is of the latter school, and his ingenious technological fix involves spraying sulphates into the atmosphere using high altitude balloons and artillery guns. The sulphates would reflect back some of the sun's rays, thus cooling the planet and counteracting the enhanced warming produced by increased greenhouse gases.[20] Crutzen argues that this approach mimics nature, since volcanic ash and other particulates from volcanoes have a demonstrable cooling effect. Thus Mount Pinatubo cooled the planet by an estimated 0.5°C in 1992 following a large volcanic eruption in 1991. A related but more costly technological fix would have tens of thousands of rockets fired into space, putting hundreds of thousands of small mirrors into orbit around the earth which would deflect a percentage of the sun's rays from the earth's atmosphere equivalent to the warming effect of greenhouse gases.[21] Another and more long-standing proposal is to fertilise the oceans with iron filings to artificially stimulate algal and phytoplankton growth; this new growth would in turn sustain more shellfish and other ocean creatures and altogether the quantity of atmospheric carbon sunk in the ocean would rise.[22]

All of these proposals involve what Dale Jamieson calls 'intentional climate change'.[23] They are attempts to geo-engineer the atmosphere to try to wrest back control over the rising temperatures that human activity is bringing about. Some are already being experimented with on a small scale. But these experiments already indicate that they involve considerable and unquantifiable kinds of risks and uncertainties. And there is another problem with these proposals. The ingenuity and range of such proposals leads some scientists and policy makers to believe that it may be possible to carry on with business as usual in terms of rising energy emissions while re-engineering the climate to compensate.

Technical fixes at the production end of fossil-fuel emissions are also widely touted as solutions to global warming. Thus it is proposed that cars could run off hydrogen fuel cells and that the electricity needed to turn water into hydrogen could be generated from renewable wind, wave or geothermal sources. A small number of vehicles already use hydrogen fuel cells, though few of these run on renewable electricity.[24] Biofuels are represented as another device for keeping the present levels of mobility and speed that industrial humans enjoy while deriving the fuel from crops rather than from fossil fuels. But this proposal, as we have seen, ignores the threats to forests and soils represented by existing agro-industrial projects. If a significant proportion of prime agricultural land is turned over to biofuel production this will only increase pressure on other species, on marginal land, and on the farms and fields of poor and small farmers. Another proposal involves carbon sequestration, where power stations bury the carbon produced when fossil fuel is burnt in subterranean locations such as oil wells. This technology is already being deployed in Norway by oil companies which use carbon gas to pressurise evacuated oil wells and so aid in the removal of more oil. Critics of this approach point out that oil wells, once evacuated of their oil, are geologically unstable and prone to leakage.[25] It is then unlikely that all the carbon buried underground in such locations would actually stay there.

These proposed engineering solutions to the climate crisis seem to present industrial societies with ways of resolving global warming while avoiding radical changes to the industrial way of life. They thus collude with the denial of global warming by corporate lobbyists, dissenting scientists, journalists, and others. They would also allow the politics of speed to continue to subvert the politics of place, and the health of other species and ecosystems. They sustain the illusion that modern humans in their quest for technological efficiency do not have the time to tend to the objective and contingent needs of the planet and its creatures. And they maintain the illusion that humanity is in control of the planet and of its own history, and that there are no biological or climatological limits to the aspirations of industrial humans for speed and wealth.

In reality these technological solutions represent a dangerous extension of existing human attempts to 'manage' the planet's ecosystems, even though we know that such attempts, for example scientific forestry and fish farming, have reduced the biodiversity and even the efficiency of the ecosystems and species they are intended to scientifically improve.[26] As Lovelock suggests, if we tie future generations into having to manage the earth's climate through such gigantic geo-engineering projects, we commit them to immeasurable costs and a serious diminishment in their relationship to the biosphere.[27] Here again we encounter the ambiguous portent of the concept of planetary stewardship as management and control.

Refusing Control by Redeeming the Time

Anabaptist theologian John Howard Yoder suggests that the modern attempt to 'grasp control of history' reflects a hubristic intent to eschew the sovereignty of the divine Creator over the earth and human history.[28] It also neglects the contingent character of biological life and evolutionary history in which human life is situated. The extreme dependence of industrial humans on fossil fuels is intimately interconnected with forgetting that the roots of the present are in the past. It involves, as we have seen, releasing the fossilised memory of sunshine drawn down beneath the crust of the earth as modern humans attempt to eschew traditional ways of life and embrace the politics of speed. This eschewal arises from an ahistorical view of human life of the kind first advanced in Thomas Hobbes' account of the 'state of nature', from which modern science and the social contract of modern politics are said to deliver industrious humans. As Wolin points out, there is something peculiarly timeless about the view of the state of nature, according to which industrious material transformation of the fruits of the earth is the only hope of deliverance from the 'savage's' bondage to natural conditions and limits.[29] As a result, the future for Hobbes, as for Marx and Smith, only becomes imaginable as hopeful when it is conceived as being politically and socially constructed by human industriousness. In this way future redemption from the brutal vicissitudes of material want becomes the guiding political project of modernity. And the artifice of a humanly-constructed social contract becomes the source of sovereign power in a world without moorings in prehistory or tradition. This way of conceiving of the political future seems to require, as Hobbes suggests, that all humans, all creatures, invest their several powers and strengths 'upon one man, or upon one assembly of men, that may reduce all their wills, by plurality of voices, unto one will.'[30]

In contrast to this fateful tying of hope in the future with material power over the 'condition of nature', and a unitary monopoly of political sovereignty which Wolin insightfully calls 'political entropy', the early Christian account of politics and time involves a conception of progressive development which arises from the Christian drama of salvation. For the prophets of industrialism the defining moment of civilisation is industrious transformation of the fruits of the earth into private property, and the construction of a state which can defend the property of one wealthy individual from that of another. Time in modernity thus becomes a human project, and ordering time towards human welfare requires economic and political artifice. By contrast, in the Christian account of redemption the future is hopeful because of the Christ events in which bondage to sin and suffering is undone by the definitive redeeming action of God in time. In the Christian era time is no longer a political project as it had been for Plato, and as it has become again in post-Christian modernity. Instead Creation, Incarnation, Resurrection are the actions of the eternal, transforming the direction and future possibilities of human existence within time from beyond time.[31]

The implications of the Christian drama for the classical account of time are first fully elaborated by Augustine, who argues that the view of time as a progressive unfolding of the beneficent purposes of God to redeem creation from futility corrects the older view of time as an endless recurring cycle of hope and despair.[32] In consequence of this new view of time, individuals in a well-ordered society are *given* time to work for peace and the common good of all while focusing their hopes not on political artifice but on the love of God who is the source of all true peace and justice, and the ultimate redeemer of history. For the early Christians, history *after* Christ is an unfolding of the new time and the new history which he inaugurates; Christ is literally 'the alpha and the omega', the beginning and end of history.[33] But this does not mean that present efforts to enact the peace and justice of God in their lives in the present are futile. On the contrary, it suggests that the effort to live peaceably on earth, and to pass it on in a better condition to the next generation, is to live in accord with the true 'grain of the universe' as revealed in the definitive character and shape of the Christ events.[34]

In the last three chapters I suggested that the Christian practices of dwelling, pilgrimage and Eucharist represent just such a form of politics. In these practices Christians take time to order their lives around the worship of God because they believe that they have been given time by the re-ordering of creation which occurs when the Creator dwells inside time in the Incarnation and so redeems time and creation from futility, and from the curse of original sin. In the shape of this apocalyptic event, Christians understand that they have seen not only the future redemption of creatureliness, but the way, the 'shape of living', that they are called to pursue between the present and the future end of time.[35] In the way of the cross Christ shows himself vulnerable to the excessive claims to power and control exercised by the imperial power of Rome and its collaborators. But in refusing to resist this power with violence, while ultimately subverting it, Christ reveals the secret of history 'hidden from before the foundation of the world', which is that love overcomes violence, and strength is made perfect in weakness. For in the resurrection Christ triumphs over the fallen powers which had made the creature subject to the futility of sin.

Hence the witness of Christians to this victory does not take the form of a triumphalistic claim to rule over the cosmos in the place of the fallen powers. Nor does it take the form of an apocalyptic pronouncement that the world is coming to an end, and that there is no time left to take care to pursue justice in human affairs, or to preserve the earth. Of course there are those who turn the prophecy of global warming into an apocalyptic pronouncement of impending doom. Some of Lovelock's less helpful reflections on global warming have this character in his *The Revenge of Gaia*. So too do films like *The Day After Tomorrow*. Warning of impending apocalyptic crisis have also played a significant part in the recent escalation of global wars and terrorism in the Middle East and the Caucasus, which, as I have suggested elsewhere, are in reality conflicts over the diminishing oil reserves of the world as production from conventional oil wells peaks and the continued fuelling of industrial mobility looks in doubt.[36] Such appeals to the apocalyptic do not, however, promote the kinds of peaceable relations between humans, and between humans and the earth, that are truly desirable if humanity is to resile from a long-term future of climate instability and ecological collapse. Instead they promote the politics of fear. And while fear of judgment, even of ecological judgment, may not be unwise, fear alone is not a beneficial political emotion. On the contrary, as Augustine suggests, fear of punishment is a much less successful motivator of change in human life than compassion and mercy.[37]

For more than sixty years the United States, often assisted by its NATO allies, and especially the United Kingdom, has devoted extensive resources to the protection of the National Security State as first promoted by the apocalyptic anti-communist rhetoric of President Truman. Fear of communism, and now of Islamist terrorism, has been utilised as the covert political tool under which the United States maintains more than 800 military bases around the world in the creation of the largest military and economic empire in human history. At the same time it has used these putative external threats to increase its surveillance of and powers over its own citizens, and to prosecute dozens of armed and covert interventions in other nations. As I hope this book has made clear, the true driver of the imperial economy in which the United States and its corporations remain the most powerful players is oil. The apocalyptic politics of fear has long been used to sustain the kinds of unjust bargains by which oil has been delivered to sustain this economy, and political power has been filleted away both from those whose lands had the oil, and those whose economies it sustained.

It would then be a tragic irony if politicians or scientists were to resort to the politics of fear to promote the need to tackle global warming. And yet some, as we have seen, have already resorted to this tactic, including Al Gore, Sir David King, and a number of other prominent voices in the climate change debate. Global warming is in one sense, of course, a greater threat to human security than terrorism. But the politics of fear has no more resolved the causes of terrorism than it will resolve the causes of global warming. In a material and geographical sense, the causes of terrorism and global warming have the same origin, and this is the death-dealing imperial power play which has seen fossil-fuel rich lands plundered in the 'great game' which has moved across the region between the Persian Gulf and Caucasus ever since oil was first discovered in the late nineteenth century. And just as terrorism is only exacerbated by the apocalyptic wars with which the twenty-first century opened, so global warming will remain unresolved so long as the Western powers remain devoted to their apocalyptic theology of individual liberty combined with economic and military coercion.[38]

The new aeon in which Christians are called to live is not so new that they can forget the past. On the contrary, it involves a kind of politics which in community and ritual and story celebrates the roots of the present in the past. This politics contrasts with the revolutionary politics of industrialism and economic progress of the fossil-fuelled era. It involves the claim that the future of the earth does not depend on giant planetary engineering projects, nor on new forms of coercive power and surveillance in which the nation-state uses the apocalyptic fear of global warming to coerce citizens into compliance with its energy-saving targets. Instead, this approach suggests a recovery of radical democracy in which individuals and local communities embrace the kinds of acknowledged dependence on the fruits of God's good earth, and on one another, which make careful use of energy and good work possible. Christians can refuse to be hurried by the politics of speed, because apocalyptic time trains them to understand that the end of history and the hope of salvation from the vicissitudes of mortal life do not depend on human effort alone, but spring instead from the grace of God.[39]

Hopeful Witness to Climate Justice

On this account then Christians, and all who care for the planet, have time to remake their lives in ways which respond hopefully to the climate crisis. And every attempt to use biopower in the form of energy more carefully in the home and the workplace, in mobility and in the food chain, has significance because it expresses solidarity with the victims of climate change and with the earth itself. The careful use of power in the home, and the redesign of dwelling and work spaces to minimise the use of power, will require billions of individual actions on a daily basis to switch off lights and machines and turn down thermostats, and so desist from the wasteful use of power. It will also require the careful redesign of tens of millions of buildings. Such is the cultural power of the promethean politics of entropy that governments refuse to invest time and energy in this kind of bottom-up energy conservation, preferring instead to bolt on new forms of lower carbon power to an inefficient grid and leaky buildings, or to attempt to sequester carbon under the earth's surface. Like the 'optics of power' of economistic development, this approach refuses the political significance of attention to myriad details by individuals and local communities. This refusal is in part responsible for the 'end of politics' and the sense of fateful fear that pervades the present political climate, where politicians and others can find no other motivator for collective action than fear of an apocalyptic climate breakdown, analogous to fear of nuclear war or terrorism.[40]

The Christian vision of politics as radically participative and grounded in an ethic of care and compassion suggests that it is unwise to construct a centralised power delivery system without attending to the uses of power that this system makes possible. Such a system is almost bound to institutionalise mass carelessness in the use of power, while at the same time fostering political apathy. The kinds of politics which are required to redress this carelessness in the uses of power will of necessity be piecemeal because they will need to engage every citizen, household, corporation, and organisation. Modern citizens are trained by industrial capitalism to imagine that only the autonomous and anonymous contractual relationships of a global market can make for freedom from coercive power.[41]

The very success of capitalism in translating all forms of political sovereignty into the sovereignty of the consumer and the corporation makes it hard to recollect the origins of democratic politics in those local communities of place whence the constitutive practices of democracy – the open public meeting, face-to-face debate, respect for every member – first emerged in Christian history.[42] Analogously, the recovery of participatory politics requires careful nurturing of face-to-face communities which are located in place, and which can resituate the human economy in the economy of the earth.[43] Christian congregations are just these kinds of community, and it should therefore be no surprise that through the eco-Congregation movement in Britain, and equivalent initiatives in other countries, many local churches are making strenuous efforts to address the climate impacts of their activities and of the lifestyles of their members.[44]

The role of Christian community in reforming global capitalism already has an exemplar in the origins of the modern fair trade movement. Fair trade represents an effort to recover the kinds of moral and relational constraints on economic behaviour once fostered in Christendom through such practices as just wage and just price, and the ban on usury.[45] There are also significant analogies between fair trade and the efforts of Christians in the nineteenth century to reform *laissez faire* capitalism. Through the witness of Christians such as William Wilberforce, laws were passed which reined in the worst excesses of *laissez faire* capitalism, gradually outlawing not only slavery but child labour, excessive hours of work, poverty wages and the poisoning of water and air by early industrialists.[46]

Multinational corporations now use global mobility and the politics of speed, to export their activities to regions of the world, such as China and India, which have fewer social and environmental controls, in order to avoid just the kinds of moral controls over industrial work that Christians campaigned for in the Victorian era. Child and adult slavery have in the process been reborn as factory owners lock workers in their factories and force them to sleep in dormitories in beds which they share on a shift basis with two other workers. Agricultural production has also become subject to the same kind of coercive work practices, as the industrialisation of food has destroyed the viability of small farms in both North and South.

The first and largest fair trade organisation in Britain, Traidcraft, began as a small cooperative in a Christian community, St John's College, Durham.[47] The principles behind fair trade are designed to right the wrongs of the global economy by reconnecting producers and consumers in a moral, instead of mechanistic, relationship of exchange which is long-term and in which producers are given a fair price for their products.[48] The movement does not eschew all international trade, and it may be that in the future the fair trade brand will need to take on carbon as another aspect of its product certification. Thus air-flown goods might be banned under such a scheme, while those which are brought by ship might still have low enough carbon inputs to make them sustainable.

Fair trade provides an important model for the global problem represented by climate change. Although it originated from the Christian commitment to addressing poverty through trade justice, the mark now is handled by a non-governmental organisation, the Fair Trade Foundation, which works to establish and guarantee the fairness of product lines ranging from bananas and coffee to wine and waistcoats. There is clearly a need for an analogous movement to establish a climate care kite mark in relation to the greenhouse gas content of all goods in the shops, of buildings and electronic and mobility devices, and of holidays and travel tickets.[49]

It may be said that fair trade or climate care kite marks are a poor sort of politics compared to the size of the problem. Fair trade still makes up less than 1 per cent of traded goods in Britain, though the proportion of fairly-traded goods has more than trebled in the last ten years, and the number of fairly-traded products has increased from 250 to 2,500 in the same period. But this is to miss the point that relationships between consumers and producers are as political as government regulations for the design of buildings, the taxation of cars and the regulation of international trade. Only through careful attendance to the moral character expressed in the myriad daily actions through which citizens sustain their lives will it be possible to truly assuage the guilt of ecological denial, and so refuse to continue to suppress moral conscience.

But this is not to suggest that climate care ends in individual actions, or even in the actions of climate-conscious congregations, consumers and corporations. If democratic governance emerged in post-Reformation Europe out of the practices of face-to-face public gatherings in Reformation churches, then Christians clearly should expect that democratic political institutions regionally, nationally and internationally should address moral concerns such as those raised by the injustices of global warming, and use the power of the law to change corporate and individual behaviours.[50]

The sad fact is that no elected politician or government official has yet shown the kind of leadership necessary to embrace the moral challenge of climate change by suggesting that the corporately-driven growth economy will have to be legally regulated and greenhouse gas emissions legally restrained if the planet is not to be wrecked. In Britain and the United States present political action on climate change relies almost exclusively on voluntary actions and agreements by corporations and individual citizens. Internationally politicians and officials have used the UNFCC to construct a system of carbon trading which draws the climate problem into the same autonomous market mechanisms, and optics of power, which are driving the global economy on its collision course with the biological health of the planet. As we have seen, legal environmental regulation has worked nationally and internationally in addressing ozone depletion, air and river pollution and threats to rare species. But corporations in the global market empire have sought ways to avoid these regulations by relocating polluting activities to low-regulation countries. There is, however, no place to hide excess greenhouse gas emissions. As the earth's carbon sinks run out of room they will end up in the atmosphere. The only solution is a global and statutory framework, under the auspices of the UNFCCC, which legally prescribes mandated emissions reductions by nation-states and by multi-state actors such as multinational corporations and international financial institutions. All the more reason, then, that concerned citizens should pioneer such reductions by embracing the kinds of practices of climate care reviewed in the last chapters. As with fair trade, and the Jubilee campaign, such witness has moral power because it is *right*, not because it is influential. And as Christians and people of goodwill have embraced change in their own lives, and engaged politicians and bankers and businesses in these moral challenges, both these movements have begun a larger change in social consciousness of the effects of global capitalism and colonialism. Climate witness has the potential to do the same.

Witness to the truth that humans live on a contingently limited planet and that industrialism is bursting the bounds of these limits may not look effective in turning the world around. It might seem that the kinds of communal, household and personal actions involved in attending to the carbon footprints of our homes, workplaces, churches and recreational spaces are infinitesimally small compared to the size of the problem. But unless those who know the truth engage in such actions, they commit themselves to the kinds of collective denial and forgetting of the truth that are the source of so much depression, hopelessness and political passivity, in late capitalist societies.

Acting rightly with respect to the earth is a source of hope, for those who so act give expression to the Christian belief that it is God's intention to redeem the earth, and her oppressed creatures, from sinful subjection to the oppressive domination of prideful wealth and imperial power. Such actions witness to the truth that the history of global warming has gradually unfolded; that those poor or voiceless human and nonhuman beings whose prospects climate change is threatening are *neighbours* through the climate system to the powerful and wealthy. And Christ's command in these circumstances is as relevant as ever: 'love your neighbour as yourself'.

Notes

Foreword

1. C. S. Lewis, *First and Second Things* (Collins, Fount paperbacks, 1985).
2. Matthew 6:33.

Introduction

1. Richard Flanagan, 'Out of control: the tradegy of Tasmania's forests', *The Monthly*, 23 (May 2007), 3–16. See also Karen McGhee, 'Logging Van Diemen's Land: Is there a green light for more sustainable forestry in Tasmania?', *Ecos* 122 (November–December 2004), 26–7, and S. M. Davey, M. Stuart, J. R. L. Hoare *et al.*, 'Science and its role in Australian Regional Forest Agreements', *International Forestry Review* 4 (2002), 39–55.
2. Robyn Hollander, 'Elections, Policy and the Media: Tasmania's Forests and the 2004 Federal Election', *Australian Journal of Political Science* 4 (December 2006), 569–84.
3. Flanagan, 'Out of control'.
4. The Wilderness Society of Australia has used Google Earth and Gunns' published logging plans to show the extent of clear felling thus far, and the planned extent over the next ten years at: http://www.wilderness.org.au/.
5. International Energy Agency, *Earth Trends Data Tables: Energy and Resources: Energy 2005*: http://earthtrends.wri.org/pdf_library/data_tables/ene1_2005.pdf.
6. The latest 'sustainable forestry' initiative reveals an extensive array of government subsidies to logging and wood chipping activities, and to the creation of new eucalyptus plantations on old growth forest land: Australian Government, Department of Agriculture, Fisheries and Forestry, *A Way Forward for Tasmania's Forests: The Tasmanian Community Forest Agreement*, 13 May 2007 at: http://www.daffa.gov.au/forestry/national/cfa/info/a_way_forward_for_tasmanias_forest.
7. Damian J. Barrett, Ian E. Galbally and R. Dean Graetz, 'Quantifying uncertainty in estimates of C emissions from above-ground biomass due to historic land-use change to cropping in Australia', *Global Change Biology* 7 (2001), 883–902.
8. Ying Ping Wangi and Damian J. Barrett, 'Estimating regional terrestrial carbon fluxes for the Australian continent using a multiple-constraint approach: Using remotely sensed data and ecological observations of net primary production', *Tellus* 55B (2003), 270–89.
9. For a fuller account of climate change in Australia see Tim Flannery, *The Weathermakers: The History and Future Impact of Climate Change* (London: Allen Lane, 2005).
10. Walter Brüggemann, *The Theology of the Book of Jeremiah* (Cambridge: Cambridge University Press, 2007), 5.
11. Brüggemann, *Book of Jeremiah*.
12. Brüggemann, *Jeremiah*, 7, n. 10.
13. Paul M. Kennedy, *The Rise and Fall of the Great Powers: Economic Change and*

Military Conflict from 1500–2000 (New York: Random House, 1987).

14. The idea of the global market as empire is elaborated in Michael Hardt and Antonio Negri, *Empire* (Cambridge, MA: Harvard University Press, 2000).
15. Dietrich Bonhoeffer, *Creation and Fall: A Theological Interpretation of Genesis 1—3*, ed. John W. De Gruchy, trs. Douglas Stephen Bax (London: SCM Press, 1959), 38.
16. Karl Polanyi, *The Great Transformation: The Political and Economic Origins of Our Time* (London: Victor Gollancz, 1945).
17. Christian Aid, *Human Tide: The Real Migration Crisis* (London: Christian Aid, May 2007), 5.
18. For a critique of this phrase and the larger imperial strategy behind it see further Michael Northcott, *An Angel Directs the Storm: Apocalyptic Religion and American Empire* (London: I. B. Tauris, 2004).
19. David Aberbach, 'Revolutionary Hebrew, Empire and Crisis: Towards a Sociological Gestalt', *The British Journal of Sociology* 48 (March 1997), 128–48.
20. The phrase 'fertile crescent' was first coined by James H. Breasted in his *The Conquest of Civilization* (New York: Harper and Brothers, 1926).
21. Breasted, *Conquest of Civilization.*
22. Brian Fagan, *The Long Summer: How Climate Changed Civilization* (New York: Basic Books, 2004).
23. H. Weiss, M.–A. Courty, W. Wetterstrom *et al.*, 'The genesis and collapse of Third Millennium North Mesopotamian civilization', *Science* 261 (August 1993), 995–1004.
24. Arie Nissenbaum, 'Sodom, Gomorrah and the other lost cities of the plain – a climatic perspective', *Climate Change* 26 (1994), 435– 46; see also T. Jacobsen, and R. M. Adams, 'Salt and silt in ancient Mesopotamian agriculture', *Science* 128 (21 November 1958), 1251–8.
25. Walter A. Brüggemann, *A Commentary on Jeremiah: Exile and Homecoming* (Grand Rapids, MI: Wm. B. Eerdmans, 1998).
26. Hannah Arendt, *The Human Condition* (Chicago: Chicago University Press, 1958), 107.
27. Abraham Heschel, *The Sabbath: Its Meaning for Modern Man* (New York: Farrer, Strauss and Giroux, 1951), 8–9.
28. See further Michael S. Northcott, *The Environment and Christian Ethics* (Cambridge: Cambridge University Press, 1996), 187–93, and also Norman Wirzba, *The Paradise of God; Renewing Religion in an Ecological Age* (Oxford: Oxford University Press, 2003), 34–41.
29. See further Northcott, *Environment and Christian Ethics*, 166–82.
30. Rémi Brague, *The Wisdom of the World: The Human Experience of the Universe in Western Thought*, trs. Teresa Fagan (Chicago: University of Chicago Press, 2003).
31. Oliver Goldsmith, *The Deserted Village* (London: W. Bulmer and Co., 1789).
32. See the article 'Consumer' in Raymond Williams, *Keywords: A Vocabulary of Culture and Society* (London: Croom and Helm, 1976). See also Kenneth R. Himes, 'Consumerism and Christian ethics', *Theological Studies* 68 (March 2007), 132–54.
33. Paul Hawken, Amory B. Lovins and L. Hunter Lovins, *Natural Capitalism: The Next Industrial Revolution* (London: Earthscan, 1999).
34. Robert Putnam, *Bowling Alone: The Collapse and Revival of American Community* (New York: Simon and Schuster, 2000); Wendell Berry, *The Unsettling of America: Culture and Agriculture* (San Francisco: Sierra Club Books, 1977).

1. Message from the Planet

1. Michelle Hunter, 'Deaths of evacuees push toll to 1577', *Picayne Times*, 19 May 2006.
2. Matthew Tiessen, 'Speed, Desire and Inaction in New Orleans: Like a Stick in the Spokes', *Space and Culture* 9 (February 2006), 35–7.
3. Holman W. Jenkins, *Wall Street Journal*, 1 February 2007.
4. Kerry Emmanuel, 'Increasing destructiveness of tropical cyclones over the past 30 years', *Nature* 436 (4 August 2005), 686–8.
5. Mary E. Davis, Lonnie G. Thompson, Tandong Yao, Ninglian Wang, 'Forcing of the Asian monsoon in the Tibetan Plateau: evidence from high-resolution ice core and tropical coral records', *Journal of Geophysical Science Research* 110 (2005), 1–13.
6. *The Climate of Poverty: Facts, Fears and Hope: A Christian Aid Report* (London: Christian Aid, 2006), 33.
7. Charles Sakar of the Christian Commission for Development in Bangladesh in *Climate of Poverty*, 34.
8. IPCC, *Climate Change 2007: Impacts, Adaptation and Vulnerability* Working Group II, pre-publication draft.
9. Christian Aid, *Facing the Inevitable: A Christian AID UNFCC Cop 12/Mop 2 Briefing* (London: Christian Aid, 2006).
10. Kevin Trenberth, 'Uncertainty in Hurricanes and Global Warming', *Science* 308 (17 June 2005), 1753–4.
11. Sydney Levitus, John I. Antonov, Julian Wang *et al.*, 'Anthropogenic warming of earth's climate system', *Science* 292 (13 April 2001), 267–72; also IPCC, *Climate Change 2007*, Working Group I, Chapter 5: 'Observations: Oceanic Climate Change and Sea Level' (draft).
12. Richard A. Anthes and Robert W. Correll *et al.*, 'Hurricanes and global warming – potential linking consequences', *Bulletin of the American Meteorological Society* 87 (May 2006), 623–8.
13. 'Global Temperature Trends: 2005 Summation', Goddard Institute for Space Studies, New York at http://data.giss.nasa.gov/gistemp/2005/.
14. 'Erratic rains, civil strife and desert locusts seriously threaten food security in sub-Saharan Africa', Food and Agriculture Organization of the United Nations (6 July 2004): (http://www.fao.org/newsroom/en/news/2004/47887/ index.html).
15. James Hansen, Makiko Sato, Reto Reudy *et al.*, 'Global temperature change', *Proceedings of the National Academy of Sciences* 103 (26 September 2006), 14288–93.
16. The double-glazing metaphor graces the cover of the first CUP edition of John Houghton's valuable *Global Warming: The Complete Briefing* (Cambridge: Cambridge University Press, 1997).
17. IPCC, *Climate Change 2007*, Working Group III, 2–6 (pre-publication draft).
18. Notable contrarian texts include Pat Michaels and Robert C. Bailing, *The Satanic Gases: Clearing the Air About Global Warming* (Washington DC: Cato Institute, 2000); Denis T. Avery and S. Fred Singer, *Unstoppable Global Warming: Every 1500 Years* (New York: Rowman and Littlefield, 2007); and Henrik Svensmark and Nigel Calder, *The Chilling Stars: A New Theory of Climate Change* (London: Icon, 2007). For a debunking of Svensmark's sunspot theory see Paul Damon and Peter Laut, 'Pattern of strange errors plagues solar activity and terrestrial climate data', *Eos, Transactions of the American Geophysical Union* 85 (2004), 370.

19. Richard Lindzen, 'Climate of fear: global-warming alarmists intimidate dissenting scientists into silence', *Wall Street Journal,* 12 April 2006.
20. Thomas Gale Moore, 'Why global warming would be good for you', *Hoover Institution Working Paper Series* (Washington DC: Hoover Institution, 1995).
21. Theodore L. Anderson, Robert J. Carlson, Stephen E. Scwartz *et al.*, 'Climate forcing by aerosols – a hazy picture', *Science,* 300 (16 May 2003), 1103–4.
22. Paul J. Crutzen and Eugence F. Stoermer, 'The Anthropocene', *IGBP Newsletter* 36 (2000) at: http://www.mpch-mainz.mpg.de/~air/anthropocene/.
23. IPCC, *Climate Change 2007,* Working Group I: 'The Physical Science Base; Summary for Policymakers' (pre-publication draft, 2006), 10.
24. David King, 'Climate change: the science and the policy', *Journal of Applied Ecology* 42 (2005), 779–83.
25. Svente Arrhenius, 'On the influence of carbonic acid in the air on the temperature on the ground', *Philosophical Magazine* 41 (1896), 237.
26. Spencer R. Weart, 'The discovery of the risk of global warming', *Physics Today* 50 (1997), 34–40.
27. Spencer R. Weart, *The Discovery of Global Warming* (Cambridge, MA: Harvard University Press, 2003).
28. Weart, 'The discovery of the risk of global warming', 34.
29. Tim Flannery, *The Weathermakers: The History and Future Impact of Climate Change* (London: Allen Lane, 2005), 57–61.
30. Personal communication, William H. Schlesinger.
31. Weart, 'The discovery of the risk of global warming', 39.
32. M. E. Mann, R. S. Bradley and M. K. Hughes, 'Northern Hemisphere temperatures during the past millennium: Inferences, Uncertainties, and Limitations', *Geophysical Research Letters* 26 (1999), 759–62.
33. On claimed anomalies in the hockey-stick graph see Stephen McIntyre and Ross McKitrick, 'Hockey Sticks, Principal Components and Spurious Significance', *Geophysical Research Letters* 32 (February, 2005).
34. IPCC, *Climate Change 2007*, Working Group I: 'The Physical Science Basis: Technical Summary', draft, 6.
35. See ch. 2 below for more detail on this.
36. IPCC, *Climate Change 2007*, Working Group III: 'Summary for Policymakers', 2.
37. Thomas Gale Moore, *Climate of Fear: Why We Shouldn't Worry About Global Warming* (Washington DC: Cato Institute, 1998), 22.
38. Thomas R. Karl and Kevin E. Trenberth, 'Modern global climate change', *Science* 302 (5 December 2003), 1719–23.
39. Prabir K. Patra, Shamil Maksutyov and Takikyo Nakazawa, 'Analysis of CO_2 growth rates at Mauna Loa using CO_2 fluxes derived from an inverse model', *Tellus B* 57 (November 2005), 357–65.
40. J. B. Miller, P. P. Tans, J. W. C. White, 'Global air sampling network reveals decreasing NH (northern hemisphere) terrestrial carbon uptake', *Geophysical Research Abstracts* 8 (2006), 10858.
41. Lesley Hughes, 'Biological consequences of global warming: is the signal already apparent', *Tree* 15. 2 (February, 2000), 56–62.
42. Camille Parmesan and Gary Yohe, 'A globally coherent fingerprint of climate change across natural systems', *Nature* (2 January 2003), 37–42.
43. Jeroen van der Sluijs, Josee van Eijndhoven, Simon Shackley and Brian Wynne, 'Anchoring devices in science policy: the case of consensus around climate sensitivity', *Social Studies of Science* 28/2 (April 1998) 291–323.

44. Stephen Bocking, *Nature's Experts: Science, Politics and the Environment* (NJ: Rutgers University Press, 2004), 112–14.
45. R. B. Alley, J. Marotzke, W. D. Nordhaus, J. T. Overpeck *et al.*, 'Abrupt climate change', *Science* 299 (28 March 2003), 2005–10.
46. Michael Crichton, *State of Fear* (London: HarperCollins, 2005). See also Thomas R. Karl and Philip D. Jones, 'Urban bias in area-averaged surface air temperature trends', *Bulletin of the American Meteorological Association* 70 (March 1989), 265–70.
47. IPCC, *Climate Change 2007*, Working Group I: 'Technical Summary', 18.
48. David Demeritt, 'The construction of global warming and the politics of science', *Annals of the Association of American Geographers* 91 (2001), 307.
49. IPCC, *Climate Change 2007*. Also John Houghton, *Global Warming: The Complete Briefing*, 3rd edn (Cambridge: Cambridge University Press, 2004), 73.
50. Robert T. Watson (ed.), *Climate Change 2001: IPCC Synthesis Report* (Cambridge: Cambridge University Press, 2001), 82.
51. IPCC, *Climate Change 2007,* 'Technical Summary'.
52. D. A. Stainforth, T. Aina *et al.*, 'Uncertainty in predictions of the climate response to rising levels of greenhouse gases', *Nature* 405 (January 2005), 403–6.
53. A. Barrie Pittock, *Climate Change: Turning up the Heat* (London: Earthscan, 2005), 102–4.
54. Nicolas Gruber, Pierre Friedlingstein, Christopher B. Field *et al.*, 'The Vulnerability of the Carbon Cycle in the 21st Century: An Assessment of Carbon-Climate-Human Interactions' in Christopher B. Field and Michael R. Raupach (eds), *The Global Carbon Cycle: integrating humans, climate and the natural world* (Washington DC: Island Press, 2004), ch. 5.
55. Chris Jones, Claire McConnell, Kevin Coleman, Peter Cox, Peter Falloon, David Jenkinson and David Powlson, 'Global climate change and soil carbon stocks; predictions from two contrasting models for the turnover of organic carbon in soil', *Global Change Biology* 11 (2005), 154–66.
56. IPCC, *Climate Change 2001*, 9–11.
57. John Houghton, 'Global warming', *Reports on Progress in Physics* 68 (2005), 1343–1403.
58. IPCC, *Climate Change 2007,* 'Technical Summary', 27.
59. James E. Hansen, 'Can we still avoid dangerous human-made climate change?', *Social Research: An International Quarterly of Social Sciences* 73 (Fall 2006), 949–74.
60. IPCC, *Climate Change 2007*, Working Group I, Chapter 6: 'Paleoclimate', 9–10 (draft).
61. Richard B. Alley, Peter U. Clark, Phillipe Huybrechts and Ian Joughlin, 'Ice-sheet and sea-level changes', *Science* 310 (21 October 2005), 456–60.
62. Robert G. Bingham, Peter W. Nienow, Martin J. Sharp and Luke Copland, 'Hydrology and dynamics of a polythermal (mostly cold) High Arctic glacier', *Earth Surface Processes and Landforms* 31 (2006), 1463–1479.
63. Eric Rignot and Pannir Kanagaratnam, 'Changes in the velocity structure of the Greenland ice sheet', *Science* 311 (17 February 2006), 986–990.
64. E. Rignot, G. Casassa, P. Gogineni, W. Krabill, A. Rivera, and R. Thomas, 'Accelerated ice discharge from the Antarctic Peninsula following the collapse of Larsen B ice shelf', *Geophysical Research Letters* 31 (22 September, 2004), L18401.
65. Paul Christoffersen and Michael J. Hambrey, 'Is the Greenland ice-sheet in a state of collapse?', *Geology Today* 22 (May–June 2006), 98–104.

66. Houghton, 'Global warming' (2005), 1343–1403.
67. IPCC, *Synthesis Report*; see also Jeremy Leggett, *The Carbon War: Global Warming and the End of the Oil Era* (New York: Routledge, 2001), 14–19.
68. Andrew Simms, 'Tuvalu and the Fate of Nations' in Simms, *Ecological Debt: The Health of the Planet and the Welfare of Nations* (London: Pluto Press, 2005), 30–49 and Mark Lynas, 'Pacific Paradise Lost' in *High Tide: How Climate Crisis is Engulfing Our Planet* (London: HarperPerennial, 2005), 81–124.
69. C. Small and R. J. Nichols, 'A global analysis of human settlement in coastal zones', *Journal of Coastal Research* 19 (2003), 584–99.
70. Nigel Arnell and Declan Conway, 'Global climate change and water resources: to 2025 and beyond', Lecture at Overseas Development Institute, 2 February 2000: http://www.odi.org.uk/speeches/water2.html, 07 February 2005.
71. IPCC,, *Climate Change 2001,* Working Group II: Impacts, Adaptation, and Vulnerability. Chapter 7: 'Human Settlements, Energy, and Industry': http://www.grida.no/climate/ipcc_tar/wg2/310.htm, 7.2.2.2
72. Zambia experienced a decline of 20 per cent of its crop output in just one year between 2004 and 2005 because of drought: presentation by Mubanga Kasakula at the GW8 conference, 'Dynamic Earth', Edinburgh, 5 July 2005. The Mediterranean countries of Spain and Portugal applied to the European Union for food aid because of severe drought in 2005.
73. David S. G. Thomas, Melanie Knight and Giles F. S. Wiggs, 'Remobilization of southern African desert dune systems by twenty-first century global warming', *Science* 435 (30 June 2005), 1218–21.
74. J. Bader and M. Latif, 'The impact of decadal-scale Indian Ocean sea surface temperature anomalies on Sahelian rainfall and the North Atlantic Oscillation', *Geophysical Research Letters* 30 (November 2003), 7.1–7.4. and Ning Zeng, 'Drought in the Sahel', *Science* 302 (7 November 2003), 999–1000. For a contrary view see S. D. Prince, E. Brown De Coulstoun and L. L. Kravitz, 'Evidence from rain-use efficiencies does not indicate extensive Sahelian desertification', *Global Change Biology* 4 (1998), 359–74.
75. *Africa – Up in Smoke? The second report from the Working Group on Climate Change and Development* (London: New Economics Foundation and International Institute for Environment and Development, 2005).
76. Jared Diamond, *Collapse: How Societies Choose to Fail or Survive* (London: Allen Lane, 2005), 157–77.
77. Albert Howard, *An Agricultural Testament* (Oxford: Oxford University Press, 1940), 8. See also Donald J. Hughes, *Ecology in Ancient Civilizations* (Albuquerque: University of New Mexico Press, 1975) and Joseph A. Tainter, *The Collapse of Complex Societies* (Cambridge: Cambridge University Press, 1988).
78. On the moral significance of embodied engagement with the world see further Albert Borgmann, *Technology and the Character of Contemporary Life* (Chicago: Chicago University Press, 1984).
79. Wendell Berry, *The Unsettling of America: Culture and Agriculture* (San Francisco, CA: Sierra Club Books, 1977), 81–4.
80. Wendell Berry, *Standing By Words* (Washington D. C.: Shoemaker and Hoard, 1983), 13.
81. Alan Durning, *How Much Is Enough? The Consumer Society and the Future of the Earth* (Washington DC: Worldwatch Institute, 1992), 34.
82. *Coming Clean: Revealing the UK's True Carbon Footprint* (London: Christian Aid, 2007) at: http://www.christian-aid.org.uk/indepth/0702_climate/missing carbon.pdf, 4–5 and *The Carbon 100: Quantifying the carbon emissions, intensities and*

exposures of the FTSE 100 (London: Henderson Global Investors and Trucost, 2006) at: http://www.trucost.com/Trucost_The_Carbon_100.pdf.

83. See further Michael Northcott, *Life After Debt: Christianity and Global Justice* (London: SPCK, 1999) and Peter Selby, *Grace and Mortgage: The Language of Faith and the Debt of the World* (London: Darton, Longman and Todd, 1997).
84. Dick Bryan, 'Financial derivatives: the new gold?', *Competition and Change* 10 (September 2006), 265–82.
85. There is an interesting correspondence between rises in CO_2 emissions in the last forty years and estimates of the rise in the size of M1, M2 and M3 in the United States. M1, 2 and 3 measure all forms of money including bank deposits and credits, paper money, coinage, stocks and shares.
86. For a fuller account of economic globalisation see my *Life After Debt.*
87. On the utopian dimensions of neoliberalism see John Gray, *Al Qaeda and What It Means to Be Modern* (London: Faber, 2003).
88. Richard Layard, *Happiness: Lessons from a New Science* (London: Penguin, 2005).
89. UNICEF, *Child Poverty in Perspective: An Overview of Child Well-Being in Rich Countries* (Florence: UNICEF, 2007).
90. R. C. N. Wit, B. E. Kampman and B. H. Boon, *Climate Impacts from International Aviation and Shipping* (Delft: CE, 2004) at http://www.ce.nl/pdf/04_4772_44.pdf.
91. Sheldon Wolin, *Politics and Vision: Continuity and Innovation in Western Political Thought*, expanded edn (Princeton, NJ: Princeton University Press, 2004), 591–4.
92. Wendell Berry, 'Thoughts in the presence of fear' in *Citizenship Papers* (Washington DC: Shoemaker and Hoard, 2004), 17–22.
93. Berry, 'Thoughts in the presence of fear', 20.
94. Berry, 'Thoughts in the presence of fear', 19.
95. Berry, 'Thoughts in the presence of fear', 22.
96. Ulrich Beck, *Risk Society: Towards a New Modernity* (London: Sage, 1992). See also Keith Spence, 'World risk society and war against terror', *Political Studies* 53 (2005), 284–302.
97. Zygmunt Bauman, *Liquid Fear* (Cambridge: Polity Press, 2006), 96.
98. Robert O'Keohane, 'The globalization of informal violence, theories of world politics, and the liberalism of fear', *Dialog-IO* (Spring 2002), 29–43.
99. Immanuel Kant, *Groundwork of the Metaphysics of Morals*, trs. Mary Gregor (Cambridge: Cambridge University Press, 1998).
100. Michele M. Moody-Adams, 'Culture, responsibility and affected ignorance', *Ethics* 104 (January 1994), 291–309; also Thomas Aquinas, *Summa Theologiae,* I, 2.6, 8.
101. Michel Foucault, *Fearless Speech* (Paris: Semiotext, 2001). See also Nancy Luxon, 'Truthfulness, risk, and trust in the late lectures of Michel Foucault', *Inquiry* 47 (2004), 464–89.
102. Michel Foucault, *History of Sexuality,* vol. 1, trs. Robert Hurley (New York: Vintage Books), 61–2, cited Luxon, 'Truthfulness, Risk and Trust', 475.
103. Berry, *Standing By Words*, 30.
104. Berry, *Standing By Words*, 30.
105. Berry, *Standing By Words*, 55.
106. Stanley Hauerwas, 'A Story-Formed Community: Reflections on *Watership Down*' in Hauerwas, *A Community of Character: Toward a Constructive Social Ethic* (Notre Dame, IN: University of Notre Dame Press, 1981), 93–5.
107. Paul Farmer, *Pathologies of Power: Health, Human Rights and the New War on the Poor* (Berkeley CA: University of California Press, 2003), 23.

108. John Howard Yoder, *For the Nations: Essays Public and Evangelical* (Grand Rapids, MI: W. B. Eerdmans, 1997), 66–70.
109. Juan Luis Segundo, *The Liberation of Theology,* trans. John Drury (Dublin: Gill and Macmillan, 1977).
110. Cornel West, *Prophetic Thought in Postmodern Times: Beyond Eurocentrism and Multiculturalism* vol. 1 (Monroe, ME: Common Courage Press, 1993), 4.
111. Bruno Latour, *Politics of Nature: How to Bring the Sciences Into Democracy,* trs. Catherine Porter (Cambridge MA: Harvard University Press, 2004).
112. Justice Peace and Creation Unit of the World Council of Churches, *Solidarity with Victims of Climate Change: Reflections on the World Council of Churches' Response to Climate Change* (Geneva: 2002), 25.
113. *Solidarity with Victims of Climate Change*, 25–6.
114. On the idea of ecological debt see Andrew Simms, *Ecological Debt: The Health of the Planet and the Wealth of Nations* (London: Pluto Press, 2005).
115. Karl Polanyi argued that modern industrial capitalism was unique in its distancing of exchange relationships from social bonds; Karl Polanyi, *The Great Transformation: The Political and Economic Origins of Our Time* (London: Victor Gollancz, 1945).
116. Among his many writings on the ethics of global warming see Lukas Vischer, 'Climate Change, Sustainability and Christian Witness', *The Ecumenical Review*, 49 (1997), 142–149.

2. When Prophecy Fails

1. Georg Kaser, Douglas R. Hardy, Thomas Molg *et al.*, 'Modern glacier retreat on Kilimanjaro as evidence of climate change: observations and facts', *International Journal of Climatology* 24 (2004), 329–39.
2. Shardul Agrawala, Annett Moehner, Andreas Hemp *et al.*, *Development and Climate Change in Tanzania: Focus on Mount Kilimanjaro* (Paris: Organisation for Economic Cooperation and Development Environment Directorate, 2003).
3. Agrawala, Moehner, Hemp *et al.*, *Development and Climate Change in Tanzania*, 32.
4. Agrawala, Moehner, Hemp *et al.*, *Development and Climate Change in Tanzania*, 37.
5. UN Development Assistance Framework Report, cited in Agrawala, Moehner, Hemp, *et al.*, *Development and Climate Change in Tanzania*, 25.
6. Archbishop Donald Mtelemela, presentation at the GW8 conference, 'Dynamic Earth', Edinburgh, 5 July 2005.
7. Agrawala, Moehner, Hemp *et al.*, *Development and Climate Change in Tanzania*, 13.
8. World Development Movement, *Climate Calendar: The UK's Unjust Contribution to Global Climate Change* (London: WDM, 2007), 10.
9. Cited WDM, *Climate Calendar,* 10.
10. Donald A. Brown, *American Heat: Ethical Problems with the United States' Response to Global Warming* (Lanham, ML: Rowman and Littlefield, 2002), table 8.2, 182.
11. Andrew Simms and Hannah Reid, *Africa Up in Smoke? The Second Report from the Working Group on Climate Change and Development* (London: International Institute for Environment and Development).
12. *The Climate of Poverty: Facts, Fears and Hope: A Christian Aid Report* (London: Christian Aid, 2006), 19.
13. Walter V. Reid, Harold A. Mooney, Angela Cropper *et al., Ecosystems and Human Well-Being: Synthesis. A Report of the Millennium Ecosystem Assessment* (Washington DC: Island Press 2005) 63.
14. Department for Environment, Food and Rural Affairs, *Experimental Statistics on Carbon Dioxide Emissions at Local Authority and Regional Level: 2004* (London:

Department for Environment, Food and Rural Affairs, 2006).

15. These are by no means undisputed figures because their calculation relies upon assumptions about the maintenance of carbon sinks as well the total output of climate heating gases. Different assumptions and a different basis for calculation indicate that the United States and Europe are responsible for a greater proportion of climate heating, and the developing world for a lower proportion than these figures suggest: see further Anil Agarwal and Sunita Narain, *Global Warming in an Unequal World: A Case of Environmental Colonialism* (New Delhi: Centre for Science and Environment, 1991) and Christian Aid, *Coming Clean: Revealing the UK's True Carbon Footprint* (London: Christian Aid, 2007) at: http://www.christian-aid.org. uk/indepth/0702_climate/missingcarbon.pdf.
16. *Coming Clean*, 4–5, and *The Carbon 100: Quantifying the carbon emissions, intensities and exposures of the FTSE 100* (London: Henderson Global Investors and Trucost, 2006) at: http://www.trucost.com/Trucost_The_Carbon_100.pdf.
17. IPCC, *Climate Change 2007: Mitigation of Climate Change,* Unpublished draft report, version 3.0 (2007), Summary for Policymakers, Table 1.
18. IPCC, *Climate Change 2007: Working Group II: Impacts, Adaptation and Vulnerability: The Summary for Policymakers*, (IPCC: Geneva, 2007).
19. Nicholas Stern, *The Economics of Climate Change: The Stern Review* (Cambridge: Cambridge University Press, 2007), ch. 3.
20. IPCC, *Climate Change 2007: Working Group III: Mitigation of Climate Change: Summary for Policymakers*, Table 4, 15. (pre-publication draft).
21. Stuart L. Pimm, *The World According to Pimm: A Scientist Audits the Earth* (New York: McGraw-Hill, 2001).
22. The presentation of EU and US emissions statistics do not seem to indicate the extent to which mobility is contributing to greenhouse gas emissions growth. More detailed local studies can be more revealing. Thus while the EIA estimates US national contributions of transport at only 25 per cent, New Jersey's greenhouse gas inventory reveals that 57 per cent of its emissions are from the transport sector: Energy Information Authority, *Annual Energy Review 2005* xxi, and New Jersey State Government, *Greenhouse Gas Emissions* (November 2006) at: http://www.nj.gov/dep/dsr/trends2005/pdfs/ghg.pdf.
23. *Inventory of US Greenhouse Gas Emissions and Sinks 1990–2004* (Washington DC: Environmental Protection Agency, 2006), 3.
24. European Environment Agency, *Greenhouse Gas Emissions and Projections in Europe 2006* (Copenhagen: 2006, EEA).
25. Mark Maslin with Patrick Austin, Alex Dickson *et al.*, *Audit of UK Greenhouse Gas emissions to 2020: will current Government policies achieve significant reductions?*, UCL Environment Institute: Environment Policy Report Number 2007: 01.
26. Pew Research Centre, *Global Warming: A Divide on Causes and Solutions*, 24 January 2007 at: http://people-press.org/reports/pdf/303.pdf.
27. Timothy Lesle, 'States take lead on mercury and global warming', *The Planet Newsletter*, July-August 2006: http://www.sierraclub.org/planet/200604/states.asp. Disputes between State and Federal governments on global warming are growing and are now entering into the judicial system: see further Justin R. Pidot, *Global Warming in the Courts: An Overview of Current Litigation and Common Legal Issues* (Georgetown, Washington DC: Georgetown Environmental Law and Policy Institute, 2006) at: http://www.law. stanford.edu/program/centers/enrlp/pdf/GlobalWarmingLit_CourtsReport.pdf.
28. Andrew C. Revkin, 'US predicting steady increase for emissions', 3 March 2007 at: http://www.nytimes.com/2007/03/03/science/03climate.html?_r=1&

oref=slogin.

29. World Energy Council, United Nations Department of Economic and Social Affairs, *World Energy Assessment Overview 2004 Update* (New York: UNDP, 2004).
30. James Lovelock, *The Revenge of Gaia: Why the Earth is Fighting Back and How We Can Still Save Humanity* (London: Allen Lane, 2006).
31. Rob Jackson, *The Earth Remains Forever: Generations at a Crossroads* (Austin: University of Texas Press, 2002).
32. A. Agarwal and S. Narain, *Global Warming: A Case of Environmental Colonialism* (New Delhi: Centre for Science and Environment, 1991). See also Joan Martinez-Alier, *The Environmentalism of the Poor: A Report for UNRISD for the WSSD*, University of Witswatersrand, 30 August 2002 at http://www.foe-scotland.org.uk/nation/ej_alier.pdf.
33. Henry Shue, 'Subsistence Emissions and Luxury Emissions', *Law and Policy* 15 (1993), 39– 59.
34. Thomas Aquinas, *Summa Theologiae* II.II, Question 66: 'Theft and Robbery', 2nd revised edn of the English translation by Fathers of the English Dominican Province, 1920, online edition edited by Kevin Knight (2006) at: http://www.newadvent.org/summa/3066.htm. See also Raymond de Roover, 'The concept of the just price: theory and economic policy', *The Journal of Economic History* 18 (December 1958), 418–34.
35. Aquinas, *Summa Theologiae* II.II, 66.2.
36. Malamo Kormetis, Dave Reay, and John Grace 'New Directions: Rich in CO_2', *Atmospheric Atmosphere* 40 (17 June 2006), 3219–20.
37. Shue, 'Subsistence Emissions and Luxury Emissions'.
38. Mark Maslin *et al.*, *Audit of UK Greenhouse Gas emissions to 2020*, and Christian Aid, *Coming Clean.* If British households are responsible for half of UK emissions, and UK corporations emit 12–15 per cent of global emissions, or 6–7 times domestic emissions, then UK corporations overseas emit three times the emissions of domestic householders.
39. Martin Khor, 'Global trends', *Third World Network* (3 November 2003) at: http://www.twnside.org.sg/title2/gtrends0304.htm. See also United Nations General Assembly, Report of the Secretary General, *International financial architecture and development, including net transfer of resources between developing and developed countries,* A 56/173, (11 July 2001).
40. UN Department of Economic and Social Affairs, *World Economic Situation and Prospects 2007* (New York: UNDESA, 2007).
41. Michael Grubb, 'Seeking fair weather: ethics and the international debate on climate change', *International Affairs* 71 (1995), 463–96.
42. Richard Leakey and Roger Lewin, *The Sixth Extinction* (New York: Doubleday, 1995).
43. Robert A. Robinson, Jennifer A. Learmouth, Anthony M. Hutson *et al.*, *Climate Change and Migratory Species* (Thetford: British Trust for Ornithology, 2005).
44. Chris Thomas, Alison Cameron, Rhys Green *et al.*, 'Extinction risk from climate change', *Nature* 427 (January 2004), 145–8.
45. Thomas, Cameron, Green, 'Extinction risk', 146.
46. Basile Michaelidis, Christis Ozounis *et al.*, 'Effects of long-term moderate hypercapnia on acid-base balance and growth rate in marine mussels *Mytilus galloprovincialis*', *Marine Ecology Progress Series* 293 (2 June 2 2005), 109–18.
47. Paul Ehrlich and Anne Ehrlich, *Extinction: The Causes and Consequences of the Disappearance of Species* (New York: Random House, 1981).
48. Robin Grove-White and Oliver O'Donovan, 'An alternative approach' in

Robin Attfield and Katherine Dell (eds), *Values, Conflict and the Environment* (Oxford: Ian Ramsey Centre and St Cross College, 1989), 117–33.

49. For a fuller account of the transition see E. P. Thompson, *Customs in Common* (Harmondsworth: Penguin, 1977).
50. Even vagrancy was outlawed in eighteenth-century England, as Thomas Hardy records in his beautiful and tragic novel *Tess of the D'Urbervilles*.
51. Hannah Arendt, *The Human Condition* (Chicago: Chicago University Press, 1958), 38–47. The erosion of genuine private property is one of the marked features both of the industrial revolution and of the present neoliberal phase of capitalism. The majority of householders in Britain and the United States do not own land – or even houses – but are instead indebted to banks through mortgages throughout their working lives, and hence tied to the job-holder economy even though most of the jobs in making have been exported to China. For an insightful account of the transition from a propertied to a debt-based economy see P-J. Proudhon, *What Is Property? An Enquiry into the Principle of Right and of Government*, trs. B. R. Tucker (New York: H. Fertig, 1966). See also John Milbank's essay *Liberality and Liberalism*, published online at http://www.theology philosophycentre.co.uk/papers.php.
52. Arendt, *Human Condition*, 38–43.
53. Arendt, *Human Condition*, 46.
54. Arendt, *Human Condition*, 43.
55. See further Charles Taylor, *The Sources of the Self* (Cambridge: Cambridge University Press, 1989).
56. Alasdair MacIntyre, *After Virtue: A Study in Moral Theory* (London: Duckworth, 1981).
57. Arendt, *Human Condition*, 165.
58. Arendt, *Human Condition*, 166.
59. On the theological origins of the cosmology of mechanism see Amos Funkenstein, *Theology and the Scientific Imagination in the Seventeenth Century* (Princeton, NJ: Princeton University Press, 1986).
60. See further Keith Thomas, *Man and the Natural World* (London: Penguin, 1986).
61. Carolyn Merchant, *The Death of Nature: Women, Ecology and the Scientific Revolution* (San Francisco: Harper and Row, 1980).
62. David Demeritt, 'The construction of global warming and the politics of science', *Annals of the Association of American Geographers* 91 (2001).
63. Bocking, *Nature's Experts*, 126.
64. John Dewey, *The Quest for Certainty: A Study of the Relation of Knowledge and Action* (London: George Allen and Unwin, 1930).
65. Stephen Toulmin, *Cosmopolis: The Hidden Agenda of Modernity* (Chicago: University of Chicago Press, 1990), 10, 29–35.
66. Aristotle, *Nicomachean Ethics*, Book 1. 7 trs. W. D. Ross (Oxford: Clarendon Press, 1908).
67. Toulmin, *Cosmopolis*, 41–2.
68. John Donne, *The First Anniversarie: An Anatomie of the World* (London: a. Matthewes, 1621).
69. William Blake, *The Chimney Sweeper* (Bushey Heath, Herts: Tauris Poems, 1969) and John Ruskin, *The Stones of Venice: Volume 3: The Fall* (London: George Allen, 1893). See also David V. Erdman, *Blake: Prophet Against Empire* (London: Courier Dover, 1991).
70. Yeats, 'The Second Coming', cited Toulmin, *Cosmopolis*, 66.
71. Toulmin, *Cosmopolis*, 67.

72. James Lovelock, *Gaia: A New Look at Life on Earth* (London: Oxford University Press, 1979).
73. For a description of the origins of the discipline see William H. Schlesinger, *Biogeochemistry: An Analysis of Global Change*, 2nd edn (San Diego, CA: Academic Press, 1997).
74. Alasdair MacIntyre, *After Virtue: A Study in Moral Theory* (London: Duckworth, 1981).
75. Philip Merowski, *More Heat than Light: Economics as Social Physics, Physics as Nature's Economics* (Cambridge: Cambridge University Press, 1989).
76. Walter B. Wriston, *The Twilight of Sovereignty: How the Information Revolution is Transforming Our World* (NY: Charles Scribner and Sons, 1992).
77. David Burrell and Stanley Hauerwas, 'From system to story: An alternative pattern for rationality in ethics' in H. T. Engelhardt and D. Callahan (eds.), *Knowledge, Value and Belief* (Hastings-on-Hudson, NY: The Hastings Center, 1977), 111–52.
78. Norman Wirzba, *The Paradise of God; Renewing Religion in an Ecological Age* (Oxford: Oxford University Press, 2003), 32.
79. Claus Westermann, *Genesis*, trs. David E. Green (Edinburgh: T. and T. Clark, 1987), 51.
80. William Ryan and Walter Pitman, *Noah's Flood: The New Scientific Discovery About the Event that Changed History* (New York: Simon and Schuster, 1999).
81. On the ecological implications of the Fall see Michael Northcott, 'Do Dolphins Carry the Cross? Biological Moral Realism and Theological Ethics', *New Blackfriars* 84 (2003), 540–553; for an alternative approach see Holmes Rolston, 'Does nature need to be redeemed?', *Zygon* 29 (1994), 205–17.
82. Wirzba, *Paradise of God*, 33.
83. Wirzba, *Paradise of God*, 33.
84. Westermann, *Genesis*, 52–3.
85. Richard B. Alley, *The Two-Mile Time Machine: Ice Cores, Abrupt Climate Change, and Our Future* (Princeton NJ: Princeton University Press, 2000).
86. On the interaction between human cultures and climate change in the last 20,000 years see Brian Fagan, *The Long Summer: How Climate Changed Civilization* (New York: Basic Books, 2004).
87. Genesis 9:12.
88. Robert Murray, *The Cosmic Covenant: Biblical Themes of Justice, Peace and the Integrity of Creation* (London: Sheed and Ward, 1992).
89. Fagan, *The Long Summer*, 143–5.
90. I am indebted to a conversation with Bryan Brock of Aberdeen University, as well as Wirzba's *Paradise of God*, for this reading of the Noah story.
91. Rowan Williams, 'On Being Creatures' in Williams, *On Christian Theology* (Oxford: Blackwell, 2000), 66–79.
92. Williams, 'On Being Creatures', 69.
93. See further Simon Conway Morris, *Life's Solution: Inevitable Humans in a Lonely Universe* (Cambridge: Cambridge University Press, 2003).
94. Metropolitan John of Pergamon, 'Proprietors or priests of creation?' Keynote address of the Fifth Symposium of Religion, Science and the Environment, 2 June 2003: http://www.orthodoxytoday.org/articles2/MetJohnCreation.php.
95. Williams, 'On Being Creatures', 70–1.
96. Williams, 'On Being Creatures', 76.
97. Hal Lindsay, 'More hot air from the UN', *Worldnet Daily*, 2 March, 2007: http://www.freerepublic.com/focus/f-religion/1794066/posts.

3. Energy and Empire

1. Mike Davis, 'A World's End: Drought, Famine and Imperialism (1896–1902)', *Capitalism, Nature, Socialism* 10 (June 1999), 3–46.
2. Barrington Moore, *The Social Origins of Dictatorship and Democracy: Lord and Peasant in the Making of the Modern World* (Hardmondsworth: Penguin, 1973).
3. Davis, 'World's End', 7–9.
4. Charles Ambler, *Kenyan Communities in the Age of Imperialism* (New Haven: Yale University Press, 1988), 3.
5. Philip Stott, 'The world's energy poor need hydrocarbon fuels', *Daily Telegraph*, 27 August 2002.
6. Bina Agarwal, 'Under the Cooking Pot: The Political Economy of the Domestic Fuel Crisis in Rural South Asia', *Institute of Development Studies Bulletin* 18 (1987), 11–22.
7. Sarah Jewitt, 'Voluntary and official forest protection committees in Bihar: solutions to India's deforestation?', *Journal of Biogeography* 22 (1995), 1003–21.
8. Bina Agarwal, 'Participatory Exclusions, Community Forestry and Gender: An Analysis for South Asia and a Conceptual Framework', *World Development* 29 (2001), 1623–48.
9. Garrett Hardin, 'The tragedy of the commons: The population problem has no technical solutions; it requires a fundamental extension of morality', *Science* 162 (15 December 1968), 1243–8. See my more extended discussion of Hardin in Michael Northcott, *The Environment and Christian Ethics* (Cambridge: Cambridge University Press, 1996).
10. See further Michael Northcott, *Life After Debt: Christianity and Global Justice* (London: SPCK 1999).
11. J. K. Galbraith, cited Andrew Simms, *Ecological Debt: The Health of the Planet and the Wealth of Nations* (London: Pluto Press, 2005), 73.
12. World Health Organization, *Climate Change and Human Health: Risks and Responses* (Geneva: WHO, 2003), 13; also Jonathan Patz, Diarmid Campbell-Lendrum, Tracey Holloway and Jonathan A. Foley, 'Impact of regional climate change on human health', *Nature* 438 (17 November 2005), 310–17.
13. IPCC, *Synthesis Report* (Cambridge: Cambridge University Press, 2001).
14. Joan Martinez-Alier, 'Ecological distribution conflicts and indicators of sustainability', *International Journal of Political Economy* 34 (Spring 2004), 13–30.
15. Friends of the Earth, 'HSBC: financing forest destruction and social conflict' (London: Friends of the Earth, May 2004) at: http://www.foe.co.uk/ resource/media_briefing/hsbc_banking_on_palm_oil.pdf.
16. Joan Martinez-Alier, 'Ecological debt and property rights on carbon sinks and reservoirs', *Capitalism, Nature, Socialism* 13 (March 2002), 115 – 119.
17. Southern Peoples Ecological Debt Creditors Alliance, *Ecological Debt: South Tells North 'Time to Pay Up'* at http://www.cosmovisiones.com/DeudaEcologica/a_timetopay.html. See also Simms, *Ecological Debt*, 86–9.
18. *Who Owes Who? Climate Change, Debt, Equity and Survival* (London: Christian Aid, 1999): http://www.christian-aid.org.uk/indepth/9909whoo/whoo1.htm 09 February 2005), 2.
19. Petter Stalenheim, Damien Fruchart, Wuyi Omitoogun and Catalina Perdomo, 'Military Expenditure', in *Stockholm International Peace Institute Yearbook 2006: Armaments, Disarmament and International Security* (Stockholm: Stockholm International Peace Institute, 2006), ch. 8.
20. Andrew Simms, *An Environmental War Economy: The Lessons of Ecological Debt and*

Global Warming (London: New Economics Foundation, 2001). On whether or not it will be possible to sue for climate damage see also Myles Allen, 'Liability for climate change: will it ever be possible to sue anyone for damaging the climate?', *Nature* 421 (27 February 2003), 891–2.

21. Christian Aid, *Human Tide: The Real Migration Crisis* (London: Christian Aid, May 2007) 28.
22. Simms, *Ecological Debt*, 23–29.
23. See further Northcott, *Life After Debt*.
24. R. Sieferle, *The Subterranean Forest: Energy Systems and the Industrial Revolution*, trs Michael Osmann (Cambridge: White Horse Press, 2001).
25. For a technical and yet readable account of the carbon cycle see further William H. Schlesinger, 'The Global Carbon Cycle' in Schlesinger, *Biogeochemistry: An Analysis of Global Change*, 2nd edn (San Diego, CA: Academic Press, 1997), 358–82.
26. For a clear graphic and numerical representation of the carbon cycle see Schlesinger, 'Global Carbon Cycle', 359, Fig. 11.1.
27. Schlesinger, 'Global Carbon Cycle', 361.
28. John Houghton, *Global Warming: The Complete Briefing*, 3rd edn (Cambridge: Cambridge University Press, 2004), 30.
29. William H. Calvin, *A Brain for All Seasons: Human Evolution and Abrupt Climate Change* (Chicago: University of Chicago Press, 2002), 99.
30. This is a summary of the fuller description given in Houghton, *Global Warming: The Complete Briefing*, 35.
31. Flannery, *The Weathermakers: The History and Future Impact of Climate Change* (London: Allen Lane, 2005), 85.
32. Michael Behrenfeld, Robert O'Malley, David Siegel *et al.*, 'Climate-driven trends in contemporary ocean productivity', *Nature* 444 (December 2006), 752–5.
33. Behrenfeld, O'Malley, Siegel *et. al.*, 'Climate-driven trends in contemporary ocean productivity', 752.
34. Houghton, *Global Warming: The Complete Briefing*, 29.
35. Stuart Pimm and Peter Raven, 'Extinction by numbers', *Nature* 43 (2000), 843–5. Also Richard Leakey, *The Sixth Extinction: Patterns of Life and the Future of Humankind* (New York: Doubleday, 1995).
36. H. G. Wells, 'History Becomes Ecology' in Wells, *The Fate of Man* (New York: Alliance Book Corporation, 1939), 27–38.
37. David S. Landes *Unbound Prometheus* (Cambridge: Cambridge University Press, 1969), G. N. von Tunzelman, *Steam Power and British Industrialization* (Oxford: Oxford University Press, 1978) and E. A. Wrigley, *Continuity, Chance and Change: The Character of the Industrial Revolution in England* (Cambridge: Cambridge University Press, 1988); also M. Flinn, *The History of the British Coal Industry 1700–1830*, Vol. 1 (Oxford: Oxford University Press, 1984).
38. Peter Brimblecombe, *The Big Smoke: A History of Air Pollution in London Since Medieval Times* (London: Methuen, 1987), 30–5.
39. Landes, *Unbound Prometheus*, 101–4.
40. D. S. L. Cardwell, *From Watt to Clausius* (London: Heinemann, 1971).
41. Watt's breakthrough insight came to him on 'a walk on a fine Sabbath afternoon': H. W. Dickson and H. Vowles, *James Watt and the Industrial Revolution* (London: Longman Green and Co., 1945), 32. See also Ben Marsden, *Watt's Perfect Engine: Steam and the Age of Invention* (Cambridge: Icon Books, 2002).
42. Wrigley, *Continuity, Chance and Change*, 5.

43. Charles Leadbetter, *Living On Thin Air: The New Economy* (London: Penguin, 2000).
44. Amory Lovins makes the same point with reference to a typewriter in *Openpit Mining* (London: Earth Island, 1973), 1.
45. E. A. Wrigley, *Poverty, Progress, and Population* (Cambridge: Cambridge University Press, 2004), 212–29.
46. Paul Ehrlich and Anne Ehrlich, *Extinction: The Causes and Consequences of the Disappearance of Species* (New York: Random House, 1981),79.
47. Mathis Wackernagel, Niels B. Shulz, *et al.*, 'Tracking the ecological overshoot of the human economy', *Proceedings of the National Academy of Sciences* 99 (June 27, 2002), 9266–71.
48. John Barrett, Jan Minx *et. al,*, *Towards a Low Footprint Scotland: Living Well, Within Ecological Limits* (Dunkeld, Perthshire: WWF Scotland and Stockholm Environment Institute, 2007).
49. James J. McKenzie, 'Energy and the environment in the 21st Century: The challenge of change', *Journal of Fusion Energy* 17 (1998), pp 141–50.
50. McKenzie, 'Energy and the environment', 142 and Willem Norde, 'Energy and entropy: a thermodynamic approach to sustainability', *The Environmentalist* 17 (1997), 57–62.
51. Simms, *Ecological Debt*, 95.
52. Energy Information Administration, *Emissions of Greenhouse Gases in the United States 2005* (Washington DC: EIA, 2005) and E-Digest Statistics, *The Global Atmosphere: UK Emissions of Greenhouse Gases, Department for the Environment and Rural Affairs* (2007) at: http://www.defra.gov.uk/environment/statistics/globatmos/gagccukem.htm.
53. DEFRA, *Greenhouse Gas Emissions from Transport* (2003) http://www.statistics.gov.uk/downloads/theme_environment/transport_report.pdf.
54. Ivan Illich, *Energy and Equity* (London: Calder and Boyars, 1974).
55. Paul Virilio, *Crepuscular Dawn* (New York: Semiotext, 2002).
56. Kenneth S. Deffeyes, *Hubbert's Peak: The Impending World Oil Shortage* (Princeton, NJ: Princeton University Press, 2003).
57. R. E. Roadifer, 'Size distribution of the world's largest known oil and tar accumulations', *Exploration for Heavy Crude Oil and Natural Bitumen*, ed. Richard F. Meyer (Tulsa, OH: American Association of Petroleum Geologists, 1987).
58. Stephane Dion, Canadian Environment Minister, as quoted in Larry Lohmann, 'Carbon Trading: A Critical Conversation on Climate Change, Privatisation and Power', *Development Dialogue* 48 (September 2006), 23.
59. P S. Wenz, *Environmental Ethics Today* (Oxford: Oxford University Press, 2001), 208.
60. On American government subsidies to American oil companies see further Robert Kennedy, *Crimes Against Nature*, (New York: Harper Collins, 2004). On British subsidies to British oil companies see Christian Aid, *Fuelling Poverty: Oil, War and Corruption* (London: Christian Aid, 2003).
61. Christian Aid, *Fuelling Poverty.*
62. Christian Aid, *Fuelling Poverty*, 8.
63. Pauline Luong and Erika Weinthal, 'Prelude to the resource curse: explaining oil and natural gas development strategies in the Soviet successor states and beyond', *Comparative Political Studies* 34 (May 2001), 367–99.
64. *Extractive Sectors and the Poor*, Michael Ross UCLA, Oxfam America, October 2001, cited Christian Aid, *Fuelling Poverty*, 10.
65. Christian Aid, *Fuelling Poverty*, 11.

66. Christian Aid, *Fuelling Poverty*, 13.
67. Richard Heinberg, *The Party's Over: Oil, War and the Fate of Industrial Societies* (Gabriola Island: New Society Press, 2003), 68–9. I draw on Heinberg's insightful account for the rest of the paragraph.
68. James Howard Kunstler, *The Long Emergency: Surviving the End of Oil, Climate Change, and Other Converging Catastrophes of the Twenty-First Century* (New York: Atlantic Monthly Press, 2005).
69. See further Michael Northcott, 'The Weakness of Power and the Power of Weakness: The Ethics of War in a Time of Terror', *Studies in Christian Ethics* 20 (April 2007), 88–101.
70. Wendell Berry, 'Thoughts in the presence of fear' in Berry, *Citizenship Papers* (Washington DC: Shoemaker and Hoard, 2004), 17–22.
71. Al Gore, 'Moment of truth', *Vanity Fair* (May 2006).
72. Francine Laden, Lucas M. Neas, Douglas W. Dockery and Joel Schwartz, 'Association of Fine Particulate Matter from Different Sources with Daily Mortality in Six US Cities', *Environmental Health Perspectives* 108 (October 2000), 941–7.
73. Friends of the Earth, 'The Exxon Files' at: http://www.foe.co.uk/resource/briefings/exxon_files.html.
74. Erik Reece, *Lost Mountain: A Year in the Vanishing Wilderness: Radical Deep Mining and the Devastation of Appalachia* (Kentucky: Riverhead Books, 2006).
75. See for example the laudatory comments on hydropower by James Lovelock in *The Revenge of Gaia: Why the Earth is Fighting Back and How We Can Still Save Humanity* (London: Allen Lane, 2006), 85.
76. Patrick McCully, *Silenced Rivers: The Ecology and Politics of Large Dams* (London: Zed Books, 1996); see also Holly Sims, 'Moved, Left No Address: Dam Construction, Displacement and Issue Salience', *Public Administration and Development* 21 (2001), 187–200.
77. Madhav A. Chitale, 'The Narmada Project', *Water Resources Development* 13.2 (1997), 169–79.
78. Arundhati Roy, *The Cost of Living* (London: Flamingo, 1999).
79. Vinod Raina, 'Why people oppose dams: environment and culture in subsistence economies', *Inter-Asia Cultural Studies* 1.1 (2000), 14–61.
80. Anil Sasi, 'Unlocking alternative sources for electricity', *Business Line*, 1 January 2007 at: http://www.blonnet.com/2007/01/02/stories/2007010201470300.htm.
81. This account of the metaphorical significance of the cedars of Lebanon was inspired by a talk given by Ched Myers at the Greenbelt Festival, Cheltenham, England in August 2004.
82. E. W. Beals, 'The Remnant Cedar Forests of Lebanon', *The Journal of Ecology* 53 (1965), 679–94.
83. Marvin W. Mikesell, 'The Deforestation of Mount Lebanon', *Geographical Review* 59 (1969), 1–28.
84. Mikesell, 'Deforestation of Mount Lebanon', 13.
85. Mikesell, 'Deforestation of Mount Lebanon', 15.
86. Beals, 'The Remnant Cedar Forests of Lebanon', 267.
87. Kenneth S. Feffeyes, *Hubbert's Peak: The Impending World Oil Shortage* (Princeton, NJ: Princeton University Press, 2001). See also Richard Heinberg, *The Party's Over: Oil, War and the Fate of Industrial Societies* (Gabriola Island: New Society Press, 2003).
88. J. Joyce Schuld, 'Augustine, Foucault and the Politics of Imperfection', *Journal of*

Religion 80 (January 2000), 1–22.

89. Schuld, 'Augustine, Foucault and the Politics of Imperfection', 19.
90. Val Plumwood, *Environmental Culture: The Ecological Crisis of Reason* (London: Routledge, 2002), 41.
91. Plumwood, *Environmental Culture*, 44.
92. Christian Aid, *Coming Clean: Revealing the UK's True Carbon Footprint* (London: Christian Aid, 2007).
93. Vijay B. Vaitheeswaran, *Power to the People: How the Coming Energy Revolution Will Transform an Industry, Change Our Lives and Maybe Even Save the Planet* (London: Earthscan, 2005), 24.
94. Sheldon Wolin, *The Presence of the Past: Essays on the State and the Constitution* (Baltimore: John Hopkins University Press, 1989), 6.
95. Wolin, *Presence of the Past*, 7.
96. Wolin, *Presence of the Past*, 24–8 and Richard Sennett, *The Corrosion of Character: The Personal Consequences of Work in the New Capitalism* (NY: Norton, 1998).
97. Wolin, *Presence of the Past*, 29.
98. Wolin, *Presence of the Past*, 183.
99. Wolin, *Presence of the Past*, 182.
100. Hannah Arendt, *The Human Condition* (Chicago: Chicago University Press, 1958), 40.
101. Ulrich Beck, *Ecological Enlightenment: Essays on the Politics of the Risk Society* trs. Marj J. Ritter (NJ: Humanities Press, 1995), 93.
102. Francis Bacon, *Organum,* 3–4 and René Descartes, *Discourse on Method*, 15–16 cited in Albert Borgmann, *Technology and the Character of Contemporary Life* (Chicago: Chicago University Press, 1984), 35–6.
103. Peter M. Vitousek, Harold A. Mooney, Jane Lubchenco, Jerry M. Melillo, 'Human domination of earth's ecosystems', *Science* 277 (25 July 1977), 494–9, and Stephen R. Palumbi, 'Humans and the world's greatest evolutionary force', *Science* 293 (7 September 2001) 1786–90.
104. Arendt, *Human Condition*, 120.
105. Arendt, *Human Condition*, 121.
106. Borgmann, *Character of Contemporary Life*, 58.
107. See also Edward Casey, *The Fate of Place: A Philosophical History* (San Francisco, CA: University of California Press, 1998).
108. William McDonough and Michael Braungart, *Cradle to Cradle: Remaking the Way We Make Things* (New York: North Point Press, 2002), 27–8.
109. Arendt, *Human Condition*, 122–5.
110. Borgmann, *Character of Contemporary Life*, 94.
111. Borgmann, *Character of Contemporary Life*, 104–7.
112. Borgmann, *Character of Contemporary Life*, 113. See also Ivan Illich, *Energy and Equity* (London: Calder and Boyars, 1974).
113. Bill McKibben, *The End of Nature* (London: Penguin, 1989).
114. Albert Borgmann, *Power Failure: Christianity in the Culture of Technology* (Grand Rapids, MI: Brazos Press, 2003), 88.
115. John Howard Yoder, *The Politics of Jesus: Vicit Agnus Noster,* 2nd edn (Grand Rapids, MI: Eerdmans, 2001).
116. Vaitheeswaran, *Power to the People*, 37. The UK Royal Commission on Energy 2004 recommended large-scale adoption of combined heat and power schemes and a radical decentralisation of the electricity grid. See also the detailed policy options explored in Shimon Awerbuch, *Restructuring Our Electricity Networks to Promote Decarbonisation, Decentralization, Mass Customization and Intermittent*

Renewables in the 21st Century, Tyndall Working Paper No. 49 (March 2004).
117. Vaitheeswaran, *Power to the People*, 225–9.

4. Climate Economics

1. Alaskans prefer Eskimo or Inupiat to Inuit, Canadians prefer Inuit to Eskimo; once again the 49th parallel divides in more ways than one.
2. Charles Wohlforth, *The Whale and the Supercomputer: On the Northern Frontier of Climate Change* (New York: North Point Press, 2004).
3. Bill Hess, 'The Gift of the Whale: The Inupiat Bowhead Hunt, A Sacred Tradition' at: http://www//hess.org.
4. For an in-depth study of the ecological impacts of climate change in the Arctic see the report of the International Arctic Science Committee at: http://www.acia.auf.edu/ (July 2005).
5. Wohlforth, *The Whale and the Supercomputer.*
6. A. S. Brierley, P. G. Fernandes, M. A. Brandon, F. Armstrong, 'Antarctic krill under sea ice: elevated abundance in a narrow band just south of ice edge', *Science* 292 (March 2002), 1890–2.
7. Fikret Berkes and Dyanna Jolly, 'Adapting to climate change: social-ecological resilience in a Canadian Western Arctic community', *Conservation Ecology* 5.2 (2002); Donald Scavia, John C. Field *et al.*, 'Climate Change Impacts on US Coastal marine Ecosystems', http://www.umces.edu/President/Scavia%20et%20al.pdf; G. A. Weller and P. A. Anderson (eds), *Implications of Global Change in Alaska and the Bering Sea Region: Proceedings of a Workshop, June 1997* (Fairbanks, Alaska: Center for Global Change and Arctic System Research, University of Alaska).
8. Wohlforth, *The Whale and the Supercomputer*, 258–9.
9. V. Loeb *et al.*, 'Effects of sea-ice extent and krill or salp dominance on the Antarctic food web', *Nature* 387 (June 1997), 897–900, and A. Clarke and C. M. Harris, 'Polar marine systems: major threats and future change', *Environmental Conservation* 2003.
10. Wohlforth, *The Whale and the Supercomputer*, 199–200.
11. Wohlforth, *The Whale and the Supercomputer*, 64.
12. Marika Holland, Ceclia Bitz and Bruno Tremblay, 'Future abrupt reductions in the summer Arctic sea ice', *Geophysical Research Letters* 33 (December 2006).
13. Fred Pearce, *The Last Generation: How Nature Will Take Her Revenge for Climate Change* (London: Transworld, 2006), 110–11.
14. Wohlforth, *The Whale and the Supercomputer.*
15. Jack D. Forbes, 'Indigenous Americans: Spirituality and Ecos', *Daedalus* 130 (Fall 2001), 283–98.
16. Peter Matheson, 'Inside the endangered Arctic Refuge', *The New York Review of Books* 53 (October 2006).
17. Tim Ingold, *The Perception of the Environment: Essays on Livelihood, Dwelling and Skill* (London: Routledge, 2000), 103.
18. Marcel Mauss and W. D. Halls, *The Gift: The Form and Reason for Exchange in Archaic Societies* (New York: W. W. Norton, 1990).
19. On the moral significance of gratitude see Bernd Wannenwetsch, *Political Worship: Ethics for Christian Citizens*, trs. Margaret Kohl (Oxford: Oxford University Press, 2004), 47–50.
20. Magnus Bergquist and Jan Ljungberg, 'The power of gifts: organizing social relationships in open source communities', *Information Systems Journal* 11 (2001), 305–20.

21. Bergquist and Ljungberg, 'The power of gifts: organizing social relationships in open source communities', 305–320.
22. Karl Polanyi, *The Great Transformation: The Political and Economic Origins of Our Time* (London: Victor Gollancz, 1945).
23. Anthony Giddens, *The Consequences of Modernity* (Cambridge: Polity Press, 1990), 25.
24. John Locke, *Second Treatise of Civil Government* (Buffalo, NY: Prometheus Press, 1986), section 4.
25. John Locke, *Second Treatise of Civil Government*, Section 46.
26. Hannah Arendt, *The Human Condition* (Chicago: Chicago University Press, 1958), 104–7.
27. Arendt, *Human Condition*, 107–8.
28. William McDonough, 'A Centennial Sermon: Design, Ecology, Ethics and the Making of Things', *Perspecta* 29 (1998), 78–85.
29. Stephen H. Schneider, 'Abrupt Non-Linear Climate Change, Irreversibility and Surprise', *Report of Working Party on Global and Structural Polices for OECD Workshop on the Benefits of Climate Policy: Improving Information for Policy Makers* (2003), 17 at: http://www.oecd.org.dataoecd/9/59/2482280.pdf.
30. Stephen Hawking, interview, *Daily Telegraph*, 16 October 2006.
31. Arendt, *Human Condition*, 1.
32. Arendt, *Human Condition*, 2.
33. Anthony Patt, 'Economists and ecologists: modelling global climate change to different conclusions', *International Journal of Sustainable Development* 2:2 (1999), 245–62.
34. R. Costanza, R. d'Arge *et al.*, 'The value of the world's ecosystem services and natural capital', *Nature* 387 (1997), 253–60. See also G. Daily (ed.) *Nature's Services* (Washington DC: Island Press, 1997), and Stephen C. Farber, Robert Costanza and Matthew A. Wilson, 'Economic and ecological concepts for valuing ecosystem services', *Ecological Economics* 41 (2002), 375–92.
35. James Gustave Speth, *Red Sky at Morning: America and the Crisis of the Global Environment* (New Haven: Yale Nota Bene, 2005), 27.
36. A. C. Pigou, *The Economics of Welfare* (London: Macmillan, 1920).
37. Gretchen C. Daily, Tore Soderqvist *et al.*, 'The value of nature and the nature of value', *Science* 289 (21 July 2000), 395–6.
38. William McDonough and Michael Braungart, *Cradle to Cradle: Remaking the Way We Make Things* (New York: North Point Press, 2002), 92–8.
39. Aubrey Meyer, *Contraction and Convergence: The Global Solution to Climate Change* (Totnes: Green Books, 2000).
40. For example in the British government's commissioned review of climate change economics carbon trading is the principal mitigating mechanism proposed: Nicholas Stern, *The Economics of Climate Change: The Stern Review* (Cambridge: Cambridge University Press, 2007) .
41. J. H. Dales, *Pollution, Property and Prices* (Toronto: Toronto University Press, 1968).
42. See also Ronald H. Coase, 'The problem of social cost', *Journal of Law and Economics* 3 (1960), 1–44, and Larry Lohmann, 'Carbon Trading: A Critical Conversation on Climate Change, Privatisation and Power', *Development Dialogue* 48 (September 2006), 57–9.
43. Jane Jacobs, *Systems of Survival: A Dialogue on the Moral Foundations of Commerce and Politics* (New York: Vintage Books, 1992).
44. See further Heidi Bachram, 'Climate fraud and carbon colonialism: the new

trade in greenhouse gases', *Capitalism, Nature, Socialism* 15.4 (December 2004), 5–20.

45. Michael Grubb, Duncan Brack and Christian Vrolijk, *Kyoto Protocol: A Guide and Assessment* (London: Earthscan, 1999), ch. 3.
46. Michael Grubb, 'International emissions trading under the Kyoto Protocol: core issues in implementation', *Implementing Trading Mechanisms* 7.2 (1998), 140–6.
47. For a much fuller description see Lohmann, 'Carbon Trading', 47–69.
48. For a fuller description see Donald McKenzie, 'Finding the ratchet: the political economy of carbon trading', *London Review of Books* 29 (5 April 2007).
49. 'The Durban Declaration on Carbon Trading', Glenmore Centre, Durban, South Africa (2004) at: http://www.sinkswatch.org/pubs/Durban%20DeclarationSeptember%202006%20leaflet.pdf; also Philippe Sands, *Lawless World: Making and Breaking Global Rules* (London: Penguin Books, 2005).
50. Australian National University Forestry, *Carbon Credit Markets: Market Report 12* (June 2000) at: http://sres-associated.anu.edu.au/marketreport/report12.pdf.
51. The Environment Audit Committee, *Inquiry into the International Challenge of Climate Change: UK Leadership in the G8 and EU: Memorandum by The Corner House, Sinks Watch and Carbon Trade Watch*, 14–16.
52. Delft Hydraulics, *Peat-CO_2: Assessment of CO_2 emissions from drained peat lands in SE Asia*, (April 2006), Wetlands International at: http://www.wetlands.org/ckpp/publication.aspx?ID=f84f160f-d851-45c6-acc4-d67e78b39699.
53. William H. Schlesinger, 'Carbon trading', *Science* 314 (24 November 2006), 1217.
54. James C. Scott, *Seeing Like a State: How Certain Schemes to Improve the Human Condition Have Failed* (New Haven, CT: Yale University Press, 1998).
55. Australian National University Forestry, *Carbon Credit Markets: Market Report 12* (June 2000) at: http://sres-associated.anu.edu.au/marketreport/report12.pdf.
56. Bachram, 'Climate fraud'.
57. Dieter Helm, 'The assessment: climate-change policy', *Oxford Review of Economic Policy* 19.3 (2003), 349–61.
58. Bachram, 'Climate fraud', 6.
59. Anil Agarwal, *Global Warming in an Unequal World* (New Delhi: Centre for Science and the Environment, 1991).
60. Anil Agarwal, 'A Southern Perspective on Curbing Global Climate' in S. H. Schneider, A. Rosencranz, and JO Niles, (eds.), *Climate Change Policy: A Survey*, (Washington DC: Island Press, 2002, 375–91, and Vandana Shiva, 'The Greening of the Global Reach' in Wolfgang Sachs (ed), *Global Ecology: A New Arena of Political Conflict* (London: Zed Books, 1993).
61. Mayer Hillman, *How We Can Save the Planet* (London: Penguin, 2004); George Monbiot, *Heat: How to Stop the Planet Burning* (London: Allen Lane, 2006) , 44–8, and Mark Lynas, 'Why we must ration the future', *New Statesman*, 23 October 2006.
62. Grubb, 'International emissions trading under the Kyoto Protocol'.
63. Increasing the price of carbon is the central recommendation of the *Stern Review*.
64. W. R. Cline, *The Economics of Global Warming* (Washington DC: Institute for International Economics, 1992).
65. Stephen H. Schneider and Lawrence H. Goulder, 'Achieving low-cost emissions targets', *Nature* 389 (4 September 1997), 13–14.
66. This assumption has long been challenged by a minority of economists such as E. J. Mishan, who in his *The Costs of Economic Growth* (London: Penguin, 1967)

argued that because of the destructive nature of the processes of capital accumulation growth creates more costs than benefits. See also Herman Daly, *Beyond Growth: The Economics of Sustainable Consumption* (Boston: Beacon Press, 1996).

67. *Stern Review: The Economics of Climate Change:* Executive Summary, x at http://www.hm-treasury.gov.uk/media/4/3/Executive_Summary.pdf.
68. This alienation was first identified by Marx in his theory of use value and alienation, and his account of 'commodity fetishism' in *Capital* Vol. 1 ch 3, trs. Samuel Moore and Edward Aveling (Moscow: Progress Publishers, 1887). However Marx's repair of the problem is inadequate as its origin is in the instrumentalism of industrial making. See Arendt, *Human Condition*, 162.
69. There is an extensive literature on this. On pathological consumption see Elizabeth C. Hirscshmann, 'The consciousness of addiction: toward a general theory of compulsive consumption', *Journal of Consumer Research* 19 (September 1992), 155–79. For an ancient reflection on the problem see John Chrysostom, *Homilies on First Timothy 12, 3–4*. See also Richard Layard, *Happiness: Lessons from a New Science* (London: Allen Lane, 2005).
70. See for example IPCC, *Climate Change 2001: Synthesis Report*, 115, figure 7.2, 'Projections of GDP losses and marginal costs in Annex II countries in the year 2010 from global models'; 113, Box 7.1 'Bottom-up and top-down approaches to cost estimates'.
71. Warwick J. McKibben and Peter J. Wilcoxen, 'The role of economics in climate change policy', *The Journal of Economic Perspectives* 16.2 (Spring 2002), 107– 29.
72. See for example Bjorn Lomborg, 'Global warming' in Lomborg, *The Skeptical Environmentalist* (Cambridge: Cambridge University Press, 2001), 258–324. See also the discussion of this issue in Stephen Gardiner, 'Ethics and global climate change', *Ethics* 114 (April 2004), 555–600.
73. William D. Nordhaus, *Warming the World: Economic Models of Global Warming* (Cambridge, MA: MIT Press, 2000).
74. McKibben and Wilcoxen, 'The role of economics in climate change policy'.
75. G. Cornelius van Kooten, *Climate Change Economics: Why International Accords Fail* (Cheltenham: Edward Elgar, 2002), 150.
76. Jason Shogren and Michael Toman, 'How much climate change is too much? An economics perspective', *Climate Changes Issues Brief* 25 (September 2000) at: http://www/rff/org/Documents/RFF-CCIB-25.pdf.
77. A similar approach is taken in Bjorn Lomborg, *The Skeptical Environmentalist* (Cambridge: Cambridge University Press, 2001), 315–19.
78. John Broome, 'Discounting the future', *Philosophy and Public Affairs* 23 (Spring 1994), 128–56.
79. Warwick J. McKibbin and Peter J. Wilcoxen, *Climate Change Policy After Kyoto: Blueprint for a Realistic Approach* (Washington DC: Brookings Institute Press, 2002), 3.
80. *Stern Review*, 278–9.
81. John Gowdy and Jon D. Erikson, 'The approach of ecological economics', *Cambridge Journal of Economics* 29.2 (March 2005), 209.
82. Emilio Padilla, 'Climate change, economic analysis and sustainable development', *Environmental Values* 13 (2004), 523–44.
83. S. Fankhauser, *Valuing Climate Change: The Economics of the Greenhouse* (London: Earthscan, 1995), 47.
84. Clive Spash, *Greenhouse Economics: Values and Ethics* (London: Routledge, 2002), 188–9.

85. Kamal Nath, cited in Spash, *Greenhouse Economics*, 190.
86. Harold A. Mooney and Paul R. Ehrlich, 'Ecosystems services: A fragmentary history' in Daily (ed), *Nature's Services*, 17. For a powerful critique of the regnant economic accounts of money see Geoffrey Ingham, *The Nature of Money* (Cambridge: Polity Press, 2004).
87. Thomas C. Shelling, 'The cost of combating global warming; facing the trade-offs,' *Foreign Affairs* 76.6 (November/December 1997), 8–14.
88. Joel Balkan, *The Corporation: The Pathological Pursuit of Profit and Power* (New York: Free Press, 2004) and McDonough and Braungart, *Cradle to Cradle*, 113.
89. Spash, *Greenhouse Economics*, 191–3.
90. Spash, *Greenhouse Economics*, 227.
91. I say putative because human longevity is a feature of many human cultures and not just industrial ones. Indeed in the early stages of the transition from agrarian to industrial human health and longevity actually markedly declines. Arendt, *Human Condition*. See also Marshall Sahlins, 'The original affluent society' in Sahlins, *Stone Age Economics* (Chicago: Aldine, 1972).
92. On this complex issue see the fuller discussion in Michael Northcott, *Life After Debt: Christianity and Global Justice* (London: SPCK, 1999).
93. Zygmunt Bauman, *Liquid Modernity* (Cambridge: Polity Press, 2000).
94. Ferdinand Hayek, *The Road to Serfdom* (Chicago IL: University of Chicago Press, 1944).
95. Richard Douthwaite, *The Growth Illusion: How Economic Growth Has Enriched the Few, Impoverished the Many and Endangered the Planet* (Gabriola Island, BC: New Society Publishers, 1999).
96. Patriarch Bartholomew, 'Address at the Oslo Sophie Prize Ceremony', 12 June 2002 at: http://www.sophieprize.org/Articles/48.html.
97. Juan Luis Segundo, *Liberation of Theology* trs. John Drury (Dublin: Gill and Macmillan,1977).
98. Pope John Paul II, *Sollicitudio Rei Socialis* (Rome: Libreria Editrice Vaticana, 1987), paragraph 36.
99. Sergei Bulgakov, 'The Lamb of God: On the Divine Humanity' in Rowan Williams, *Sergei Bulgakov: Towards a Russian Political Theology* (Edinburgh: T. and T. Clark, 1999), 214.
100. Bulgakov, 'The Economic Ideal', in Williams, *Sergei Bulgakov*, 28.
101. Bulgakov, 'The Economic Ideal', 47.
102. Bulgakov, 'The Economic Ideal', 35.
103. Bulgakov, 'The Economic Ideal', 49.
104. Arendt, *Human Condition*, 133.
105. Bulgakov, 'The Economic Ideal', 49.
106. Arendt, *Human Condition*, 127.
107. Arendt, *Human Condition*, 132.
108. Arendt, *Human Condition*, 132.
109. Arendt, *Human Condition*, 135.

5. Ethical Emissions

1. Teferra Haile-Selassie, *The Ethiopian Revolution 1974–1991: From a Monarchical Autocracy to a Military Oligarchy* (London: Kegan Paul International, 1977).
2. C. T. Agnew and A. Chappell, 'Drought in the Sahel', *Geojournal* 48 (1999), 299–311.
3. Ben Parker, *Ethiopia: Breaking New Ground* (Oxford: Oxfam, 1995).
4. A. Giannini, R. Saravanan, P. Chong, 'Oceanic forcing of Sahel rainfall on

interannual to interdecadal time scales', *Science* 302 (2003), 1027–30.

5. J. Bader and M. Latif, 'The impact of decadal-scale Indian Ocean sea surface temperature anomalies on Sahelian rainfall and the North Atlantic Oscillation', *Geophysical Research Letters* 30 (November 2003), 7.1–7.4. Also James Verdin, Chris Funk *et al.*, 'Climate science and famine early warning', *Philosophical Transactions of the Royal Society B* (2005), 360, 2155–68.
6. M. Hulme, R. Doherty *et al.*, 'African Climate Change: 1900–2100', *Climate Research* 17 (2001), 2145–68.
7. Pari *et al.*, cited Bill Hare, 'Relationship between increases in global mean temperature and impacts on ecosystems, food production, water, and socio-economic systems', *Proceedings of Avoiding Dangerous Climate Change,* February 1–3 2005, Exeter, Met Office, at http://www.stabilisation2005.com/programme2.html.
8. Sari Kovats, 'The global burden of disease due to climate change', *Proceedings of Avoiding Dangerous Climate Change,* February 1–3 2005.
9. *Presentation of Robert Watson at the Sixth Conference of the Parties: Proceedings of COP 6* (2000) cited Jekwu Ikeme, 'Climate Change Adaptational Deficiencies in Developing Countries: The Case of Sub-Saharan Africa', *Mitigation and Adaptation Strategies for Climate Change* 8 (2003), 29–52.
10. Rachel Warren, *Impacts of Global Climate Change at Different Annual Mean Temperature Increases,* Working Paper, Tyndall Centre for Climate Change Research, 2004.
11. Interview with Pius Ncube quoted in *Africa Up in Smoke? The Second Report from the Working Group on Climate Change and Development* (London: New Economics Foundation, 2005), 11.
12. Wangari Maathai, *The Greenbelt Movement* (Nairobi: Wangari Maathai, 1985). See also Dianne E. Rocheleau, 'Gender ecology and the science of survival: stories and lessons from Kenya', *Agriculture and Human Values* (1991), 156–65.
13. Archbishop Donald Mtemela, Keynote Address to the GW8 conference, 'Dynamic Earth', Edinburgh, 8 July 2005.
14. Jekwu Ikeme, 'Equity, Environmental Justice and Sustainability: Incomplete Approaches in Climate Change Politics', *Global Environmental Change* (2003), 195–206.
15. Ikeme, 'Equity, Environmental Justice and Sustainability', 195–9.
16. Bhaskar Vira, 'Trading with the Enemy? Examining North-South Perspectives in the Climate Change Debate' in Daniel W. Bromley and Jouni Paavola (eds), *Economics, Ethics and Environmental Policy: Contested Choices* (Oxford: Blackwell, 2002), 164–80.
17. See further Northcott, *The Environment and Christian Ethics* (Cambridge: Cambridge University Press, 1996), 169–78.
18. There is an excellent account of the developing universalism of the Israelite sense of calling in the Old Testament, and its further elaboration in the New Testament, in Donald Senior and Carroll Stuhlmueller, *The Biblical Foundations for Mission* (Maryknoll, NY: Orbis Books, 1983).
19. Allan C. Wilson, and Rebecca L. Cann, 'The Recent African Genesis of Humans', *Scientific American* 266 (1992), 68–73.
20. H. Gamoran, 'The biblical laws against loans on interest', *Journal of Near Eastern Studies* 30 (1971), 127–34.
21. John Rawls, *A Theory of Justice* (Oxford: Oxford University Press, 1973).
22. Rawls, *A Theory of Justice.*
23. Hannah Arendt, *The Human Condition* (Chicago: Chicago University Press,

1958), 157.

24. Arendt, *Human Condition*, 157.
25. Arendt, *Human Condition*, 157.
26. Arendt, *Human Condition*, 166.
27. Arendt, *Human Condition*, 166.
28. See further Michael S. Northcott, 'The theology of making: concept art, clones and co-creators', *Modern Theology* 21 (April 2005), 219–36.
29. See the extensive exposition of this problem in Sheldon Wolin, *Politics and Vision: Continuity and Innovation in Western Political Thought*, expanded edn (Princeton, NJ: Princeton University Press, 2004), 248–97.
30. Liam B. Murphy, 'Institutions and the demands of justice', *Philosophy and Public Affairs* 27 (1998), 251–91.
31. Arendt, *Human Condition*, 159.
32. Jon Garthoff defends this Rawlsian division of labour in his essay 'Zarathrustra's dilemma and the embodiment of morality', *Philosophical Studies* 117 (2004), 259–74.
33. Leo Strauss, *Liberalism Ancient and Modern* (Chicago: University of Chicago Press, 1995).
34. See the fuller account of this development in Northcott, *An Angel Directs the Storm: Apocalyptic Religion and American Empire* (London: I. B. Tauris, 2004), 80–7.
35. David Adams and Rob Evans, 'New figures reveal scale of industry's impact on climate', *Guardian*, 16 May 2006.
36. Richard Bauckham, *God and the Crisis of Freedom: Biblical and Contemporary Perspectives* (Louisville, KY: Westminster John Knox Press, 2002), 33–6.
37. Stephen Bocking, *Nature's Experts: Science, Politics and the Environment* (NJ: Rutgers University Press, 2004).
38. *Human Development Report 2005* (Geneva: United Nations Development Programme, 2006).
39. Martha C. Nussbaum, *Frontiers of Justice: Disability, Nationality, Species Membership* (Cambridge, MA: Belknap Press of Harvard University Press, 2006), 74. The reference to 'truly human functioning' is from Karl Marx's 1844 *Economic and Philosophical Manuscripts.*
40. Nussbaum, *Frontiers of Justice*, 89.
41. Rawls, *Theory of Justice*, 8.
42. On the imperial nature of present American foreign and economic policies see Northcott, *An Angel Directs the Storm.*
43. Nussbaum, *Frontiers of Justice*, 236.
44. Hugo Grotius, *The Rights of War and Peace,* Book 1, edited by Richard Tuck (Indianapolis: Liberty Fund, 2005).
45. Oliver O'Donovan, *Ways of Judgment* (Grand Rapids, MI: Wm. B. Eerdmans, 2006), 218–19.
46. Grotius, *Rights of War and Peace*, 201.
47. Grotius, *Rights of War and Peace*, 347–8.
48. O'Donovan, *Ways of Judgment*, 225–8.
49. Stanley Hauerwas makes a similar criticism of Nussbaum's *Frontiers of Justice* in his paper 'The politics of gentleness: random thoughts for a conversation with Jean Vanier' given at a public 'Conversation with Jean Vanier' at the University of Aberdeen in September 2006.
50. Michael Cowen and Robert W. Shenton, *Doctrines of Development* (London and New York: Routledge, 1996).
51. For a fuller account of the many disastrous development projects sponsored by

the World Bank and other official aid agencies see further Michael Northcott, *Life After Debt: Christianity and Global Justice* (London: SPCK, 1999).

52. James C. Scott, *Seeing Like a State: How Certain Schemes to Improve the Human Condition Have Failed* (New Haven, CT: Yale University Press, 1998), 286.
53. Scott, *Seeing Like a State*, 253–4.
54. John Gray, *False Dawn: The Delusions of Global Capitalism* (London: Granta, 1998).
55. Val Plumwood, *Environmental Culture: The Ecological Crisis of Reason* (London: Routledge, 2002).
56. Monbiot, *Heat*, xii.
57. As I write this 1.6 million British citizens have just signed the largest petition in British history calling on the government to abandon proposals to enact road-pricing procedures designed to restrict car driving.
58. S. Seacrest, R. Kuzelka and R. Leonard, 'Global climate change and public perception: the challenge of translation' cited in Thomas Lowe, Katrina Brown *et al.*, *Does Tomorrow Ever Come? Disaster Narrative and Public Perceptions of Climate Change,* Tyndall Centre for Climate Change Research Working Paper 72 (March 2005), 7.
59. Wooter Pourtinga, Nick Pidgeon and Irene Lorenzoni, *Public Perceptions of Nuclear Power, Climate Change and Energy Options in Britain: Summary Findings of a Survey Conducted During October and November 2005*, Understanding Risk Working Paper (2006) 2, School of Environmental Sciences, University of East Anglia, 13.
60. Pourtinga, Pidgeon and Lorenzoni, *Public Perceptions of Nuclear Power*, 15.
61. Maurice Merleau-Ponty, *Phenomenologie de la Perception,* 493–4, trs. and cited in Richard M. Zaner, *The Problem of Embodiment: Some Contributions to a Phenomenology of the Body* (The Hague: Martinus Nijhoff, 1971), 191.
62. Zaner, *Problem of Embodiment*, 191.
63. Marcel Mauss, 'Les Techniques du Corps', cited in Thomas J. Csordas, 'Introduction: the body as representation and being-in-the-world', in Csordas (ed.), *Embodiment and Experience: The Existential Ground of Culture and Self* (Cambridge: Cambridge University Press, 1994), 1–24.
64. David Bohm, *Wholeness and the Implicate Order* (London and Boston: Routledge and Kegan Paul, 1980); Gregory Bateson, *Steps to an Ecology of Mind: Collected Essays in Anthropology, Psychiatry, Evolution and Epistemology* (St Albans: Paladin, 1973); J. J. Gibson, *The Ecological Approach to Visual Perception* (Boston, MA: Houghton Mifflin, 1979). See also Ian Burkitt, *Bodies of Thought: Embodiment, Identity and Modernity* (London: Sage, 1999).
65. On the growing role of embodiment in Artificial Intelligence and cognitive science see A. Clark, *Being There – Putting Brain, Body and World Together Again* (Cambridge MA: MIT Press, 1997).
66. See further Stanley Hauerwas and David Burrell, 'From system to story: an alternative pattern for rationality in ethics', 158–190 in Stanley Hauerwas and L. Gregory Jones (eds), *Why Narrative: Readings in Narrative Theology* (Grand Rapids, MI: William B Eerdmans, 1989); see also Alasdair MacIntyre, *After Virtue: A Study in Moral Theory* (London: Duckworth, 1981).
67. The significance of narrative is beginning to emerge as a key theme in cognitive science as researchers in the fields of artificial intelligence and robotics attempt to chart and describe, so that they may ultimately be able to replicate, all that makes human mental and physical agency possible; see further Chrystopher L. Nehaniv and Kerstin Dautenhahn, 'Embodiment and

Memories: Algebras of Time and History for Autobiographic Agents' at: http://homepages.feis. herts.ac.uk/~comqkd/em6pp.ps.

68. Jean-Pierre Dupuy argues that the collective modern commitment to the ethereal economy of market relations, despite its relative detachment from the real relations and welfare of bodies in society, arises from an idealist strain at the heart of modern economics which he exposes through an examination of inconsistencies between anarcho-capitalist and morally conservative tendencies in the work of Ferdinand Hayek: Jean-Pierre Dupuy, 'Intersubjectivity and Embodiment', *Journal of Bioeconomics*, 6 (2004), 275–94.
69. Alistair McFadyen, *Bound to Sin: Abuse, Holocaust and the Christian Doctrine of Sin* (Cambridge: Cambridge University Press, 2000), 27–8.
70. McFadyen, *Bound to Sin*, 37.
71. Val Plumwood, 'Nature, self and gender: feminism, environmental philosophy, and the critique of rationalism' in Robert Elliot (ed.), *Environmental Ethics* (Oxford: Oxford University Press, 1995), 160.
72. McFadyen, *Bound to Sin*, 161.
73. On the significance of the words 'on earth' in the second petition of the Lord's Prayer see James Jones, *Jesus and the Earth* (London: SPCK, 2003), 20.
74. I have described the significance of the Trinitarian divine nature for an appreciation of ecological and social participation more fully in Northcott, *Environment and Christian Ethics.*
75. Oliver O'Donovan, *Common Objects of Love* (Grand Rapids, MI: Eerdmans, 2002), 13.
76. McFadyen, *Bound to Siné* 161–2.
77. O'Donovan, *Common Objects of Love*, 16–17.
78. On the analogy of scripture and nature as the two books of God see R. J. Berry, *God's Book of Works: The Nature and Theology of Nature* (London: Continuum, 2003).
79. Arendt, *Human Condition*, 15.
80. Augustine, *City of God*, trs. O'Donovan in *Common Objects of Love,* 20.
81. Esther de Waal, *A World Made Whole: The Rule of St Benedict* (London: Darton, Longman and Todd, 1995).

6. Dwelling in the Light

1. The Kogi story was taken up by an anthropologist filmmaker, Alan Ereira. The resulting film *From the Heart of the World: The Elder Brother's Warning* (London: BBC, 1992) was made with the active collaboration of the Kogi and is a remarkable piece of environmental filming. See further Peter Loizos, 'First Exits from Observational Realism: Narrative Experiments in Recent Ethnographic Films', in Marcus Banks and Howard Morphy (eds), *Rethinking Visual Anthropology* (New Haven: Yale University Press, 1997), 85.
2. Remi Brague, *The Wisdom of the World: The Human Experience of the Universe in Western Thought*, trs. Teresa Fagan (Chicago: University of Chicago Press, 2003).
3. George Mackay Brown, *Winter Tales* (London: Polygon, 1995).
4. J. B. Harris, 'Electric lamps, past and present', *Engineering Science and Education Journal* 2 (August 1993), 161–70.
5. Asheville, North Carolina is one such place where the beauty of the Appalachians is marred by commercial light, commodified spaces and forests of tall ugly neon signs. The makers of the film *Cold Mountain* had to go to Romania to find mountain environments which looked like those of America at the time of the Civil War. Cold Mountain itself is still there but its lower

slopes and surrounding hills are crowded with golf courses, mobile homes, motels and the other paraphernalia of industrialised recreation.

6. Bryan Brock, *Singing the Ethos of God: On the Place of Scripture in Christian Ethics* (Grand Rapids, MI: Wm. B. Eerdmans, 2007), 342.
7. See further A. N. Williams, *The Ground of Union: Deification in Aquinas and Palamas* (Oxford: Oxford University Press, 1999).
8. Paul Halsall, 'Hesychasm' in *Internet Medieval Sourcebook,* Fordham University Center for Medieval Studies (1997) at: http://www.fordham.edu/halsall/source/hesychasm1.html.
9. Leonid Ouspenski and Vladimir Lossky, *The Meaning of Icons,* revised edn (Crestwood, NY: St Vladimir's Seminary Press, 1982).
10. St Symeon the New Theologian, *The Catechetical Discourses* XXII trs. C. J. De Cantazaro (New York: Paulist Press, 1980).
11. Rudolf Clausius, 'On the motive power of heat and on the laws which can be deduced from it for the theory of heat' in E. Clapeyron and R. Clausius, *Reflections on the motive power of fire by Sadi Carnot and other papers on the Second Law of Thermodynamics,* edited with an introduction by E. Mendoza (London: Dover Press, 1960).
12. Teresa Brennan, *Exhausting Modernity: Grounds for a New Economy* (London: Routledge, 2001), 78–9.
13. Hans U. Fuchs, 'Thermodynamics: A Misconceived Theory' in J. D. Novak (ed.), *Second International Seminar on Misconceptions and Educational Strategies in Science and Mathematics, July 26–29, 1987: Proceedings,* Volume III (Ithaca, NY: Cornell University Press, 1987), 160–7.
14. Oliver O'Donovan, *Resurrection and Moral Order: An Outline for Evangelical Ethics* (Leicester: Apollos, 1986), 125.
15. David Hallman, 'The WCC Climate Change Programme: History, Lessons and Challenges' in Martin Robra (ed.), *Climate Change* (Geneva: World Council of Churches Justice Peace and Creation Team, 2005), 13.
16. See further *Solidarity With Victims of Climate Change: Reflections on the World Council of Churches' Response to Climate Change* (Justice Peace and Creation Team, Geneva: WCC, 2002) at: http://www.wcc-coe.org/wcc/what/jpc/ climatechange.pdf.
17. Lukas Vischer, 'Climate Change, Sustainability and Christian Witness', *The Ecumenical Review* 49 (1997), 142–9.
18. International Energy Agency, *Earth Trends Data Tables: Energy and Resources: Energy 2005* http://earthtrends.wri.org/pdf_library/data_tables/ene1_2005.pdf.
19. See for example the Canadian interfaith programme 'Greening Sacred Spaces: Living faithfully – Living Green' at: http://www.faith-commongood.net/greenspaces/index.asp. See also 'Sustainable Housekeeping through Environmental Management Systems' European Christian Environmental Network, 2003, at http://www.ecen.org/cms/index.php?mact=News,cntnt01,detail,0&1cntnt01articleid=6&cntnt01dateformat=%25e%20%25B%20%25G&cntnt01returnid=44.
20. Press Release: 'Church launches "Shrinking the Footprint" Campaign', Church of England Communications Office, Church House, 2 June 2006.
21. Sigurd Bergmann, 'Theology in its Spatial Turn: Space, Place and Built Environments Challenging and Changing the Images of God' in *Religion Compass* 1 (2007), 1–25.
22. Interview with Anthony Lawlor, *The Monthly Aspectarian,* (1998) http://www.meditatie.net/sthapatya/lawlor.html.

23. Kenneth Clark, *Civilisation: A Personal View* (Harmondsworth: Penguin, 1982), 36–41.
24. Anthony Lawlor and Jeremy P. Tarcher, *The Temple in the House: Finding the Sacred in Everyday Architecture* (East Rutherford, NJ: Putnam, 1994).
25. Victor Papanek, *The Green Imperative: Ecology and Ethics in Design and Architecture* (London: Thames and Hudson, 1995), 124 – 6.
26. Lawlor and Tarcher, *The Temple in the House*, 121–7.
27. Le Corbusier, *Towards a New Architecture,* trs. Frederick Etchells (London: Architectural Press, 1946).
28. William McDonough and Michael Braungart, *Cradle to Cradle: Remaking the Way We Make Things* (New York: North Point Press, 2002).
29. Stephen Wiel, Nathan Martin *et al.*, 'The role of building energy efficiency in managing atmospheric carbon dioxide', *Environmental Science and Policy* 1 (1998), 27–38.
30. 'What is a passive house?' at: http://www.passivehouse.com/English/Passiveh.html.
31. Brenda Boardman, Gavin Killip *et al.*, *The 40% House Report* (Oxford: University of Oxford Environmental Change Institute: 2005) at: http://www.eci.ox.ac.uk/lowercf/40house.html.
32. Monbiot, *Heat*, 66–8.
33. 'Households living in fuel poverty' in *Sustainable Development: The Government's Approach* (London: Sustainable Development Unit, Department for Environment, Food and Rural Affairs, 2005): http://www.sustainable-develop ment.gov.uk/regional/summaries/63.htm.
34. Department of Trade and Industry, *The Carbon Challenge*, Executive Summary (2006).
35. 'Clean Energy: The European Deficit', *Power Economics* (5 May 2005), 7–10.
36. Catherine Casebolt, 'Home alone: living off the grid', *Home Energy Magazine Online May/June 1993,* http://homeenergy.org/archive/hem.dis.anl.gov/eehem/93/930509.html.
37. M. A. White and F. A. Barata, 'State-of-the-art mountain top removal and contour mining', *International Journal of Rock Mechanics, Mining Sciences and Geomechanics* (July 1996), 232–43, and Erik Reece, 'Moving Mountains', *Orion* magazine (January/February 2006) at: http://www.oriononline.org/pages/om/06-1om/Reece.html.
38. Heather Lovell, 'Supply and Demand for Low Energy Housing in the UK: Insights from a Science and Technology Studies Approach', *Housing Studies* 20 (September 2005), 815–29.
39. David Hallman, 'Climate change and ecumenical work for sustainable community' in Dieter Hessell and Larry Rasmussen (eds), *Earth Habitat: Eco-Injustice and the Earth's Response* (Indiana: Fortress Press, 2001), 125–34.
40. Vijay B. Vaitheeswaran, *Power to the People: How the Coming Energy Revolution Will Transform an Industry, Change Our Lives and Maybe Even Save the Planet* (London: Earthscan, 2005).
41. David Reay, *Global Warming Begins at Home* (London: Macmillan, 2005).
42. Ellen Davis, 'The Tabernacle is Not a Storehouse: Building Sacred Space', *Sewanee Theological Review*, 49 (Pentecost 2006), 305–19.
43. Davis, 'The Tabernacle is Not a Storehouse', 311.
44. Timothy Gorringe, *A Theology of the Built Environment* (Cambridge: Cambridge University Press, 1998).
45. Wendell Berry, *The Unsettling of America: Culture and Agriculture* (San Francisco:

Sierra Club Books, 1996), 219.
46. See the excellent biography of Fiona McCarthy, *William Morris: A Life for Our Time* (London: Faber and Faber, 1994).
47. Gorringe, *Theology of the Built Environment*, 90–101.
48. Gorringe, *Theology of the Built Environment*, 99–100.
49. Parts of what follows occur in fuller form in my essay 'The Word in Time and Space' in John Vincent and Peter Francis (ed.), *Faithfulness in the City* (Hawarden, Wales: Monad Press, 2003), 244–65.
50. For an account of the Morris Workshops see MacCarthy, *William Morris*.
51. John Ruskin, *Sesame and Lilies*, cited Richard Sennett, *The Conscience of the Eye: The Design and Social Life of Cities* (London: Faber and Faber, 1990), 20.
52. John K. Galbraith, *The Affluent Society*, 4th British edn (London: Penguin, 1987).
53. Sennett, *Conscience of the Eye*, 29.
54. See further Northcott, 'The Word in Time and Space', 252–4.
55. See further Michael S. Northcott, 'The theology of making: concept art, clones and co-creators', *Modern Theology* 21 (April 2005), 219–36.
56. Metropolitan John of Pergamon, 'Proprietors or priests of creation?' Keynote address of the Fifth Symposium of Religion, Science and the Environment, 2 June 2003: http://www.orthodoxytoday.org/articles2/MetJohnCreation.php.
57. John Milbank, 'On Complex Space' in John Milbank, *The World Made Strange: Theology, Language and Culture* (Oxford: Blackwell, 1997), 277.
58. Sennett, *Conscience of the Eye*, 22–7 and Milbank, 'On Complex Space', 279.
59. John Muir, *My First Summer in the Sierra* (Edinburgh: Canongate Classics, 1988).

7. Mobility and Pilgrimage

1. Friedrich Engels, *The Condition of the Working Classes in England*, 1845.
2. E. A. Wrigley, 'The supply of raw materials in the Industrial Revolution' in J. Hoppit and E. A. Wrigley (eds): *The Industrial Revolution in Britain* (Oxford: Blackwell, 1994).
3. Henry Ford, in collaboration with Samuel Crowther, *My Life and Work* (New York: Garden City Publishing, 1922).
4. World Council of Churches, Justice, Peace and Creation, *Mobile – but not Driven: Towards Equitable and Sustainable Mobility and Transport* (WCC: Geneva, 2002): http://www.wcc-coe.org/wcc/what/jpc/mobile.pdf.
5. Commission for Global Road Safety, *Make Roads Safe: A New Priority for Sustainable Development* (London: CGRS, 2007).
6. Department of Trade and Industry, *Energy Trends 2005* (London: DTI, 2005), Table 1.3a, 'Supply and use of fuels', 33.
7. US Federal Government Energy Information Administration, *Energy Perspectives 2005*, figure 7: http://eia.doe.gov/emeu/aer/ep/ep_text.html.
8. One life-cycle estimate of energy use in automobiles estimates energy used in manufacture at 10 per cent of the total used in a 14-year vehicle life, 15 per cent in maintenance, servicing, insuring and financing and 75 per cent in petroleum: Heather L. MacLean and Lester B. Lave, 'A life-cycle model of an automobile', *Environmental Policy Analysis* 3 (1988), 322–30A.
9. Energy Information Administration, *Energy Perspectives*.
10. Joyce E. Penner, David H. Lister *et al.*, *IPCC Special Report: Aviation and the Global Atmosphere: Summary for Policymakers* (IPCC, 1999), 8 at: http://www.earthscape.org/p1/ipc01/ipc01.pdf.
11. J. M. W. Turner, *Rain, Steam and Speed: The Great Western Railway* (1844) at: http://www.j-m-w-turner.co.uk/artist/turner-rain-steam.htm.

12. John Urry, 'Automobility, Car Culture and Weightless Travel: A discussion paper', Department of Sociology, Lancaster University (1999) at: http://www.comp.lancs.ac.uk/sociology/papers/Urry-Automobility.pdf.
13. Paul Virilio, 'Speed and Information: Cyberspace Alarm!', *Ctheory.net* (August 1995) at: http://www.ctheory.net/articles.aspx?id=72.
14. Paul Virilio, 'The Museum of Accidents', *International Journal of Baudrillard Studies* 3 (July 2006) at: http://www.ubishops.ca/baudrillardstudies/vol3_2/virilio.htm.
15. Sandy Baldwin, 'On Speed and Ecstasy: Paul Virilio's "Aesthetics of Disappearance" and the Rhetoric of Media', *Configurations* 10 (2002), 130.
16. Milan Kundera, *Slowness,* trs. Linda Asher (London: Faber and Faber, 1996), 4.
17. Informant quoted in Pauline Garvey, 'Driving, Drinking and Daring in Norway' in Daniel Miller (ed.), *Car Cultures* (Oxford: Berg, 2001), 133–52.
18. Eileen Spencer, *Car Crime and Young People on a Sunderland Housing Estate,* Police Research Group: Crime Prevention Unit Series, Paper No. 40 (London: Home Office Police Department, 1992) at: http://www.homeoffice.gov.uk/rds/prgpdfs/fcpu40.pdf.
19. On the emotional phenomenology of car driving see further Mimi Sheller, 'Automotive Emotions: Feeling the Car', *Theory, Culture and Society* 21 (2004), 221–42.
20. Oliver O'Donovan, 'The Loss of a Sense of Place' in Oliver O'Donovan and Joan Lockwood O'Donovan, *Bonds of Imperfection: Christian Politics Past and Present* (Grand Rapids, MI: Eerdmans, 2004), 306.
21. Romand Coles, 'The Wild Patience of John Howard Yoder: Outsiders and the Otherness of the Church', *Modern Theology* 18 (July 2002), 306–31.
22. Donald B. Kraybill, *The Riddle of Amish Culture* (Baltimore: John Hopkins University Press, 1989), 165–8.
23. Donald B. Kraybill, 'Introduction: The Struggle to be Separate' in Donald B. Kraybill and Marc A. Olshan (eds.), *The Amish Struggle With Modernity* (New Haven, CT: University Press of New England, 1994), 7.
24. Wendell Berry, *The Unsettling of America: Culture and Agriculture* (San Francisco: Sierra Club Books, 1996), 176.
25. John M. Theilmann, 'Medieval Pilgrims and the Origins of Tourism', *Journal of Popular Culture* 20 (Spring 1987), 94.
26. See further David Frankfurter, 'Introduction' in Frankfurter (ed.), *Pilgrimage and Holy Space in Late Antique Egypt* (Leiden: Brill, 1998), and R. A. Markus, 'How on earth could places become holy? Origins of the Christian Idea of Holy Places', *Journal of Early Christian Studies* 2 (1994), 257–71.
27. Letters of Saint Jerome, 46. 10, at: http://www.newadvent.org/fathers/3001046.htm.
28. John C. Olin, 'The Idea of Pilgrimage in the Experience of Ignatius of Loyola', *Church History* 48 (December 1979), 387–97.
29. Eamon Duffy, *The Stripping of the Altars: Traditional Religion in England c.1400–c.1580* (New Haven, CT: Yale University Press, 1992), 191.
30. Victor Turner and Edith Turner, *Image and Pilgrimage in Christian Culture: Anthropological Perspectives* (Oxford: Basil Blackwell, 1978), 6–8.
31. Blake Leyerle identifies a heightened role of healing in records of pilgrimages to the Holy Land between the fourth and fifth centuries: Blake Leyerle, 'Landscape as Cartography in Early Christian Pilgrimage Narratives', *Journal of the American Academy of Religion* 64 (Spring 1996), 119–43.
32. Turner and Turner, *Image and Pilgrimage,* 11.

33. Turner and Turner, *Image and Pilgrimage*, 1 and 33–4.
34. John Urry's concept of 'meetingness' bears strong analogies with the Turners' account of *communitas*: Urry, 'Social networks, travel and talk', *The British Journal of Sociology* 54 (June 2003), 155–75.
35. Analogously in modernity, Urry suggests that mobility is an important source of social networks beyond the local: see Urry, 'Social networks, travel and talk', 162–3.
36. *La Voz Guadalupana* (July 1940) in Turner and Turner, *Image and Pilgrimage*, 95.
37. Richard Niebuhr, unpublished lecture, as cited in Corelyn F. Senn, 'Journeying as Religious Education: The Shaman, the Hero, the Pilgrim and the Labyrinth Walker', *Religious Education* 97 (Spring 2002), 124.
38. O'Donovan, 'The Loss of a Sense of Place', 306.
39. Senn, 'Journeying as Religious Education', 129–130.
40. Extract from Stanley Hauerwas, *The Peaceable Kingdom* in John Berkman and Michel Cartwright (eds), *The Hauerwas Reader* (Durham, NC: Duke University Press, 2001), 126–7.
41. Blake Layerle, 'Landscape as Cartography in Early Christian Pilgrimage Narratives', *Journal of the American Academy of Religion* 64 (Spring 1996), 119–43.
42. Layerle, 'Landscape as Cartography', 126–7.
43. Arne Naess, *Ecology, Community and Lifestyle: Outline of an Ecosophy*, trs. David Rothenberg (New York: Cambridge University Press, 1989).
44. Val Plumwood, 'Nature, self and gender: feminism, environmental philosophy, and the critique of rationalism' in Robert Elliot (ed.), *Environmental Ethics* (Oxford: Oxford University Press, 1995), 160.
45. Val Plumwood, 'Prey to a crocodile', *Aisling Magazine* 30 (2002): http://www.aislingmagazine.com/aislingmagazine/articles/TAM30/ValPlumwood.html.
46. See further Simone Fullagar, 'Desiring Nature: Identity and Becoming in Narratives of Travel', *Cultural Values* 4 (2000), 58–76.
47. Thomas Eriksen, *Tyranny of the Moment: Fast and Slow Time in the Information Age* (London: Pluto Press, 2001), 53.
48. *The Way of a Pilgrim*, trs. R. M. French (NY: Ballantine Books, 1979).
49. Paul Call, 'The Way of a Pilgrim and the Pilgrim Continues His Way', *Slavic Review* 22 (March 1963), 172.
50. Kelly Winters, *Walking Home: A Woman's Pilgrimage on the Appalachian Trail* (Los Angeles: Alyson., 2001).
51. Nancy Humpel, Alison Marshal *et al.*, 'Changes in Neighborhood Walking Are Related to Changes in Perceptions of Environmental Attributes', *Annals of Behavioral Medicine* 27 (2004), 60–7.
52. Richard Bauckham, *God and the Crisis of Freedom: Biblical and Contemporary Perspectives* (Louisville, KY: Westminster John Knox Press, 2002), 36–7.
53. Submission by Dr Alan Storkey to the Transport Committee Inquiry into Bus Services across the UK, Select Committee on Transport, House of Commons, 5 May 2006 at: http://www.publications.parliament.uk/pa/cm200506/cmselect/cmtran/1317/1317am04.htm. See also Monbiot, *Heat*, 148–54.
54. Kundera, *Slowness*, 34.

8. Faithful Feasting

1. Brian Fagan, *The Long Summer: How Climate Changed Civilization* (New York: Basic Books, 2004).
2. Colin Tudge, *So Shall We Reap: What's Gone Wrong With the World's Food – And How to Fix It* (London: Penguin, 2003), 52–7.

3. Marshall Sahlins, 'The original affluent society' in Sahlins, *Stone Age Economics* (Chicago: Aldine, 1972).
4. Tudge, *So Shall We Reap*, 54.
5. William J. Burroughs, *Climate Change in Prehistory: The End of the Reign of Chaos* (Cambridge: Cambridge University Press, 2005), 192.
6. Ched Myers, 'Cultural/Linguistic Diversity and Deep Social Ecology (Genesis 11:1–9)' at: http://www.thewitness.org/agw/myers.032802.a.html.
7. Evan Eisenberg, *The Ecology of Eden* (New York: Alfred A. Knopf, 1998), 83.
8. Colin M. Turnbull, *The Forest People* (London: Chatto and Windus, 1961) and Peter Bellwood, *First Farmers: The Origins of Agricultural Societies* (Oxford: Blackwell, 2005).
9. IPCC, *Fourth Assessment Report*, Working Group II: 'Climate Change Impacts: Technical Summary' (Geneva: IPCC, 2007).
10. Paula A. Harrison, 'Climate Change and Agriculture in Europe: Assessment of Impacts and Adaptation' in O'Riordan (ed.), *Agriculture and Climate Change*, 20–6.
11. J. Reilly, F. Tubiello, B. McCarl *et al.*, 'US Agriculture and Climate Change: New Results', *Climatic Change* 57 (2003), 43–69.
12. David B. Lobell and Christopher Field, 'Global scale climate-crop yield relationships and the impacts of recent warming', *Environmental Research Letters 2* (March 2007), 014002.
13. M. L. Parry, C. Rosenzweig, A. Iglesias, M. Livermore, G. Fischer, 'Effects of climate change on global food production under SRES emissions and socio-economic scenarios', *Global Environmental Change* 14 (2004), 53–67.
14. Fred Pearce, *The Last Generation: How Nature Will Take Her Revenge for Climate Change* (London: Transworld, 2006), 248–9.
15. On the ambiguous ecological and social outcomes of the green revolution see P. B. R. Hazell, C. Ramasamy and P. K Aiyasamy, *The Green Revolution Reconsidered: the Impact of High-Yielding Rice Varieties in South India* (New Delhi; Oxford University Press, 1993). On the climatological implications see Pekka Kaupi and Roger Sedjo *et al.*, 'Technological and Economic Potential of Options to Enhance, Maintain, and Manage Biological Carbon Reservoirs and Geo-engineering' in B. Metz *et al.*, *IPCC Third Assessment Report: Climate Change 2001: Mitigation* (Cambridge: Cambridge University Press, 2001), 303–53.
16. Jules Pretty and Andrew Ball, 'Agricultural Influences on Carbon Emissions and Sequestration: A Review of Evidence and the Emerging Trading Options', *Centre for Environment and Society and Department of Biological Sciences, University of Essex,* UK Centre for Environment and Society Occasional Paper, March 2001, University of Essex, 5.
17. David Pimentel and Marcia Pimentel, 'Sustainability of meat-based and plant-based diets and the environment', *American Journal of Clinical Nutrition* (2003) 78 (supplement), 660–3.
18. Annika Carlsson-Kanyama, 'Climate change and dietary choices – how can emissions of greenhouse gases from consumption be reduced?', *Food Policy* 23 (1998), 277–93.
19. Jeremiah 4:23 as cited in Margaret Barker, *The Great High Priest: The Temple Roots of Christian Liturgy* (London: T. and T. Clark, 2003), 45.
20. Mary Douglas, cited Barker, *Great High Priest,* 45.
21. *Jewish Encyclopaedia*, cited Barker, *Great High Priest*, 47.
22. Barker, *Great High Priest*, 51.
23. Isaiah 52:15 and 53:4–5.
24. Isaiah 53:5.

25. Barker, *Great High Priest*, 55.
26. On early Christian attitudes to animals see Richard Bauckham, 'Jesus and Animals II: What did he practise?' in Andrew Linzey, ed., *Animals on the Agenda: Questions About Animals for Theology and Ethics* (London: SCM Press, 1998), 49–60.
27. Prayer of Basil the Great quoted in John Passmore, 'The treatment of animals', *Journal of the History of Ideas* 36 (April–June 1975), 195–218.
28. Andrew Linzey, *Animal Theology* (London: SCM Press, 1994).
29. The classic work on this is Peter Singer, *Animal Liberation*, 2nd edn (London: Jonathan Cape, 1990), though Singer presumes erroneously that modern ethical attitudes to, and treatment of, animals represent an ethical advance over predecessor cultures, which is quite contrary to the historical evidence.
30. Danielle Nierenberger, 'A fowl plague', *World Watch Magazine* (January/February 2007) at: http://www.worldwatch.org/node/4779 and 'Bird flu crisis: Small farms are the solution not the problem', *Grain* (July 2006) 24–8 at: http://www.grain.org/seedling_files/seed-06-07-11.pdf.
31. B. A. Swinburn, I. Caterson, J. C. Seidell, and W. P. T. James, 'Diet, nutrition and the prevention of excess weight gain and obesity', *Public Health Nutrition* 7 (February 2004), 123–46, and S. A. Bingham, 'High meat diets and cancer risk', *Proceedings of the Nutrition Society* 58 (1999), 243–8.
32. Michael Pollan, 'Our national eating disorder', *New York Times* (17 October 2004) and Tim Lang and Geof Rayner, 'Obesity: a growing issue for European health policy?', *Journal of European Social Policy* 15 (2005), 301–27.
33. Industrially produced sugars and sweeteners are also novel foods with equally problematic health and ecological implications; Pollan, 'Our national eating disorder'.
34. Mike Fitzpatrick, 'Soy formulas and the effects of isoflavines on the thyroid', *New Zealand Medical Journal* 113 (11 February 2000), 24–6; Clinton D. Allred, Kimberley F. Allred, Young H. Ju *et al.*, 'Soy Diets Containing Varying Amounts of Genistein Stimulate Growth of Estrogen-dependent (MCF-7) Tumors in a Dose-dependent Manner', *Cancer Research* 61 (1 July 2001), 5045–50; R. M. Sharpe, B. Martin, K. Morris *et al.*, 'Infant feeding with soy formula milk: effects on the testis and on blood testosterone levels in marmoset monkeys during the period of neonatal testicular activity', *European Society of Human Reproduction and Embryology* 17 (2002), 1692–1703.
35. L. Chang, 'Review article: epidemiology and quality of life in gastrointestinal disorders', *Alimentary Pharmacology and Therapeutics* 20 (November 2004), 31–9.
36. Simona Noaghiul and Joseph Hibbeln, 'Cross-National Comparisons of Sea-Food Consumption and Rates of Bipolar Disorders', *American Journal of Psychiatry* 160 (2003), 2222–7 and J. R. Hibbeln, L. R. Nieminen, and W. E. Lands, 'Increasing homicide rates and linoleic acid consumption among five Western countries, 1961-2000', *Lipids* 39 (December 2004), 1207–13.
37. Greenpeace, *Eating Up the Amazon* (Washington DC: Greenpeace USA, 2006): http://www.greenpeace.org.uk/MultimediaFiles/Live/FullReport/7555.pdf.
38. Greeanpeace, *Eating Up the Amazon*, 16.
39. Friends of the Earth, *Greasy Palms – Palm Oil, the Environment and Big Business* (London: Friends of the Earth, 2005) at: http://www.foe.co.uk/resource/reports/greasy_palms_summary.pdf.
40. 'Indonesia government to build 11 biodiesel plants in 2006', *endonesia.biz*, 21 July 2006 at: http://www.endonesia.biz/mod.php?mod=publisher&op=viewarticle&cid=17&artid=728.
41. Joan Martinez-Alier, 'Ecological distribution conflicts and indicators of sustain-

ability', *International Journal of Political Economy* 34 (Spring 2004), 13–30.

42. M. L. Khandekar, T. S. Murty, D. Scott and W. Baird, 'The 1997 El Niño, Indonesian Forest Fires and the Malaysian Smoke Problem: A Deadly Combination of Natural and Man-Made Hazard', *Natural Hazards* 21 (2000), 131–44.
43. Prabir K. Patra, Shamil Maksyutov *et al.*, 'Analysis of atmospheric CO_2 growth rates at Mauna Loa using CO_2 fluxes derived from an inverse model', *Tellus* B 57 (2005), 357–65.
44. Vaclav Smil, 'Detonator of the population explosion', *Nature* 400 (29 July 1999), 415.
45. James N. Galloway, Ellis B. Cowling, Sybil P. Seitzinger, Robert H. Socolow, 'Reactive nitrogen: too much of a good thing?', *Ambio: A Journal of the Human Environment* 31 (2002), 60–3.
46. Vaclav Smil, *Enriching the Earth: Fritz Haber, Carl Bosch and the Transformation of World Food Production* (Cambridge, MA: MIT Press, 2001).
47. Willem Norde, 'Energy and entropy: a thermodynamic approach to sustainability', *The Environmentalist* 17 (1997), 57–62.
48. Andrew Whitley, *Bread Matters* (London: Fourth Estate, 2006).
49. 'Letter: Caries caused by industrially produced bread', *Medizinische Klinik* 5 (1975), 1442.
50. R. Lal, 'Soil Carbon Sequestration Impacts on Global Climate Change and Food Security', *Science* 34 (11 June 2004), 1623–6.
51. N. N. Rabelais, R. E. Turner, W. J. Wiseman, 'Gulf of Mexico hypoxia, a.k.a. 'The Dead Zone', *Annual Review of Ecology and Systematics* 33 (2002), 236–63.
52. Wes Jackson, *New Roots for Agriculture* (San Francisco: Friends of the Earth, 1980); also Wes Jackson, 'Natural systems agriculture: a truly radical alternative', *Agriculture, Ecosystems and Environment* 88 (2002), 111–17.
53. Sidney W. Mintz and Christine M. Du Bois, 'The Anthropology of Food and Eating', *Annual Review of Anthropology* 31 (2002), 99–119.
54. Mary Douglas, 'Food as a system of communication' in Douglas, *In the Active Voice* cited A. McGowan, *Ascetic Eucharists: Food and Drink in Early Christian Ritual Meals* (Oxford: Clarendon Press, 1998), 4.
55. E. P. Thompson, 'The Moral Economy of the English Crowd in the Eighteenth Century', *Past and Present* 50 (1971), 79.
56. Sean Freyne, 'Herodian Economics in Galilee: Searching for a Suitable Model' in Philip F. Esler (ed.), *Modelling Early Christianity: Social Scientific Studies of the New Testament in its Context* (London: Routledge, 1995).
57. John Dominic Crossan, *The Historical Jesus: The Life of a Mediterranean Jewish Peasant* (Edinburgh: T. and T. Clark, 1991), 122–6.
58. This is the central claim of John Koenig, *The Feast of the World's Redemption: Eucharistic Origins and Christian Mission* (Harrisburg, Penn: Trinity Press International, 2000). See also Eugene LaVerdiere, *Dining in the Kingdom of God: The Origins of the Eucharist According to Luke* (Collegeville, Minn: Liturgy Training Publications, 1994).
59. See further Crossan, *The Historical Jesus*, 128.
60. For a fuller version of what follows see Michael Northcott, 'Eucharistic eating: The moral ecology of food' in Lukas Vischer (ed.), *The Eucharist and Creation Spirituality* (Geneva: Centre Internationale Reformé John Knox, 2007).
61. Dennis E. Smith, *From Symposium to Eucharist: The Banquet in the Early Christian World* (Minneapolis: Fortress Press, 2003), 2.
62. In addition to Crossan and Smith, Richard Horsley, Marcus Borg, Gerd

Theissen, N. T. Wright and William Herzog offer interpretations of the meals of Christ which are analogous to this approach.

63. Smith, *From Symposium to Eucharist*, 3–9.
64. N. T. Wright, 'One God, One Lord, One People: Incarnational Christology for a Church in a Pagan Environment', *Ex Auditu* 1 (1985): http://www.northpark.edu/sem/exauditu/papers/wright.html.
65. McGowan, *Ascetic Eucharists*, 272.
66. Smith, *From Symposium to Eucharist*, 32–48.
67. Novatian, *On Jewish Foods*, 3 and 4, trs. Russell De Simone, *The Fathers of the Church: A New Translation* (Washington DC: Catholic University Press of America, 1974), 147–51.
68. Titus 1:15 and 1 Tim. 4:4 cited Novatian, *On Jewish Foods*, 5.
69. 1 Tim. 6:10 cited Novatian, *Jewish Foods*, 6.
70. Novatian, *Jewish Foods*, 7.
71. See further Koenig, *Feast of the World's Redemption.*
72. Michael Northcott, 'A place of our own', in Peter Sedgwick (ed.), *God in the City* (London: Mowbray, 1995), 135.
73. Words of offering from the 1982 Eucharistic Order of the Scottish Episcopal Church.
74. See further Michael Northcott, *The Environment and Christian Ethics* (Cambridge: Cambridge University Press, 1996).
75. C. S. Lewis, *The Abolition of Man* (London: Geoffrey Bles, 1943).
76. C. S. Lewis, *The Discarded Image: An Introduction to Medieval and Renaissance Literature* (Cambridge: Cambridge University Press, 1967): see also Remi Brague, *The Wisdom of the World: The Human Experience of the Universe in Western Thought*, trs. Teresa Fagan (Chicago: University of Chicago Press, 2003).
77. James Lovelock, *The Revenge of Gaia: Why the Earth is Fighting Back and How We Can Still Save Humanity* (London: Allen Lane, 2006).
78. Wendell Berry, *The Unsettling of America: Culture and Agriculture* (San Francisco: Sierra Club Books, 1986), 13.
79. Miriam Weinstein, *The Surprising Power of Family Meals: How Eating Together Makes Us Smarter, Stronger, Healthier and Happier* (Hanover, NH: Steerforth Press, 2005).
80. Karl Marx, *Capital*, Vol. 1, ch. 3. Marx is the spectre at the feast/famine of the modern food economy; while Marxism advanced forms of collectivised farming in the twentieth century as wicked as the enslaving agriculture economy of the Pharaohs, nonetheless Marx's analysis of the moral and spiritual genius of capitalism remains unsurpassed.
81. B. L. Smith, 'Organic versus supermarket foods: element levels', *Journal of Applied Nutrition* 45.1 (1993) offers a laboratory-based investigation demonstrating greater nutrient content in organic foods, though not in a peer reviewed journal. See also Kirsten Brandt and Jens Peter Molgard, 'Organic agriculture: does it enhance or reduce the nutritional value of foods?', *Journal of the Science of Food and Agriculture* 81 (June 2001), 924–31.
82. Frederick Kirschenmann, 'Ecological Morality: A New Ethic for Agriculture', in D. Rickerl and C. Francis (eds), *Ecological Morality: A New Ethic for Agriculture* (Madison, WI: American Society for Agronomy, 2004), 167–76.
83. For a fuller account see Northcott, 'Eucharistic Eating: The Moral Ecology of Food'.
84. Vigen Guroian, *Ethics After Christendom: Toward an Ecclesial Christian Ethic* (Grand Rapids, MI: Eerdmans, 1994), 160.
85. Henri de Lubac, *Corpus Mysticum: L'Eucharistie et l'Eglise au Moyen Age*, 2nd edn

(Paris: Aubier, 1949). See also Paul McPartlan, *The Eucharist Makes the Church: Henri de Lubac and John Zizioulas in Dialogue* (Edinburgh: T. & T. Clark, 1993).

86. John D. Rempel, *The Lord's Supper in Anabaptism: A Study in the Christology of Balthasar Hubmaier, Pilgram Marpeck, and Dirk Philips* (Scottdale, Penn: Herald Press, 1993), 32–3.
87. John Howard Yoder, *Täufertum und Reformation in Gespräch*, cited Rempel, *The Lord's Supper in Anabaptism*, 33.
88. John Zizioulas supports an analogous position when he opposes the practice of private priestly Mass and suggests that the celebration of the Eucharist requires the actual personal presence of the people of God: John Zizioulas, 'The ecclesiological presuppositions of the holy eucharist', cited Paul McPartlan, *The Eucharist Makes the Church: Henri de Lubac and John Zizioulas in Dialogue* (Edinburgh: T. and T. Clark, 2003), 178.
89. Wendell Berry, 'The use of energy' in Berry, *Unsettling of America*, 95.
90. Duffy, *Stripping of the Altars*, 95–102.
91. Jualynne E. Dodson and Cheryl Townsend Gilkes, '"There's Nothing Like Church Food": Food and the Afro-Christian Tradition: Re-Membering Community and Feeding the Embodied S/spirit(s)', *Journal of the American Academy of Religion* 63 (1992), 519–38.

9. Remembering in Time

1. Miroslav Volf, *The End of Memory: Remembering Rightly in a Violent World* (Grand Rapids, MI: Willam B. Eerdmans, 2006).
2. James Lovelock, *The Revenge of Gaia: Why the Earth is Fighting Back and How We Can Still Save Humanity* (London: Allen Lane, 2006).
3. See further Michael Northcott, 'Soil, Stewardship and Spirit in the Era of Chemical Agriculture' in R. J. Berry (ed.), *Environmental Stewardship: A Critical Reader* (London: T. and T. Clark, 2006), 213–19.
4. Bob Holmes, 'Imagine earth without people', *New Scientist* 2573 (12 October 2006), 36–41: http://www.newscientist.com/channel/life/mg19225731.100.
5. Ross McKintrick, Joseph D'Alco, Mhadav Khandekar *et al.*, *Independent Summary for Policymakers: IPCC Climate Change 2007* (Vancouver, CA: Fraser Institute, 2007).
6. David Bellamy, 'Glaciers are cool', *New Scientist* 2495 (16 April 2005).
7. Senator James Inhofe, 'Hot and Cold Media Spin: A Challenge To Journalists Who Cover Global Warming', United States Senate, 25 September 2006.
8. Internal memo cited in *Smoke and Mirrors: How ExxonMobil Uses Big Tobacco's Tactics to Manufacture Uncertainty on Climate Science* (Cambridge, MA: Union of Concerned Scientists, 2007), 9.
9. Pat Michaels and Robert C. Bailing, *The Satanic Gases: Clearing the Air About Global Warming* (Washington DC: Cato Institute, 2000), 75–91. See also Roger Bate and Julian Morris, *Global Warming: Apocalypse or Hot Air?* (London: Institute of Economic Affairs, 1994), and Myron Ebell, 'Global Warming and Energy' in Angela Logomasini (ed.), *The Environmental Source* (Washington DC: Competitive Enterprise Institute, 2004), 67–80: http://www.cei.org/pdf/2317.pdf, 31 January 2005.
10. Henrik Svensmark, 'Cosmic rays and the biosphere over 4 billion years', *Astronomische Nachrichten* 327 (2006), 871–5 and Nigel Marsh and Henrik Svensmark, 'Solar influence on earth's climate', *Space Science Reviews* 107 (2003), 317–25.
11. Hanspeter Holzhauser, Michel Magny and Heinz J. Zumbuhl, 'Glacier and

lake-level variations in west-central Europe over the last 3500 years', *The Holocene* 15.6 (2005), 789–801.

12. Paul Damon and Peter Laut, 'Pattern of strange errors plagues solar activity and terrestrial climate data', *Eos, Transactions of the American Geophysical Union* 85 (2004), 370.
13. Natalia G. Andranova and Michael E. Schlesinger, 'Causes of global temperature changes during the 19th and 20th centuries', *Geophysical Research Letters* 27 (July 2000), 2137–40.
14. Nigel Calder, 'An experiment that hints we are wrong on climate change', *Sunday Times*, 11 February 2007. See also Henrik Svensmark and Nigel Calder, *The Chilling Stars: A New Theory of Climate Change* (London: Icon, 2007).
15. Michael Crichton, *State of Fear* (London: Harper Collins, 2005) and Richard Lindzen, 'Climate of fear: global-warming alarmists intimidate dissenting scientists into silence', *Wall Street Journal*, 12 April 2006.
16. Michael Crichton, *State of Fear*. See also Amy Laura Hall's forthcoming *Conceiving Parenthood: The Protestant Spirit of Biotechnological Reproduction* (Grand Rapids, MI: Eerdmans).
17. Blaise Pascal, *Pensées*, trs. A. J. Krailsheimer (Harmondsworth: Penguin, 1968).
18. Stephen F. Haller, *Apocalypse Soon? Wagering on Warnings of Global Catastrophe* (Montreal: McGill-Queen's University Press, 2002).
19. Wendell Berry, *Standing By Words* (Washington D. C.: Shoemaker and Hoard, 1983), 24–7.
20. Paul Crutzen, 'Albedo enhancement by stratospheric sulfur injections: a contribution to resolve a policy dilemma?', *Climate Change* 77 (2006), 211–19.
21. Robert E. Dickenson, 'Climate engineering: a review of aerosol approaches to changing the global energy balance', *Climatic Change* 33 (1996), 279–290.
22. Ken O. Buesseler, John E. Andrews, Steven M. Pike, Matthew A. Charette, 'The effects of iron fertilization on carbon sequestration in the Southern Ocean', *Science* 304 (16 April 2004), 414–17.
23. Dale Jamieson, 'Intentional Climate Change', *Climatic Change* 33 (1996), 323–36.
24. There is to my knowledge one such car which does run on hydrogen derived from renewable sources in the Shetlands Islands: Michael Magee, 'Catching the wind: hydrogen car fuel cell breakthrough from remotest Scottish island', *Sunday Herald*, 12 March 2006.
25. I am grateful to an anonymous peer reviewer for this suggestion. See also Stephen W. Pacala, 'Global constraints on reservoir leakage', Princeton University Carbon Mitigation Initiative, Working Paper at: http://www.-princeton.edu/~cmi/research/kyoto02/pacala%20kyoto%2002.pdf.
26. Jamieson, 'Intentional climate change'. On fish farming see also Michael Northcott, 'Farmed Salmon and the Sacramental Feast: How Christian Worship Resists Global Capitalism' in William F. Storrar and Andrew Morton (eds), *Public Theology for the 21st Century* (London: T. and T. Clark, 2004), 213–30.
27. Lovelock, *Revenge of Gaia*.
28. John Howard Yoder, *The Politics of Jesus: Vicit Agnus Noster* (Grand Rapids, MI: William B. Eerdmans, 1972), 234, and Michael Northcott, *An Angel Directs the Storm: Apocalyptic Religion and American Empire* (London: I. B. Tauris, 2004), 166.
29. Sheldon Wolin, *Politics and Vision: Continuity and Innovation in Western Political Thought*, expanded edn (Princeton, NJ: Princeton University Press, 2004), 237–8.
30. Thomas Hobbes, *Leviathan*, 17.

31. Wolin, *Politics and Vision*, 112–13.
32. Augustine, *City of God*, 11.6. My interpretation here is indebted to Wolin, *Politics and Vision*, 112–13.
33. Oscar Cullman, *Christ and Time* (Philadelphia: Westminster Press, 1950).
34. Stanley Hauerwas, *With the Grain of the Universe: The Church's Witness and Natural Theology* (London: SCM Press, 2002).
35. David F. Ford, *The Shape of Living: Spiritual Directions for Everyday Life* (London: Fount, 1997).
36. Northcott, *Angel Directs the Storm.*
37. Augustine, *Letter 153: Augustine to Macedonius* in Margaret Atkins, *Augustine: Political Writings* (Cambridge: Cambridge University Press, 2001, 71–86.
38. Again see Northcott, *Angel Directs the Storm* for a fuller account.
39. Stanley Hauerwas, 'Democratic Time: Lessons Learned from Yoder and Wolin,' *Crosscurrents* (Winter 2006), 534–52.
40. On the politics of fear see further Corey Robin, *Fear: The History of a Political Idea* (Oxford: Oxford University Press, 2004).
41. Wolin, *Politics and Vision*, 261.
42. See further Northcott, *Angel Directs the Storm*, 39–54.
43. Val Plumwood, *Environmental Culture: The Ecological Crisis of Reason* (London: Routledge, 2002), 93–6.
44. Eco-Congregation is an ecumenical initiative enabling local churches to address their ecological footprint, and to reconceive their mission and worship in the light of the ecological crisis: http://www.ecocongregation.org/ and http://www.ecocongregation.org/scotland/.
45. See further Michael Northcott, 'The World Trade Organisation, Fair Trade and the Body Politics of Saint Paul' in John Atherton (ed.) *Through the Eye of a Needle: Theology, Ethics and Economy,* (London: Epworth Press, 2007).
46. Garth Lean, *God's Politician: William Wilberforce's Struggle* (London: Darton, Longman and Todd, 1980); see also the film *Amazing Grace: The Story of William Wilberforce* (Michael Apted, 2006).
47. Chris Sugden, *Fair Trade as Christian Mission* (Bramcote, Nottingham: Grove Books, 1999).
48. Geoff Moore, 'The Fair Trade Movement: Parameters, Issues, and Future Research', *Journal of Business Ethics* 53 (2004), 73–86.
49. The World Wildlife Fund is urging the UK government to adopt a climate care kite mark scheme but given the compromises over carbon trading it is doubtful that a government scheme will be as trustworthy as the fair trade mark: Mark Rice-Oxley, 'Britain helps citizens atone for emissions', *Christian Science Monitor,* 8 February 2007 at: http://www.csmonitor.com/2007/0208/p06s01 woeu.html/.
50. Northcott, *Angel Directs the Storm*, 17–20. See also A. David Lindsay, *The Modern Democratic State* (London: Royal Institute of International Affairs, 1943).

Index